UFO HOSTILITIES
And The Evil Alien Agenda

Conspiracy Journal
P R O D U C T I O N S

UFO HOSTILITIES
And The Evil Alien Agenda

Lethal Encounters With Ultra-Terrestrials Exposed

Timothy Green Beckley and Sean Casteel
with
Dr. James McDonald, George D. Fawcett. Brad Steiger, Arthur Crockett,
Nigel Watson, Dr. Olavo Fontes, Bob Pratt, Scott Corrales, B. Ann Slate,
George Andrews, Tim R. Swartz, Nomar Slevik, Hercules Invictus,
Allen Greenfield, William Kern, Dr. Karla Turner

Copyright 2018 by Timothy Green Beckley
dba Global Communications/Conspiracy Journal

First Edition
Published in the United States of America By
Global Communications/Conspiracy Journal
Box 753 · New Brunswick, NJ 08903

Staff Members
Timothy G. Beckley, Publisher
Carol Ann Rodriguez, Assistant to the Publisher
Sean Casteel, General Associate Editor
Tim R. Swartz, Graphics and Editorial Consultant
William Kern, Editorial and Art Consultant

Sign Up On The Web For Our Free Weekly Newsletter
and Mail Order Version of Conspiracy Journal
and Bizarre Bazaar
www.Conspiracy Journal.com
Order Hot Line: 1-732-602-3407
PayPal: MrUFO8@hotmail.com

Contents

Reality in the human mind is simply a matter of perception, therefore, one must sometimes believe what one cannot actually see without succumbing to the futility of trying to find meaning in events that may have no meaning.

Anon

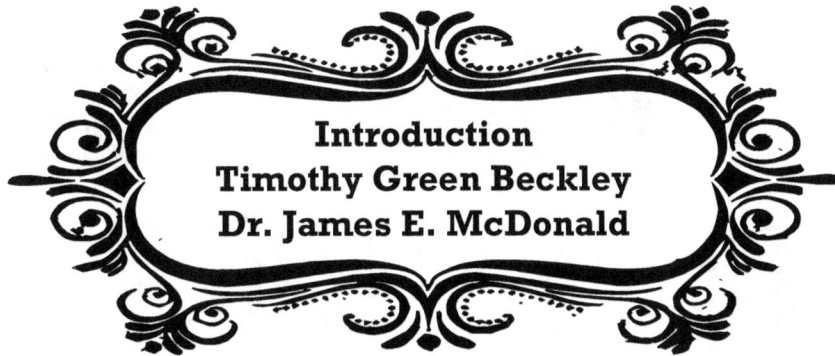

Introduction
Timothy Green Beckley
Dr. James E. McDonald

WORLD NEWS

★ ★ ★ ★ ★ SPECIAL EDITION FINAL

WORLD THREATENED WITH DESTRUCTION BY DISTANT PLANET

Story Inside

Army Meets Strange Foe! Machine guns and tanks make no impression on 9-foot indestructible robot on a death-dealing rampage in Washington, D. C. Menacing visitor arrives in weird space ship accompanied by strange companion. Demands audience with the President. Threatens to destroy universe if we don't see things his way.

UFO HOSTILITIES AND THE EVIL ALIEN AGENDA

Introduction

UFOS – "SPACEMEN" – ARE THEY FRIEND OR FOE?
By Timothy Green Beckley

What are their overall intentions?

Are their occupants benevolent? Sympathetic? Non-committal? Or downright hostile toward humankind?

From the testimony of a good percentage of eyewitnesses and those claiming close encounters – on the ground or in the air – it is obvious that the intentions of the "visitors," from wherever they might originate – is not all "squeaky clean."

They have been on many occasions hassled, harassed, burned, poisoned, pummeled, beaten, laser beamed, shocked, and perhaps – as extreme a term as it may seem – murdered those who have come in close proximity to these unknown manifestations.

Pilots have likewise been terrorized, had their engines stalled – even melted – and "shot down" out of the sky, most often without any provocation. Dozens of planes – and their crews and passengers – have vanished into thin air.

Motorists, likewise, often do not have an inkling of what is about to hit them between the proverbial eyes on a lonely road in the dead of night, their car engines idling or conking out altogether, with these unsettling events perhaps leading up to an abduction or molestation at the hands of some extraterrestrial "monsters."

Ample exposure has also been given to cases in which UFOs have brought to a complete halt automobiles and other mechanical devices. Several well-seasoned UFO researcher-writers have penned lengthy papers on this so-called EM (electromagnetic) effect. What of other even more bizarre and irrational side effects, traceable to disturbingly close confrontations with UFOs and their many-faceted crew members?

Over a period of years I have kept track of existing patterns – some of them very strange – which might not be obvious to those who do not have access to an extensive UFO data base. Ours goes back over 50 years!

UFO HOSTILITIES AND THE EVIL ALIEN AGENDA

STRANGE GROWTH PATTERNS

One of our major concerns in this work is the possibility of radioactivity being unleashed or spread by these objects, effecting humans and our earthly surroundings alike. The Cash-Landrum incident, as detailed in this work by Sean Casteel, is the leading example among cases that demonstrate the possibility that a potentially lethal dose of radioactive emissions from a UFO can cause undue harm to those who don't know any better than to keep out of the way (not that you're going to have much of a choice!)

Scientific evaluation of the Bikini atom bomb experiments some years ago established the fact that existing life forms on and near this closely-guarded South Pacific coral reef had undergone a drastic transformation following the testing of two atomic warheads there in June of 1946. For years after these top secret blasts (whose mushroom clouds rose 30,000 feet), sea creatures – from the most microscopic to larger varieties – were found washed ashore, their bodies mutated in any number of ways. Marine navigators in the area even reported seeing a monstrous 66-foot shark, much larger than any ever previously seen. On the island itself, insects were reported to have grown to tremendous proportions, six and seven times their normal size. Often they were found to have additional limbs and tentacles.

Flying Saucer 'Passenger' Declares A-Bomb Blasts Reason For Visits

By LEN WELCH

Fasten your safety belt, Buster, and take a firm grip on your chair for we are about to take off on a story to end all stories about flying saucers.

Woven into this incredible tale is what was reported to be probably the first person-to-person conversation with a man in a flying saucer, an explanation why flying saucers are flitting about our skies, a beautiful woman from another planet, and mysterious footprints in the desert sands.

Few questions about flying saucers are left unanswered by this story that has its beginning on a lonely spot on the California desert between Parker, Ariz., and Desert Center, Calif.

ITS PRINCIPALS are four Arizonans out to get a look at a flying saucer, a Valley Cen-

Is this a flying saucer or a

fred C. Bailey, 38, of Winslo for 12 years an employe of t Santa Fe Railway and n "braking" on passenger tra and Mrs. Bailey.

Williamson's interest in f ing saucers was intensified stories of saucers he uncov ed among Indian legends wh doing independent resear among the Chippewas.

"I had corresponded w Prof. George Adamski, f merly of Palomar Observat near San Diego, and learn that he had made pictures flying saucers," Williams said. "We (my wife and t Baileys) decided to go on picnic lunch with Profes Adamski in the hope that would see a flying saucer."

THE GROUP in addition the Williamsons, the Bail and Professor Adamski cluded Alice K. Wells a Lucy R. McGinnis, the lat the professor's secretary b

4

UFO HOSTILITIES AND THE EVIL ALIEN AGENDA

Of course the causes of these abnormalities are easily traced, scientifically. We know, for instance, that excessive radiation, especially "close-control" radiation, where atoms and molecules undergo initial changes, causes mutations to occur in nature. And certainly the radiation swept into the atmosphere by these two tremendous atomic weapons were among the most deadly ever.

However, what of instances where radioactivity cannot be pinpointed as having set off the phenomenal and rapid growth of plant life, insects – and, yes, even human beings? Let us consider cases in which UFOs can be associated with such mutations.

Several years ago a story was brought to my attention which, at the time, seemed incredible. So unusual was it that I neatly creased the report – which had come to me in the form of two personal letters – down the middle and placed them in the back of an unmarked file folder out of sight and for a long time out of memory. It was not until recently, with the reporting of several similar cases, that I vaguely recalled the rejected episodes and retrieved them from their resting place. During the summer of 1967, 18-year-old Jerry James, his father, mother and younger sister were driving toward California and a long-cherished vacation from their Colorado home. It was during the first night of their trip that a series of peculiar events began to transpire.

Jerry, scheduled to enter college in the fall, had taken turns with the driving chores. Finally, after being on the road for more than 18 continuous hours, the family pulled their new Chevrolet pickup, behind which they towed a house trailer, onto a service station lot to refresh themselves.

As they proceeded to stretch their limbs and catch a breath of fresh desert air, their eyes were attracted to a peculiar-looking object moving about overhead. Later, Jerry was to give me the following description of what the family had witnessed. "The object – a definite aircraft of a completely unknown type – was shaped similar to a child's top – round at the top and pointed at the bottom. It was encircled by a ring of brightly-colored lights, and from out of the lower portion of the 'ship' came a beacon similar to a searchlight. This light beam passed over the area as if in search of something. We became quite shaken, moments later, when it shone directly on our trailer."

Jerry states that the object appeared to be about the size of an automobile. "Before disappearing from our view, the UFO hovered only a few feet above the top of our truck and then, quite noiselessly, zoomed off into the distance and went out of sight." None of those present found it easy to believe what they had seen.

The next day, the James family camped off the main highway, miles from where their observation of the previous evening had taken place.

"We thought we were once again going crazy when we saw a silver-looking vessel floating in from over a nearby mountain peak." This time, Jerry revealed,

UFO HOSTILITIES AND THE EVIL ALIEN AGENDA

"The UFO got so close – within 75 feet – that we could even see rivets visible in the metallic gray structure. At one point, as the orb made a direct pass at us, my mom fainted dead away. We had to use smelling salts to revive her!"

Entering the back of the trailer they immediately noticed that something truly strange – well beyond their mysterious traveling companion – had come about. "Everything inside the trailer compartment was in perfect order, although, when we left it, the bunks had been unmade and the sink area strewn with silverware and unwashed dishes and utensils. Our beds had been put together and everything was as neat as it could be – though not as my Mom would have done it." It was then, looking around them, that Jerry noticed the most unusual feature. "On the kitchen cabinet we found a small dish with a sprouting carrot which had been placed there sometime during the last 24 hours. But something drastic and truly puzzling had happened to it, for the carrot did not resemble a carrot any longer. It had spread out, as if having grown for weeks in a tempered, professional hot house. Roots were everywhere, running outside the dish and reaching almost to the floor."

Examining the vegetable closely, Jerry and his parents and sister noticed a slimy green-colored substance on the cabinet and dish. "It was a vile smelling substance," Jerry observed, "and other than that I can't tell you much about it. One thing strange, though. When I touched it with a pencil it 'ate up' the wood and lead in a matter of seconds."

Soon after this, Jerry telephoned the police, who they hoped would investigate the series of phenomena. En route, their truck began to accelerate against the will of its driver. "Only a fractional depressing of the accelerator was enough to send the vehicle – truck and trailer – speeding along at 40 miles per hour."

But this was not all, as Jerry later confirmed. Upon arriving at the nearest gas station, the attendant proceeded to remove the gas cap. As he did, out rushed an odorous and gaseous material, causing a hissing noise and bringing with it a strong smell like that of sulfur dioxide, or rotten eggs.

Even later, they discovered that the part of the truck seat where the driver sat had rotted out and the ignition key itself now glowed in the dark and had gotten soft and rubbery.

While the law enforcement officials they contacted refused to believe that the James family had actually encountered a strange visitor from the depths of the universe, their insurance company was remarkably less skeptical. "They bought us a brand new truck to replace the one that had been wrecked so badly," Jerry reported. "They informed us that they had never seen anything quite like it before. What happened really had them stumped – almost as much as it had us frightened."

Jerry remarked in a later communication that he was well aware that a good

The Bakersfield Californian

AIRLINER, 44 ABOARD, MISSING

U.S. Tries to Catch Up With Russia

Jupiter Missile Will Be Used for Launching

700 Attend Quarterly Convention

Special Policy Talks Highlight Program Today

Big Plane Presumed Down at Sea

Last Reported Past Halfway Mark to Hawaii

Ike to Go to Hospital for Checkup

State Dept. Can't Get Its Top Scientist

FLASHES

Troops Guard Against Riots

Notable Persons on Lost Airliner

Young Killer of Two Sent to Hospital

Sputnik II Goes Silent; No News of Dog Given

British Claim Clean Bomb

portion of what transpired could have been caused by a harmful dose of radioactivity. Yet he also added that no one in his family showed the slightest symptoms of radiation poisoning. "None of us became ill afterwards. Even if the amount of radioactivity had been negligible, we should have become sick. In addition, it seems very probable that if the UFOs which passed directly overhead had carried with them some sort of contamination – enough to make the carrot root grow so wildly, or the ignition key to glow and turn soft –surely it would have made us ill enough to need medical treatment. Probably we would have been quarantined in a hospital for observation. Thus I have my doubts that radioactivity was what caused these peculiar things to happen."

But if radiation wasn't the cause of this family's problems, what was? It seems more than coincidental that two UFOs were seen while all this was taking place. A careful examination of other close sightings reveals numerous instances of a similar rapid "growth pattern."

A good deal of excitement and tension was generated in the out-of-the-way community of Galt, Ontario, Canada, where, in July, 1957, a group of frightened teenagers came within inches of a "spaceship," as they readily referred to the object on the ground in front of them.

Jack Stephens, truly among the trustworthy, according to parents and teachers, said the party of five had accidentally stumbled upon the object, complete with portholes, as it rested on two ball-shaped "landing gear" in a field outside town.

Additional credibility came after an analysis was made of the scorched black earth discovered on the spot. Paul Hartman, a stringer for the "Galt Reporter," says the change in soil composition at this location was incredible. Top-soil dug up from the site glowed in the dark. Also, when grain samples from the burnt-out region were studied under a microscope, it was found that they were healthier and sturdier than the grain samples taken from elsewhere in the field. Finally, the insects here had undergone a certain instability and peculiarity in character. Ants where the saucer had touched down were larger and stronger-looking than their exact counterparts located in untouched areas. The ant hills themselves, according to Mr. Hartman, were much higher than ordinary, and a spider, which had by chance found its way into a jar containing soil from the Gait UFO landing site, had grown to about ten times the usual size for this particular species.

Strange? Unexplainable? All a coincidence? During the summer of 1973, a mysterious encompassing "ooze," dubbed the "blob," began sprouting up nightmarishly in yards throughout Texas. Although never directly related to UFO activity, speculation that the fungus was an extraterrestrial manifestation remained on the tip of everyone's tongue. While news of the "blob" spread far and wide, other mutations in the world of biology and botany have not been publicly discussed, let alone explained.

5½ Inches in a Month

Girl Who Saw UFO Is Growing Taller

UFO HOSTILITIES AND THE EVIL ALIEN AGENDA

In Corvallis, Oregon, on November 12th, 1973, a woman telephoned the "Gazette Times" after having located a hillside where a large unidentified object allegedly crash-landed. Police at the site confirmed that there was a sizable burned area and that here – and only here – a strange wild-growing fungus seemed to be thriving in the charred soil. Investigation determined they weren't any species of toadstool or mushroom known to the area – because they were much too large!

Similarly, from Argentina, Guillermo Aldunati, President of the Association de Observadores de los Astros, a private UFO investigating body in Rosario, confirmed that mushrooms of enormous size had been found in the provinces of Santa Fe and Necochea (north and south of Buenos Aires). The latter was discovered on a supposed UFO landing strip and the soil around the mushrooms was scorched for a radius of 25 feet.

In the previous instances we have seen how UFOs apparently are able – either accidentally or for some unknown purpose – to influence the growth of plants and insects. But what of humans? Have they had such a devastating impact on unsuspecting observers?

Back in the spring of 1964, one of the biggest, most productive UFO flaps of that decade took place in the Southwest corner of the United States. Literally thousands of individuals in New Mexico, Texas, Oklahoma, Missouri and Nevada huddled in groups nightly as an Armageddon of whirling discs and penetrating lights cascaded across a starry backdrop.

Without a doubt, the major highlight of this majestic wave of sightings came when Lonnie Zamora, a Socorro, New Mexico, police officer, came to within 50 feet of a landed UFO and its small-sized occupants.

While this flap was well publicized, one sighting did not receive the attention it rightfully deserved.

On April 29th of that year, 10-year-old Sharon Stull stood outside the Lowell schoolyard, where she was attending the fourth grade. With her on the day in question was her sister, Robin, age eight, and another classmate.

As they went about their recess games, they noticed an egg-shaped "thing" high in the sky hovering above their elementary school. No one paid much attention to it, though Sharon kept looking up at it from time to time for roughly 30 minutes.

After their noon break, the children returned to class. A half hour later, Sharon asked to be excused so she could wash out her eyes, which had gotten red and were causing her considerable discomfort. Noting that the irritation did not disappear after a while, Sharon's teacher told her to see the school doctor. He in turn called Sharon's mother, Mrs. Max Stull, and recommended that the child be taken at once to the Batton Hospital for a thorough examination.

Sharon was treated at the infirmary and was told not to allow any sunlight to touch upon her face, which had turned a blue-red color. "They instructed me to keep her blinds drawn and to protect her inflamed eyes and eyelids from light," Mrs. Stull told newsmen. "Part of my daughter's face and nose appeared to be puffy and red. She continued to complain of burning pains."

Seeking an explanation, Sharon told the attending physician what she had seen in the clear, blue noonday sky. "The doctor seemed interested," her mother revealed, "and I remember him remarking about how such a short exposure to the sun, even at the brightest time of day, would under normal circumstances be insufficient to cause such burns and inflammation."

Subsequent newspaper coverage revealed that Sharon was getting better, although it was reported that she would have to wear dark glasses for a while, in order to protect her sensitive eyes.

Nothing more was to be heard about Sharon's unfortunate luck until four months later, when a brief update appeared in The Albuquerque Tribune. What had been a major front page story only weeks before was now only worthy of two paragraphs, and these were neatly tucked away somewhere between the comic section and the sports pages. Yet, despite the apparent lack of interest on the part of the Tribune's editorial staff, Sharon had undergone a continuing series of traumatic changes, all of which had a near-disastrous effect upon her emotional and physical wellbeing.

According to Mrs. Stull, her little girl had grown 5 and a half inches and gained 25 pounds in the month following the sighting of the mysterious egg-shaped craft which had hung over the school yard. "A while ago she was just a child who liked to play with dolls and paper cut-outs," Sharon's mother said. "Now she is suddenly mature and grown-up, cooks meals by herself, cleans house and takes care of the younger children."

At the time she saw the strange object in the sky, Mrs. Stull confirmed that Sharon had been 4 feet, 8 inches tall and weighed about 85 pounds. "Now Sharon is 5 feet, 2 inches tall and weighs 110 pounds – and is still growing. My daughter had outgrown all her clothes and quickly outgrows new garments and shoes. I'm so confused I don't know what to believe," she sadly observed. "I just feel funny" was all Sharon herself would say when questioned about her UFO experience.

"I know she definitely saw something in the sky that day, but I don't know what," Mrs. Stull said. "It has been a nightmare ever since. I wish I had let her play hooky that day."

While this is the only case of phenomenal human growth attributable to the close approach of a flying saucer, there remains a strong possibility that similar effects have transpired – and may be taking place at this very moment.

Girl Says She Was Burned While Watching UFO Here

By JERRY SMOTHERS

Concern over unidentified flying objects continued at a high pitch here today.

Latest incidents include:

—A report from a 11-year-old girl that she was burned while watching a strange egg-shaped thing in the sky yesterday afternoon.

—A report from a truck driver and his wife that they saw two mysterious objects in the sky between here and Socorro.

The rash of incidents started Friday when a policeman saw a UFO take off south of Socorro. UFO experts from different parts of the country are in New Mexico to investigate.

Suffering eye, nose and face burns yesterday afternoon was Sharon Stull, one of five children of Mr. and Mrs. Max L. Stull of 1900 Wilmore SE.

On Lowell School Ground

She said she gazed at a UFO for five or 10 minutes shortly before 12:30 p.m. on the Lowell School ground at 1600 Ross SE after returning there from lunch at home.

After 1 p.m., while in class, she noticed burning sensations about the eyes and face. She and her mother believe her burns were inflicted by the thing in the sky.

Sharon said she was not facing the sun while watching the thing, slightly smaller than an airplane and having no windows, speed across the sky and disappear in the distance. Her mother said the child has never had a sunburn and couldn't have gotten one in the few minutes she was outside.

Her sister, Robin, 8, says she, too, saw the thing, but paid little attention to it and didn't continue to watch it. Sharon said two other children also glanced at it briefly.

Sharon was treated by a doctor at Bataan Hospital last night and was kept in bed at home today with her blinds drawn to protect her inflamed eyes and eyelids.

Complains of Pains

Part of her face and nose appeared to be puffy and red. She continued to complain of burn pains.

Police interviewed her at

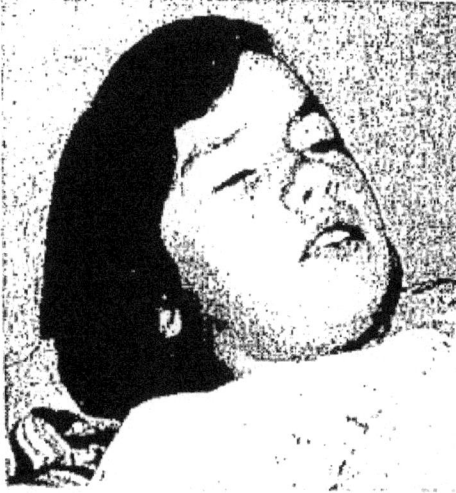

length about the incident and had her draw the silvery object.

Sharon's doctor said his diagnosis is conjunctivitis (membrane inflammation) of both eyes and first - degree sunburn under the eyes and on the nose, adding:

"The sun exposure she had yesterday would usually be considered insufficient to cause the burns and inflammation. It appears to be the type of burns and inflammation caused by longer exposure to the sun."

He said he believes there is no permanent eye damage. Mrs. Stull was extremely concerned and frightened over the incident today. She also kept Robin home from school.

"I'm just scared to death," Mrs. Stull said.

Truck driver Napoleon Green, 37, and his wife, 35, of 1805 Walter SE, said they sighted two UFO's traveling at about the speed of jet planes at 5:30 p.m. yesterday as they were driving toward Albuquerque on U. S. 85, 17 miles north of Socorro.

Shine Brightly

"I wasn't going to report it because I didn't want to believe my eyes," Mr. Green said today. He said his wife dived beneath the truck's dashboard after seeing the egg-shaped things, about half the size of jet planes and shining brightly in the setting sun.

Green said he saw them possibly two minutes before they disappeared over the eastern horizon.

Federal Aviation Agency officials said they have seen nothing unusual in the sky or on their radar screens. Airplanes appear only as small dots on the screens and it is unlikely a UFO could be distinguished from an airplane on radar.

One official said it would be possible in some areas to track a UFO to see where it goes, if they received a report from a responsible agency immediately after a sighting. Radar sightings have to be in a direct line and can be blocked by mountains, making low - altitude radar tracking difficult sometimes, he advised.

The Weather Bureau, FAA and Kirtland AF Base said there were no balloons or other strange - appearing objects in the sky here, to their knowledge, at the time Sharon saw what she did.

Sharon Stull in Bed Today

After all, how many persons would report such a peculiar change in their physical stature to authorities? Also, it is an established fact that once they have been stricken from the pages of newspapers, most UFO observers return to a life of anonymity. Researcher John Keel has stated that UFO percipients who have established a seeming rapport with these phantoms by way of a close sighting frequently undergo a series of physical and emotional changes.

We admit we cannot pinpoint the precise nature of every UFO. What we may deem as hostile may be irrelevant to those "on the other side" of our reality. Perhaps they have taken some action or reaction of ours too seriously and have deemed it necessary to protect themselves against an unmerciful army of humans. Some have suggested this negative attitude toward humans started at the close of WWII and the development of the A-bomb, but that's not necessarily true, as a review of history can determine.

But rest assured that the number of cases of UFO hostility make it almost certain that you will not undergo the wrath of a rebellious space commander. Such personal attacks, thank God, are few and far be-

UFO HOSTILITIES AND THE EVIL ALIEN AGENDA

tween. But percentage-wise, if you do encounter a UFO at close range, we cannot presently determine or predict what the outcome is likely to be.

Caveat Emptor – and let the observer beware!

SUGGESTED READING
ROUND TRIP TO HELL IN A FLYING SAUCER
UFOS – WICKED THIS WAY COMES
EVIL EMPIRE OF THE ETS AND ULTRA TERRESTRIALS
CURSE OF THE MEN IN BLACK – RETURN OF UFO TERRORISTS

TIMOTHY GREEN BECKLEY
UFO & Paranormal Pioneer

Tim Beckley has had so many careers that even his own girlfriend doesn't know what he does for a living...

Timothy Green Beckley has been described as the Hunter Thompson of UFOlogy by the editor of UFO magazine, Nancy Birnes.

Since an early age his life has more or less revolved around the paranormal. At the age of three his life was saved by an invisible force. The house he was raised in was thought to be haunted. His grandfather saw a headless horseman. Beckley also underwent out of body experiences starting at age six.And saw his first of three UFOs when he was but ten, and has had two more sightings since – including an attempt to communicate with one of these objects.

Photo by April Troiani

12

Chapter 02

A SCIENTIST REPORTS ON THE POSSIBLE HOSTILITY OF UNEXPLAINED AERIAL PHENOMENA
By Dr. James McDonald

Editor's Note: Though he is no longer with us, Dr. McDonald was among the most astute researchers in the UFO field. He studied the subject meticulously and is the only phenomenologist to have appeared before Congress to air his – sometimes spectacular – views on this most controversial of all topics.

James E. McDonald (1920-1971) was Associate Director at the University of Arizona's Institute of Atmospheric Physics from 1954 until 1956, and professor until his death in 1971. His areas of research included cloud physics, weather modification, and micrometeorology. He was the author of *Physics of Cloud Modification*, in the fifth volume of Advances in Geophysics. His scientific interests also led him to contribute writings and testimony to contemporary issues of his time such as the supersonic transport debates, placement of Titan II missiles around the Tucson basin, and scientific dismissal of unidentified flying objects. His hypothesis that UFOs were extraterrestrial instruments on information gathering missions, and his beliefs that current scientific reports, such as the Condon Report, were superficially done, led him to investigate sightings and combat governmental impediments into his research topics.

This statement has been submitted by Dr. James E. McDonald, Senior Physicist, Institute of Atmospheric Physics, and professor, Department of Meteorology, The University of Arizona, Tucson, Arizona, to the House Committee on Science and Astronautics at July 29, 1968, Symposium on Unidentified Flying Objects, Rayburn Bldg., Washington, D.C.

IS THERE ANY EVIDENCE OF HAZARD OR HOSTILITY IN THE UFO PHENOMENA?

Official statements have emphasized, for the past two decades, that there is no evidence of hostility in the UFO phenomena. To a large degree, this same conclusion seems indicated in the body of evidence gathered by independent inves-

A young, very astute, Dr. James McDonald UFO researcher and scientist.

tigators. The related question as to potential hazard is perhaps less clear. There are on record a number of cases (I would say something like a few dozen cases) wherein persons whose reliability does not seem to come into serious question have reported mild, or in a very few instances, substantial injury as the result of some action of an unidentified object. However, I know of only two cases for which I have done adequate personal investigation, in which I would feel obliged to describe the actions as "hostile." That number is so tiny compared with the total number of good UFO reports of which I have knowledge that I would not cite "hostility" as a general characteristic of UFO phenomena.

One may accidentally kick an anthill, killing many ants and destroying the ants' entrance, without any prior "hostility" towards the ants. To walk accidentally into a whirling airplane propeller is fatal, yet the aircraft held no "hostility" to the unfortunate victim. In the UFO phenomena, we seem to confront a very large range of unexplained, unconventional phenomena and if among them we discern occasional instances of hazard it would be premature to adjudge hostility. Yet, as long as we remain so abysmally ignorant of the over-all nature of the UFO problem, it seems prudent to make all such judgments tentative. If UFOs are of extraterrestrial origin, we shall need to know far more than we now know before sound conclusions can be reached as to hazard-and-hostility matters. For this reason alone, I believe it to be urgently important to accelerate serious studies of UFOs.

In the remainder of this section, I shall briefly cite a number of types of cases that bear on questions of hazard:

1. Car-stopping Cases:

In a two-hour period near midnight, November 2-3, 1957, nine different vehicles all exhibited ignition failures, and many suffered headlight failures as objects described as about 100-200 ft. long, glowing with a general reddish or bluish glow, were encountered on roads in the vicinity of the small community of Levelland, Texas. This series of incidents became national headline news until officially explained in terms of ball lightning and wet ignitions. However, on checking weather data, I found that there were no thunderstorms anywhere close to Levelland that night, and there was no rain capable of wetting ignitions. Although I have not located any of the drivers involved, I have interviewed Sheriff Weir Clem of Levelland and a Levelland newspaperman, both of whom investigated the incidents that night. They confirmed the complete absence of rain or lightning activity. The incidents cannot be regarded as explained.

This class of UFO effect is by no means rare. In France in the 1954 wave of UFO sightings, Michel has described many such cases involving ignition-failure in motorbikes, cars, etc. Similar instances were encountered in my checks on Australian UFO cases. There are probably of the order of a hundred cases on record. In only a very few cases has there been any permanent damage to the vehicle's electrical system. In the Levelland case, for example, as soon as the luminous object receded from a given disturbed vehicle, its lights came back on automatically (in instances where the switches had been left on), and the engines were immediately re-startable. The latter point in itself makes the "wet ignition" explanation unreasonable, of course.

It is unclear how such effects might be produced. One suggestion that has been made as to ignition-failure is that very strong magnetic fields might so saturate the iron core of the coil that it would drive the operating point up onto the knee of the magnetization curve, so that the input magnetic oscillations would produce only very small output effects. Only a few oersteds would have to be produced right at the coil to accomplish this kind of effect, but when one back-calculates, allowing for shielding effects and typical distances, and assumes an inverse-third-power diple field, the requisite H-values within a few feet of the "UFO diple" end, to speak here somewhat loosely, come out in the mega-gauss range. Curiously, a number of other back-calculations of magnetic fields end up in this same range; but obviously terrestrial technologies would not easily yield such intensities. Clear evidence for residual magnetization that might be expected in the foregoing hypothesis does not exist, so far as I know. The actual mechanism may be quite unlike that mentioned.

How lights are extinguished is even less clear, although, in some vehicles, relays in the lighting circuits might be magnetically closed. The lights pose more mystery than the ignition. Such cases do not constitute very disturbing questions of hazard or hostility. One might argue that highway accidents could be caused by lighting and ignition failures; however, more serious highway-accident dangers are implicit in other UFO cases where no electrical disturbance was caused. Many motorists have reported nearly losing control of vehicles when UFOs have swooped down over them; this hazard is distinctly more evident than hazard from the car-stopping phenomenon. Indeed, the number of instances of what we might term "car-buzzing" instances that have involved road-accident hazards is large enough to be mildly disturbing, yet I know of no official recognition of this facet of the UFO problem either. An incident I learned of in Australia involved such fright on the part of the passengers of the "buzzed" vehicle that they jumped out of the car before it had come to a stop, and it went into a ditch. A similar instance occurred not long ago in the U.S. For reasons of space-limitations, I shall not cite other such cases, though it would not be difficult to assemble a list that would run to perhaps a few dozen.

2. Mild Radiation Exposure:

By "radiation" here, I do not mean exposure to radioactivity or to other nuclear radiations, but skin irritations comparable to sunburn, etc. I have interviewed a number of persons who have experienced skin-reddening from exposure to (visible) radiations near UFOs. Rene Gilham, of Merom, Indiana, watched a UFO hovering over his home-area on the evening of Nov. 6, 1957, and received mild skin-burns, for example. I found in speaking with him that the symptoms were gone in a matter of days, with no after-effects. The witnesses in a car-stopping incident at Loch Raven Dam, Md., on the night of Oct. 26, 1958, who were close to a brightly luminous, blimp-sized object after getting out of their stopped car, experienced skin-reddening for which they obtained medical attention. Without citing other such instances, I would say that these cases are not suggestive of any serious hazard, but they warrant scientific attention.

Dr. McDonald being interviewed by the press, circa 1969.

3. More Serious Physical Injuries:

James Flynn, of Ft. Myers, Fla., in a case that has been rather well checked by both APRO and NICAP investigators, reportedly suffered unusual injuries and physical effects when he sought to check what he had taken to be a malfunctioning test vehicle from Cape Canaveral that had come down in the Everglades, March 15, 1965. I have spoken with Flynn and others who know him and believe that his case deserved much more than the superficial official attention it received when he reported it to proper authorities. He was hospitalized for about a week, treated for a deep hemorrhage of one eye (without medical evidence of any blow), and suffered loss of all of the principal deep-tendon reflexes for a number of days, according to his physician's statement, published by APRO.

An instance of more than mere skin-reddening, associated with direct contact with a landed unidentified object, reportedly occurred in Hamilton, Ontario, March 29, 1966. Charles Cozens, then age 13, stated to police and to reporters (and recounted to me in a telephone interview with him and his father) that he had seen two rather small whitish, luminous objects come down in an open field in Hamilton that evening- He moved towards them out of curiosity, and states that he finally moved right up beside them, and touched the surface of one of them to see what it felt like. It was not hot, and seemed unusually smooth. One of the two small (8 ft. by 4 ft. plan form, 3-4 feet high) bun-shaped objects had a projection on one end that the boy thought might have been some kind of antenna, so he touched it, only to have his hand flung back as a spark shot out from the end of the projection

DR. JAMES E.
MCDONALD'S
FIGHT FOR
UFO SCIENCE

Firestorm

Foreword by
Dr. Jacques Vallée

Ann Druffel

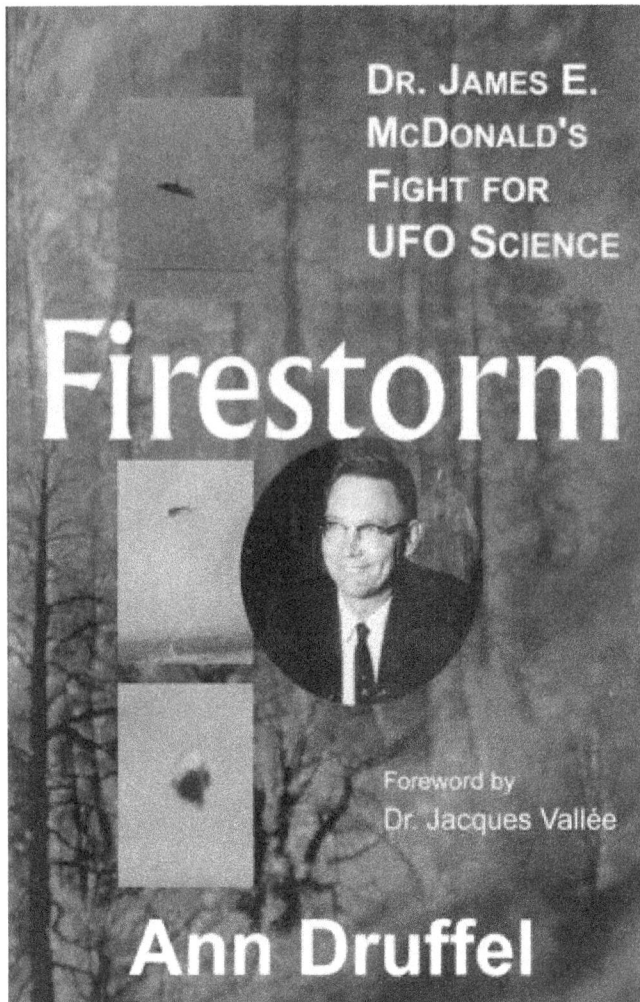

Those seeking to do further research on the work of Dr. McDonald will find this book by Ann Druffel to be very useful.

UFO Buff Found Dead Near Tucson

TUCSON (AP) — Dr. James E. McDonald, well known for his controversial stands on unidentified flying objects and the supersonic transport planes, is dead at age 51.

Dr. McDonald, a University of Arizona physics professor, was found dead in the desert north of here. Pima County sheriff's deputies said he was the victim of an apparent suicide.

McDonald was an outspoken critic of an Air Force report that claimed almost all reports of UFOs could be explained as natural phenomena or government projects. He maintained that some could be from other inhabited worlds.

into the air.

He ran, thinking first to go to a nearby police substation. But, on looking over his shoulder after getting to the edge of the field and seeing no objects there, he decided the police might not believe him and ran to his home. His parents, after discussing the incident at some length with the frightened boy, notified police, which is how the incident became public knowledge. Two others in Hamilton saw that night seemingly similar objects, but airborne rather than on the ground. Cozens was treated for a burn or sear on the hand that had been in contact with the projection at the moment the spark was emitted. On questioning both the boy and his father, I was left with the impression that, despite the unusual nature of the report, it was described with both straightforwardness and concern and that it must be given serious consideration. Clearly one would prefer a number of adult witnesses to an individual boy; yet I believe the case will stand close scrutiny.

There are a few other such reports of moderate injury reportedly sustained in direct physical contact with landed aerial objects for which I do not yet feel satisfied with the available degree of authentication. It would be very desirable to conduct far more thorough investigations of some foreign cases of this type, to check the weight of the evidence involved. That only a very small number of such cases is on record should be emphasized.

4. Instances Suggesting Overt Hostility:

In my own investigative experience, I know of only two cases of injuries suffered under what might be describable as overt hostility, and for which present evidence argues authenticity. There are other reports on record that might be construed as overt hostility, but I cannot vouch for them in terms of my own personal investigations.

In Beallsville, Ohio, on the evening of March 19, 1968, a boy suffered moderate skin burns in an incident of puzzling nature. Gregory Wells had just stepped out of his grandmother's house to walk a few tens of yards to his parents' trailer when his grandmother and mother heard his screams, ran out and found him rolling on the ground, his jacket burning. After being treated at a nearby hospital, he described to parents, sheriff's deputies, and others what he had seen. Hovering over some trees across the highway from his location, he had seen an oval-shaped object with some lights on it. From a central area of the bottom, a tube-like appendage emerged, rotated around, and emitted a flash that coincided with ignition of his jacket. He had just turned away from it and so the burn was on the back of his upper arm. In the course of checking this case, I interviewed a number of persons in the Beallsville area, some of whom had seen a long cylindrical object moving at very low altitude in the vicinity of the Wells' property that night. There is much more detail than can be recapitulated here. My conversations with persons who know the boy, including his teacher, suggest no reason to discount the story, despite its unusual content.

After checking the Beallsville incident, I checked another report in which burn-injuries of a more serious nature were sustained in a context even more strongly indicative of overt hostility. I prefer not to give names and explicit citation of details here, but I remark that there appears to me, on the basis of my present information and five interviews with persons involved, to be basis for accepting the incident as real. Partly because of its unparalleled nature, and partly because some of the evidence is still conflicting, I shall omit details and state only that the case, taken together with other scattered reports of injuries in UFO encounters, warrants no panic response but does warrant far more thorough investigation than any that has been conducted to date.

5. UFOs and Other Electromagnetic Disturbances:

UFO HOSTILITIES AND THE EVIL ALIEN AGENDA

There are so many instances in which close-passage of an unidentified flying object led to radio and television disturbance that this particular mode of electromagnetic effect of UFOs seems incontrovertible. One would require nothing more than broad-spectrum electromagnetic noise to account for these instances, of course.

There is a much smaller number of instances, some of which I have checked, in which power has failed only within an individual home coincident with nearby passage of a UFO. Magnetic saturation of the core of a transformer might conceivably account for this phenomenon.

Then there are scattered instances in which substantial power distribution systems have failed at or very near the time of observation of aerial phenomena similar, broadly speaking, to one or another UFO phenomenon. I have personally checked on several such instances and am satisfied that the coincidence of UFO observation and power outage did at least occur. Whether there is a causal connection here, and in which direction it may run, remains quite uncertain. Even during the large Northeast blackout, November 9, 1965, there were many UFO observations, several of which I have personally checked. I have inquired at the Federal Power Commission to secure data that might illuminate the basic question of whether these are merely fortuitous, but the data available are inadequate to permit any definite conclusions. In other parts of the world, there have also been reports of system outages coincident with UFO sightings. Again, the evidence is quite unclear as to causal relations.

There is perhaps enough evidence pointing towards strong magnetic fields around at least some UFOs that one might hypothesize a mechanism whereby a UFO might inadvertently trigger a power outage. Perhaps a UFO, with an accompanying strong magnetic field, might pass at high speed across the conductors of a transmission line, induce asymmetric current surges of high transient intensity, and thereby trip circuit breakers and similar surge-protectors in such a way as to initiate the outage. There are some difficulties with that hypothesis, of course; but it could conceivably bear some relation to what has reportedly occurred in some instances.

I believe that the evidence is uncertain enough that one can only urge that competent scientists and engineers armed both with substantial information on UFO phenomena and with relevant information on power-system electrical engineering, ought to be taking a very close look at this problem. I am unaware of any adequate study of this potentially important problem. Note that a problem, a hazard, could exist in this context without anything warranting the label of hostility.

"Four Star" Hostile Encounters

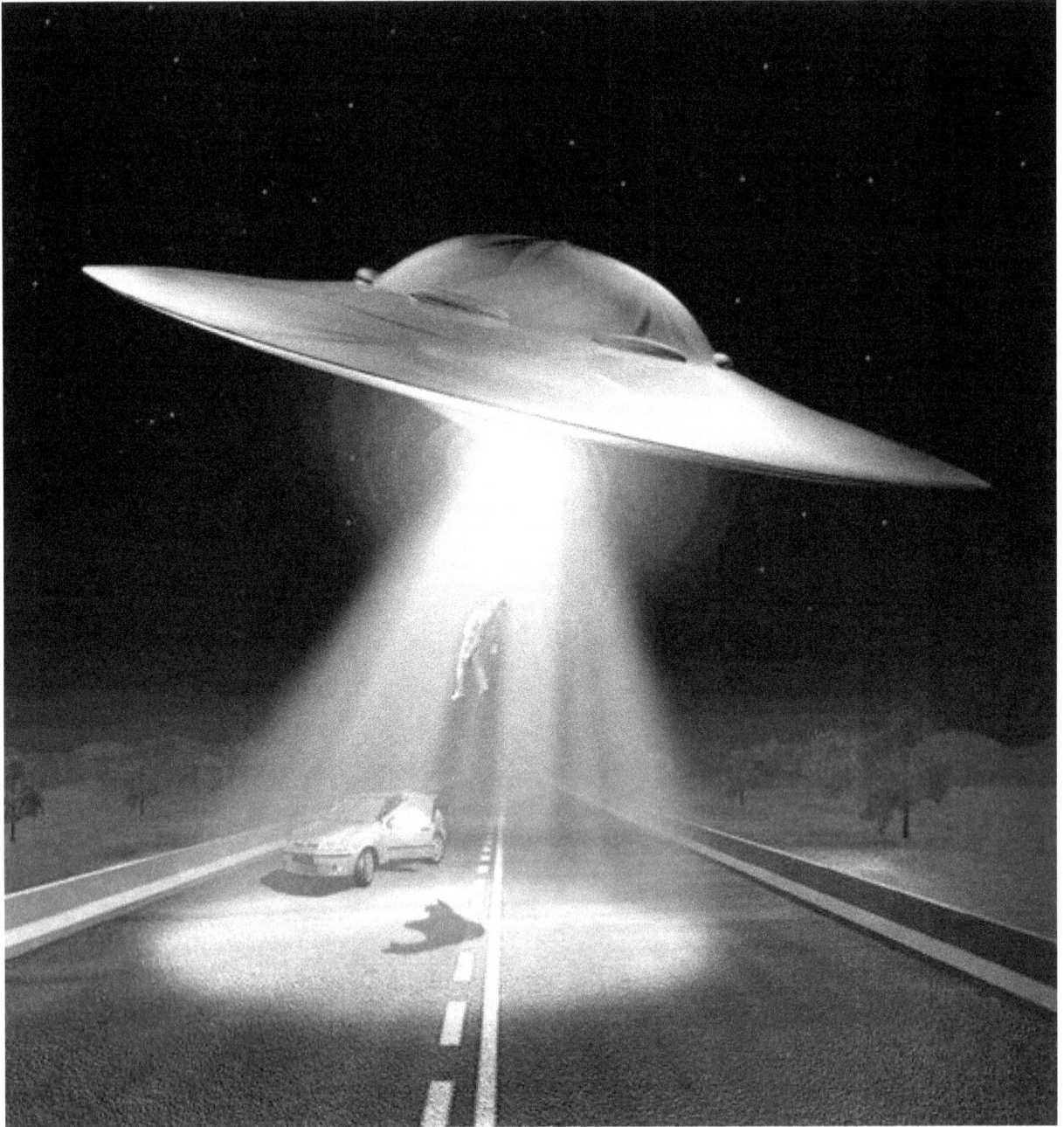

Chapter 03

THE HOSTILE UFO UNIVERSE OF GEORGE D. FAWCETT
Appraised and Abridged by Timothy Green Beckley

One of my all-time favorite songs is "Juke Box Hero" by Foreigner. It's a tune from the Eighties about a star struck kid who hears a song standing out in the rain with his ears pressed to the building's auditorium wall because it's a sold out concert and he can't get inside. The lyrics tell the heroic story:

Standing in the rain, with his head hung low

Couldn't get a ticket, it was a sold out show

Heard the roar of the crowd, he could picture the scene

Put his ear to the wall, then like a distant scream

He heard one guitar, just blew him away

He saw stars in his eyes, and the very next day

Bought a beat up six string, in a secondhand store

Didn't know how to play it, but he knew for sure

That one guitar, felt good in his hands,

Didn't take long to understand

Just one guitar, slung way down low

Was a one way ticket, only one way to go

So he started rockin', ain't never gonna stop

Gotta keep on rockin', someday gonna make it to the top

As a teenager I had my own "juke box hero," but it wasn't a rock band, it was a serious, highly trumpeted UFOlogist by the name of George D. Fawcett who had a huge flying saucer photo collection and lived about 80 miles away in another state. But I just had to meet him, even if I had to beg, borrow or steal a ride to Easton, PA. I had read a couple of articles by Fawcett (whose full time career was working as a YMCA Director for over 40 years) in the pulp saucer magazines of the day and couldn't believe that here was a researcher who was up on the entire topic of UFOs. He knew all the important cases so well, and devoted every spare hour to cataloging and listing some of the most impressive, unexplained aerial events that had occurred from the end of World War II up until he took it upon himself to type up all the information he had retrieved in a coherent fashion and present it to his fellow researchers.

I knew I wanted to follow in Fawcett's footsteps and once I "started rockin'" I was "never gonna stop" and someday I was "gonna make it to the top" and present my own unique opinions and exploits in this crazy universe of men in black stories, alien abductions, close encounters and probably most important CASES WHERE UFOS CAUSED UNDUE TRAUMA, HARM AND HARDSHIPS to those who came in contact or close proximity with them.

To set the record straight, GF was born July 21, 1929 (we are both Cancers) in Mount Airy NC. He served with the U.S. Army in the Panama Canal Zone, and somewhere in between acted in an editorial capacity for several newspapers. He was a Field Investigator for MUFON and worked with the late Dr. J Allen Hynek on several important UFO cases.

As we visited in his home, George opened up scrapbook after scrapbook to show me that he had been collecting clippings on the subject starting with World War II and the sighting of the Foo Fighters. Said Fawcett, "I noticed during our bombing raids over in Germany and Japan later that our pilots were reporting head-on crashes, near collisions, wing-tip formations, with what the Associated Press at the time called 'shining silver balls in the sky.' Later LATEST NAZI DEVICES, was the headline. And that caught my eye on December 28, 1944." GF was just 15 and already he knew there was something utterly devastating about these sightings that needed our immediate, and utmost, attention!

In February 1961 Fawcett stunned thousands when Ray Palmer published a controversial article titled FLYING SAUCERS ARE HOSTILE which GF had penned for RP's "Flying Saucers From Other Worlds Magazine." It caused quite a hubbub in the field and drew both criticism and praise from those who devoured its content. George told me years later that he was undecided as to whether or not he should publish his findings because he realized they would be most unpopular with a large segment of the audience who were more interested in sweeping the true nature of the phenomena under the proverbial rug and would have to put aside their personal view that the saucers were piloted by friendly Nordic-looking alien types who were supposedly visiting here to show us the evils of our ways, which were decidedly non-cosmic when it came to the proliferation of nuclear weapons and our continuing warrior-like nationalistic attitudes toward one another.

In the end, Fawcett knew he should come forward with his findings and let the chips fall where they may.

"I felt like we had both the good guys in the white hats and the bad guys in the dark hats, whereas we had some that were beneficial and we also had some that were hostile and (possibly) demonic."

FLYING SAUCERS ARE SERIOUS BUSINESS
By George Fawcett

(Published with the specific request of the author that no editing or cutting be done; the editors guarantee that this is a word for word presentation of Mr. Fawcett's manuscript.)

In September 1947, the Air Force concluded that the flying saucers were real. In 1948, a top secret Air Force conclusion was that these "Unidentified Flying Objects" (Air Force name for "flying saucers") were spaceships. In 1950, Air Force Intelligence analysis showed that these UFOs demonstrated "controlled maneuvers." In 1953 however a secret Pentagon scientists' panel urged more proof, asking the Air Force to quadruple its UFO project and to give the American people all

The FLYING SAUCERS ARE HOSTILE!

By George D. Fawcett

(Published with the specific request of the author that no editing or cutting be done, the editors guarantee that this is a word for word presentation of Mr. Fawcett's manuscript.)

UFO HOSTILITIES AND THE EVIL ALIEN AGENDA

UFO information, including the Air Force's startling conclusions.

Since that date, through official censorship, ridicule and misinformation, the public has constantly been told that the "flying saucers" do not exist, despite the continued radar and visual sightings by qualified observers, UFO movies and photographic evidence, jet pursuits and the recovery of "fallen fragments."

But now the SITUATION has changed, for on December 24, 1959, in a TIG AF Brief issued by the Inspector General to Operations and Training Commands, the UFO's are called "SERIOUS BUSINESS" after almost 12 years of telling the public repeatedly that there were no such things as flying saucers. The press and news media for the most part have mirrored the publicly expressed opinions of the Air Force over the same period, but now were quick to see the "sudden change" in the Air Force policy. On Capitol Hill, many Senators and Congressmen were beginning to ask for "open hearings" on the subject. Armed Forces Censorship Regulations JANAP 146-B, CIRVIS REPORTS, and AFR 200-2, which carry heavy penalties for servicemen reporting UFOs to the General Public, were keeping the military mum on the subject, at least until they were discharged or retired from active duty.

THE AMERICAN LEGION MAGAZINE   December, 1945

The Foo Fighter
MYSTERY

In the meantime, the CIA carried on its censorship of the strange flying objects for the Air Force. The FBI, the Armed Services, the Civilian Defense Directors, the Moon-watch Teams and the Ground Observers Corp have all been asked to relay UFO sightings to the Air Force for urgent action.

In JANAP 146-B, issued in March, 1954, it was interesting to note that Unidentified Flying Objects were listed among aircraft, missiles, submarines and surface vessels and should be reported if engaged in suspicious activity or observed in any unusual location or course.

These many strange "whatniks" were in the skies long before the "Sputniks

and Explorers" of today. And long before the Russians. Canadians, Norwegian, French and American began to develop circular craft using a new "air cushion" principle, the "flying saucers" had already demonstrated the superior speeds and maneuvers over any manmade aircraft or missiles of this Earth. The true "flying saucers," I decided in 1952, were real, ageless, worldwide, appearing in periodic cycles of time, with hundreds of landings recorded in history and having an outer space origin.

And even since the advent of the so-called "Space Age" in 1957, I have seen no reasons to change my former beliefs, except with one new addition. It is my present belief, in keeping with the Air Force announcement that the UFOs are SERIOUS BUSINESS, that they have a right to be, because the FLYING SAUCERS are HOSTILE. The FEAR of PANIC is the chief reason behind the present international censorship and conspiracy on the subject. In addition, these objects are a threat to the defense of our country and others of the world and these "Unidentified Flying Objects" have already contributed to the technical and scientific knowledge of the world in the present and future SPACE RACE. That is why "flying saucers" are being censored, chased, and imitated and investigated by many of the major countries of the world at the present time. I honestly believe that the people of the world have a right to know about the dangers that they must face and are being educated, but too slowly, to the fact that someone, somewhere many years ago accomplished SPACE TRAVEL, while we all were in the stage of just talking about it and perhaps ignorantly joking about the subject. Now for the STORY behind the HOSTILE FLYING SAUCERS . . . and let the readers draw their own conclusions!!

In recent months, the U-2 Plane Incident has died down somewhat, but instead much has been made of seven or eight "Mystery Moons," in orbit around the Earth. Both radar screens and professional and amateur astronomers around the globe have reported picking them up on the screens and observing them with telescopes. With some of them, neither the Americans nor the Russians are willing to admit that they belong to either of them. Right away this recalled to my mind a statement by General Douglas McArthur in 1955, that "the nations of the world would have to unite against attack by people from other planets. For the next war will be an interplanetary one!"

Without further delay, here is the up-to-date story of the ***HOSTILE FLYING SAUCERS***!

1944 - Army Pilot Charles F. Lane reported that his plane motors stopped and the instruments froze when he was approached by one UFO over the Himalaya Mountains in India.

August 28, 1945 -Passenger Leonard J. Stringfield reported that his C-46 developed engine trouble when approached by three luminous blobs of light near

UFO HOSTILITIES AND THE EVIL ALIEN AGENDA

Iwo Jima.

December 5, 1945 - Five TBM Avenger Bombers and One PBM Martin Mariner with a total of 27 men disappeared without a single trace during a training mission, in good weather and with good radio contact. UFOs were being reported by the news during that period of time, near Fort Lauderdale, Florida.

June 21, 1947 - Lt. Frank M. Brown and Captain William Davidson of Air Force Intelligence died when their plane crashed near Tacoma, Washington State. Their mission: bringing back UFO fragments that crashed to earth at Tacoma.

July 23, 1947 -Pilot John H. Janseen reported that his plane motors conked off, but his plane remained suspended in the air, during a close approach of 2 UFOs over Morristown, New Jersey. Janseen reported an electric-like pricking sensation during his hair raising experience.

September, 1947 - ATIC conclusion that the flying saucers were real.

January 7, 1948 - Captain Thomas F. Mantell, Jr. killed while chasing a "flying disk" over Godman Air Force base in Kentucky. His F-51 fighter plane's wreckage was found scattered over a wide area.

July, 1948 - Captain C. S. Chiles and Co-Captain John B. Whitted reported that a cigar-shaped UFO almost collided with their Eastern Airlines DC-3, that it rocked their plane over Montgomery, Alabama.

1948 - A top secret ATIC conclusion that the UFOs were spaceships.

October 12, 1949 - "Captured Saucers" first reported in New Mexico.

December, 1949 -Air Force Jets Fired on UFO over New Jersey.

December 39, 1949 - Major Jeremiah Boggs. Air Force Intelligence said that some pilots had fired at UFOs to bring them down for closer examination.

March 9, 1950 - "Flying Disk" Reported to Have Landed at Mexico City, Mexico.

April 10, 1950 - At Amarillo, Texas, David Lightfoot, 12, reported that he had touched a landed "flying saucer," but that the object released a gas or spray of something when it took off, which turned his arms and face bright red, causing welts. A younger boy confirmed his story.

April 20, 1950 - At Lufkin, Texas, pharmacist Jack Robertson reported that a globe-like UFO, which emitted a red glow, hovered above his car about 20 feet, then made a "whooshing sound" and threw off sparks that burned his face.

1950 - Pilot Chet Swital reported that his plane and another plane were circled by UFOs, which paralyzed the plane motors and instruments over the Muroc Air Force Base in California and a New Mexico Air Force Base.

1950 - An Air Force Intelligence Analysis Showing "Controlled UFO Maneuvers." That the "Flying Saucers" Were Real and Spaceships.

July, 1952 - In Atlanta, Georgia, pilot Fred Reagan's Piper Cub Plane Collided with Bright, Dazzling Lozenge Shaped UFO. Reagan died in May, 1953, because of a degeneration of brain tissue due to extreme atomic radiation, for which authorities there were unable to offer an explanation.

July 24-25, 1952 - An Italian named Carlos was fired at by hovering UFO, which used some sort of green ray near the Sercho River in Italy.

July 26, 1952 - The U.S. Air Force Issued a "Shoot 'Em Down" Order to Pilots Encountering UFOs. This was mentioned in AF press release.

1952 - An Air Force F-8Q Jet Pilot Opened Fire At a Hovering UFO in Ohio.

1952 - In Dublin, Ireland, a 10-inch "red hot" disk landed and burned a child there.

August 19, 1952 - At West Palm Beach, Florida, a Scoutmaster named J. D. "Sonny" Desvergers reported his encounter with a hovering UFO, which shot a "Ball of Fire" at him, burning his arms, face and head and rendering him unconscious. Charred roots indicated possible electric induction.

September 13, 1952 - At Sutton, West Virginia, Mrs. Kathleen May, a National Guardsman and 5 other boys were investigating a landed globe-like UFO object on a hillside.

All of the witnesses were repelled from the hilltop by a noxious sickening odor of the landed object. They all suffered choking spasms and vomiting, because of the incident.

September 13, 1952 - At Frametown, West Virginia, Mr. and Mrs. George Snitowsky and baby were gassed and felt electric-like shocks after their car stopped near a soft, violet-colored landed UFO. This happened on the same night of the Sutton incident at a spot not far from the same location.

September 14, 1952 - At Belle Glade, Florida employee Floyd Brown of the Everglades Experiment Station noticed his eyes and nostrils began to smart and burn, after a circling UFO passed several times over the station. It smelled like an acid, ammonia-like odor. The cows were badly frightened by the incident, and there was a 33% reduction of milk production on the following morning.

September 22, 1952 - At Centerville, Virginia, Mrs. E. L. Hazelwood reported that she watched several brightly illuminated UFOs and became ill for several days because of the odor given off by the strange objects.

October 23, 1955 - At Indianapolis, Indiana, farmer John Hobner was forced by a blinding bluish-white lighted UFO in a nearby field close to the highway to crash his car into a tree. Hobner suffered minor injuries.

December 4, 1952 - At Laredo AFB in Texas, a Lt. in the Air Force reported a near collision with his F-51 and a strange flying blue-lighted UFO. The pilot turned off his plane lights in order to avoid further pursuit by the UFO, and frightened he made a hasty landing.

UFO HOSTILITIES AND THE EVIL ALIEN AGENDA

January 29, 1953 - Merchant Lloyd C. Booth near Conway, South Carolina, reported that he shot at a flying egg-shaped UFO that hovered over his farm for 45 minutes. The bullet bounced back with a metallic thud and the next morning 20 of his livestock were found dead on his farm from "unknown causes."

January, 1953 - A Secret Report by a Scientists' Panel at the Pentagon, which urged (1) that the AF quadruple its UFO investigation project, and (2) that they give the American people all UFO information, including the AF conclusions. This in 1960 has yet to be done!!

October 19, 1953 - Near Philadelphia, Pennsylvania, a National Airlines DC-6 Pilot put his plane into a dive to avoid being rammed by a shiny disc-shaped UFO. Passengers were hospitalized.

October, 1953 - Near Washington, D.C., Captain J. L. Kidd put his American Airlines DC-6 passenger ship into a dive to avoid collision with a lighted UFO. Passengers were thrown into the aisles by the unexpected maneuver.

November 23, 1953 - Kinross Air Force Base, Michigan, Pilot Lt. Felix Moncla, Jr. and Radar Office Lt. R. R. Wilson and their F-89 plane was being tracked on radar when it merged on the screen with another UFO, 70 miles off Keweenaw Point.

The plane and its pilots never returned from the mission and up to this date, no trace has been found of either.

January-March, 1954 - Over the Tyrrhenian Sea, Near Malta, and in the Mediterranean Sea area, four RAF Planes and Pilots Disappeared. Wreckage of One Was Found, which indicated that an Explosion Was Followed by Fire and Intense heat. Flying saucer- and cigar-shaped UFOs were being reported in large numbers at that time in Italy, France, England, New Zealand and the United States.

January 4, 1954 - At Marigname Airport in France, Fragments Were Left by a reported cigar-shaped UFO with windows, which landed on the airport runway.

1954 - On Heligoland Island, off Germany, Scientists Were Reported Investigating a UFO which crashed to the earth there.

April 20, 1954 - A Part of a Hovering UFO that fell upon a farmhouse was reported under close investigation.

May, 1954 - An F-94 Jet Fighter Pursued a UFO above Utica, New York. As the gleaming pulsating object moved towards the F-94, a mysterious blast of heat filled his cockpit. Unable to breathe and feeling the plane about to melt, the pilot and radar officer bailed out. The jet itself plunged into the town of Walesville, New York, and killed 4 persons.

August, 1954 - Australian Sheep Grower W. C. Hall saw six petrol tank-shaped UFOs land on his ranch in North Queensland, Australia. His chickens, cattle and even the jack rabbits on his farm were affected by the oddly colored exhaust fumes of the UFOs, which he believed brought about a change in the genes of animal life on his ranch, due to Atomic Radiation, as various freaks were born afterwards.

August 27, 1954 - At Woodside, California, fragments from a hovering UFO landed on the open highway. Now under investigation by authorities.

October 5, 1954 - Beaumont, near Paris, France, as a Luminous UFO approached to within 150 yards of eyewitnesses they reported that they felt a strange sensation "as if nailed to the ground." In addition to the temporary paralysis, the witnesses noticed the UFO gave off a peculiar smell.

October 21, 1954 - A silvery UFO object hovered above the Rubber Plant near Pozzuoli, Italy. As it rose vertically with a piercing whistle the observers noticed that a Pekinese dog watching the UFO barked and fell dead.

October 28, 1954 - Near Monza, Italy, a large group of townspeople watched a "silvery disc-shaped UFO" land on a sports field. Several hundred witnessed the event. For several days after the object landed and took off, many of the population of eyewitnesses still had red eyes as aftereffects of the blinding light.

April 8, 1955 - Air Force Jets fired at a UFO circling a balloon, but destroyed the balloon and missed the UFO, over Rockfield, Illinois. This was a story about the fish, er, I mean, the "saucer" that got away,

August 6, 1955 - Cincinnati, Ohio, a Witness, Not Wanting Name Revealed, Reported he saw an oval-shaped UFO resting near his driveway. While watching the pulsating of the object, his eyes suffered severe irritation and he had to consult the doctor for relief.

1955 - Government Investigators announced that they had found an Increase in Background Radiation in the Skies after UFO sightings had been reported in the vicinity.

August 2, 1956 - In Amarillo, Texas, pilot J. G. Kirby photographed a UFO with a long exhaust trail of some sort near it. Hensley Air Force Base Officers told Kirby that the Glow Was Caused by Radiation Vapor. He was instructed to keep silent about his photograph and his sighting by the Air Force, which he did for over a year, before breaking secrecy.

1956 - Above the Atlantic Ocean, Off Gander, Newfoundland, a Navy R7V-2 Transport, a Four Engine Super Constellation, Had to Dive to Avoid a Collision Course with a giant disc-shaped UFO "Like Gigantic Dishes On top of one another," which appeared to be approximately 350 and 400 Feet in Diameter, with Lighted Edges. Confirmed by Ground Radar.

September, 1956 - At Great Falls, Montana, the Air Force Base sent 3 jets to investigate hovering UFO over Missoula, Montana. As they approached the strange flying object, their plane instruments went crazy (magnetic field?) and became useless. One jet went into a dive for 15,000 feet. The nose and wing of his plane were slightly damaged.

UFO HOSTILITIES AND THE EVIL ALIEN AGENDA

October 2, 1956 – Night watchman Harry J. Sturdevant of the Hebeert Elkins Construction Firm of Trenton, New Jersey, reported while on duty he saw a red-lighted cigar shaped UFO near the Delaware River. The smell of the object made him nauseated and he couldn't swallow. He collapsed in pain, but managed to crawl to his parked car. Six weeks later his sense of taste and touch were impaired and he had only been able to work a few days. Leonard B. Willits, a New Jersey State Workmen's Compensation referee awarded Medical Compensation to the night watchman, who had encountered a UFO, which made him ill and affected his senses of smell, taste, and touch.

March 10, 1957 - Near Puerto Rico, Captain Matthew A. Van Winkle of Pan American Airlines was forced off his course by a "green circular object UFO," which was witnessed by other pilots in the vicinity also. Due to his sudden maneuver, some of the plane passengers were hospitalized.

July 17, 1957 - An American Airlines DC-6 Aircoach had to swerve to avoid collision with a UFO at Salt Flats, Texas. Some of its passengers were hospitalized, among the 85 on board the ship.

July 24, 1957 - Near Amarillo, Texas, a TWA Constellation Airliner had to dive to avoid collision with a moving UFO and some of its passengers were treated for injuries.

July 28, 1957 - At Knoxville, Tennessee, an American Airlines plane was struck by a "Ball of Fire UFO," which knocked a small puncture in the tail of the plane.

August 14, 1957 - At Joinville, Brazil, Commander Jorge Campos Araujo of a VARIG Airlines C-47, the Cargo Ship PP-UCC reported that the engines of the Airliner began acting up, coughing and missing and that all the cabin lights dimmed and also went dead after the near approach of a strange luminous UFO, which had a dome on the top of it.

August 21, 1957 - Parana City, Argentina, fragments from a UFO landed on the ground.

September 7, 1957 - Near Ubatuba, Brazil (located in Sao Paulo State), an exploding UFO showers fragments to the beach, witnessed by many.

November 4, 1957 - At Orcgrande, New Mexico, Many car motors stopped and headlights went out when a "Luminous Egg-shaped UFO" hovered above the highway. Rocket Engineer James Stokes received burns on his face from the object that gave off a heat wave and created electrical interference with the so- called "hallucinated machines" on the ground.

November 4, 1957 - Brazilian Fortress, Itaipu, near Santos, Brazil, this incident tells of a Luminous Orange UFO with a humming sound and a wave of heat that overcame two sentinels on duty, rendering them unconscious and burning them on various places of the body. The UFO caused lights all over the fortress to collapse. Turrets, heavy cannons, and elevator electric systems also collapsed. Intercommunications were disrupted and alarms on electric clocks began to ring without reason.

November 4, 1957 - At Ararangua, Santa Catarina, Brazil, Captain De Beyssac of VARIG Airlines reported as he closed in on a large hovering red UFO that his ADF, his right generator, and his transmitter receiver all burned out at the same time.

November 6, 1957 - Near Merom, Indiana, ironworker Rene Gilham saw a brightly luminous hovering UFO and suffered burns on his face and eyes.

November 6, 1957 - Near Montville, Ohio, plasterer Olden Moore saw a hovering circular UFO and heard a "whirring sound" as an object landed in the field 500 feet away. A Geiger counter later showed a deadly radioactivity level existed after the sighting in the area.

November 8, 1957 - Pacific Ocean, Panamerican Stratocruiser "Romanee of the Skies" crashed in the Pacific Ocean. UFOs had been reported in the vicinity.

November 9, 1957 - In Hillsborough, California, fragments from a UFO landed, now under investigation.

November 19, 1957 - In Madison, Ohio, housewife Mrs. Leita Kuhn saw an acorn-shaped UFO above her garage. She noticed that she had to keep covering her face in order to view the strange object. As a result of the incident, she suffered body rash, failing of vision and burns on her arms and legs, similar to "Radiation Burns" noted her doctor.

November 13, 1957 - Crownsville, Maryland, a strange metallic UFO exploded over the area and left fragments on the ground, which are under investigation by the Air Force.

December 7, 1957 - Dallas, Texas, a man there is reported dying from "Radioactive Burns" received by a close approach of a luminous UFO.

December 15, 1957 - Langdon, North Dakota, UFO fragments found on the ground, now under investigation by the Air Force.

February 17, 1958 - Santa Fe, New Mexico, Mrs. Leroy Evans and Mrs. Fred McIntosh each suffered burns while watching a hovering UFO near the ground.

May 5, 1958 - Near Montevideo, Uruguay, pilot Ale Jo Rodriguez, flying in his Piper Aircraft, reported a bright silvery UFO, metallic top-shaped object with a slight vapor trail which hovered over Uruguay and created such intense heat that he was forced to open the windows and the plane door and remove his jacket.

October 26, 1958 - Above the Lock Raven Reservoir, Near Baltimore, Maryland, Mr. Philip Small and Mr. Alvin Cohen watched a hovering luminous egg-shaped UFO, felt a wave of heat, heard an explosion as the object left the area and suffered burns from their encounter. Noted electrical interference with car lights and motor.

December 10, 1958 - Rio Negro, Uruguay, persons watched a luminous egg-shaped UFO, felt heat coming from the object, and found fragments which fell to the ground during the incident.

December 16, 1958 - Tampa Bay, Florida, a B-47 Stratojet crashed and exploded, UFOs reported seen in the vicinity.

December 21, 1958 - Juyjuy, Argentina — UFO fragments land, under investigation.

1958 - Pacific Ocean, an Air Force pilot indicated that his plane was "shot at" in a UFO encounter, and crew opinions that the strange "bursts" or explosions near the transport were linked with previous disappearances of AF planes in this Pacific area. More on this case later!

February 22, 1959 – Utah, PFC Bernard G. Irwin of Ft. Bliss, Texas, reported the sighting of a "Ball of Light" Circular UFO in the skies. The object, whatever it was, rendered him unconscious for 24 hours.

April 1, 1959 - McChord Air Force Base, California, Air Force C-118 Transport Crashes, 3 Persons Killed, UFOs seen in vicinity.

April, 1959 - Seattle, Washington, Air Force acknowledges a "Frantic Radio Report" that "we have hit something or something has hit us," which preceded a mysterious crash of a C-118. No survivors. Eyewitnesses on ground reported sighting 3 or 4 parachute-like glowing UFOs following the plane prior to its crash.

September 7, 1959 - Wallingford, Kentucky, Fragments Fall to Ground From Hovering UFO. Under Investigation.

September 24, 1959 - Redmond, Oregon, a UFO with Tentacle-Like Projections, Glowing with Red and Greenish Colors, moved through "Controlled Maneuvers." Reported from the Ground observers, picked up on Air Force radar for over an hour. After the object moved away, an active search for the UFO was conducted by 7 jet fighters, a B-47 bomber, and a Tripacer, which later was asked to check the area for signs of "RADIOACTIVITY."

September 29, 1959 - Texas— A Braniff Airways Turbo-Prop Airliner crashed in Texas, claiming 34 lives. UFOs were reported in the vicinity.

October 19, 1959 - Near the "Big Marsh," outside of Poquoson, Virginia, two schoolboys, Mark Muza and Harold Moore, "shot at" a hovering UFO. The bullet struck a metallic surface and rebounded to the ground.

1959 - A Brazilian Air Force B-26 Commander asked permission to take off without lights on his bomber in order to avoid further pursuit by a strange UFO that had followed his plane for almost an hour before he reached the Airport.

December 24, 1959 - In private instructions briefing Air Force for Operations and Training Commands, Major General Richard E. O'Keefe, Acting Inspec-

UFO HOSTILITIES AND THE EVIL ALIEN AGENDA

tor General, recently warned that UFO sightings would increase. The TIG brief was headed "UFOs-SERIOUS BUSINESS." The AF he said was chiefly concerned with defense and technical aspects of the phenomena. Air Force Investigators should use binoculars, cameras, Geiger counters, magnifying glasses and containers for any "fragments or ground samples" recovered. This order came as the Air Force was again telling the public that the "flying saucers" just don't exist.

March 2, 1960 - Near Labuttendorf, Austria, newspaper man Edgar Schedelbauer of Vienna, Austria, for newspaper "WIENER MONTAG," suffered "great red spots" on his face and hands after his close encounter with a milky-white spider-shaped hovering UFO. It made a low humming sound and then a jet-like roar as it hovered about 50 feet above a road. Photographer took pictures of the object. Now under investigation.

March 12, 1960 - The AERIAL PHENOMENA RESEARCH ORGANIZATION at Alamogordo, New Mexico, announced that its organization possessed fragments of an "Extraterrestrial Flying Saucer" that fell from an exploding UFO in Brazil in 1957 and photographs of an "Extraterrestrial Flying Saucer" taken at Trinidad Island in 1958.

May, 1960 - The NATIONAL INVESTIGATIONS COMMITTEE ON AERIAL PHENOMENA in Washington, D.C. announced that the "UFOs or FLYING SAUCERS" are unknown devices which have been reported by scores of reliable observers, and are "intelligently controlled machines from outer space." Vice Admiral R.H. Hillenkoetter, former head of the Central Intelligence Agency, stated that the "unknown objects are operating under intelligent control." He added "I know that neither Russia nor this country has anything even approaching such high speeds and maneuvers." Speaking as a NICAP Board Member, Admiral Hillenkoetter said it was high time the secrecy was ended.

June, 1960 - The U.S. Air Force is currently busy investigating films and movies of "UFOs and FLYING SAUCERS" taken over Michigan in February, over Austria and Iowa in March, over Wisconsin in April and over New York in August, all of this year 1960. They were photographed by a newspaper man, a restaurant owner, a pilot, a housewife and an astronomer. (A point of information.) I have listed over 171 different photographs or movies taken worldwide since 1905 in my own private "SAUCERIANA COLLECTION."

September 6, 1960 - Near Hartford, Connecticut, fragments Fell to Ground from Bluish-Green speeding UFO and landed with a hissing sound. They are now under investigation by the National Aeronautics and Space Administration in Washington, D.C.

This article has been printed in hopes that it will be given consideration as an accurate research paper into acts of UFO and FLYING SAUCER hostility. It is far from being a complete list, as time does not permit me to do a better job

UFO HOSTILITIES AND THE EVIL ALIEN AGENDA

It should serve as a warning that many of the UFOs are dangerous and hostile. So before you go running to meet those so-called peaceful visitors from Venus, should you encounter a UFO, you should think twice, because they may not be peaceful or from Venus!

I became interested in these objects in 1947, witnessed them in 1951 and 1954, was convinced of their reality in 1952, ran into Army censorship in Panama in 1955, concluded with their hostility in 1960. 1961 and the years that lie ahead should bring us many other "visitors from space," even as we cross over the threshold into visiting space ourselves.

My thanks for information for this article go out to FATE, TRUE, and LIFE magazines, the Air Force, and IFSB, FSI, CSI. NICAP, and APRO, all of which are made up of civilian and military researchers and investigators. As of late, my thanks go to APRO, the Woonsocket, R I. YMCA "Project Saucer" Committee, and the Massachusetts and Rhode Island TWO-STATE UFO STUDY GROUP.

POSTSCRIPT

I hope that the enclosed items regarding UFO hostility will prove to be food for thought for those thousands, perhaps millions, who are searching for the many answers to the riddle of the "flying saucers." I hope that the 90 incidents regarding the dangerous and hostile UFOs will help others in preparing a defense against attack by such objects, and will provide information, which will help mankind's efforts in achieving space travel for himself and for peaceful purposes only: to the benefit of mankind. For the thousands of other groups and individuals

who made this pamphlet possible. I also extend my thanks and appreciation.

Since completing this pamphlet, the following case has been reported to me. November 6, 1957 (note causes for this date alone already mentioned) near San Antonio, Texas, civilian cook Lon Yarborough at Lackland Air Force Base reported electric interference with his car motor and headlights when approached by bright egg-shaped UFO which landed on ground. Yarborough heard a "whistling sound" and felt a blast of hot air, but suffered no burns.

STATEMENTS

Among the many eyewitnesses to the phenomena are such persons as Claire Booth Luce, U.S. Lady Ambassador; Navy Secretary Dan Kimball, Astronomers Clyde W. Tombaugh, American who discovered the Planet PLUTO, and British Lunar Expert H. P. Wilkins.

AIR CHIEF MARSHALL Lord Dowding (England) - "The saucers exist and they are not manufactured on the Earth."

DR. HERMAN OBERTH (Germany) - V-2 Rocket Fame - "The saucers are 'Vikings' from another Planetary System."

LT. COLONEL JOHN O'MARA (United States) - Admitted in Effect "The saucers are interplanetary, and we also have saucer-shaped crafts ourselves."

WILBUR B. SMITH (Canada) Head of the Project Magnet-Flying Saucer Detector Station, "I'm convinced they're real, that they're machines of some kind."

******

"From the weight of evidence, I believe THE FLYING SAUCERS ARE HOS-

UFO HOSTILITIES AND THE EVIL ALIEN AGENDA

TILE!"

That is the title of a lengthy report prepared by George D. Fawcett, based on his more than ten years of gathering UFO evidence. We don't quite agree with George's title, which may give a conclusion stronger than his evidence justifies; but we want to give him a stage where this little-circulated research report can reach the many readers of FLYING SAUCERS for their consideration.

The incident above, which took place at the Brazilian Army Fortress Itaipu, along the coast of Sao Paulo state, near Santos, on Nov. 4, 1957, is only one of 90 reports he lists as involving hostility on the part of saucers.

While other researchers have published magazines, Fawcett has devoted his time to collecting and organizing UFO information into useful reference units. We first heard of him when we joined Albert K. Bender's ill-fated organization, The International Flying Saucer Bureau, back in 1952. At that time, Fawcett was on Bender's International Council. Since the closing of that important organization, Fawcett has carried on his research independently, as he did even then, though remaining constantly in touch with the writer.

WERE THESE ACCIDENTS?

It may be unfortunate that Fawcett does not editorialize in his reports, and that the reader, as a consequence, may not reach a valid conclusion.

Were these near-misses and occasional injuries the result of deliberate attacks by saucers — or were they only accidents, without bad intentions by the saucerians?

With the superior technology demonstrated in these encounters, the saucers surely could have inflicted more destruction than actually occurred. Yet, with such a superior technology, and the advanced intelligence we tend to connect with such, could not the operators of such craft have avoided the damaging contacts with terrestrials?

In the case of the "attack" on the fortress, the business of the saucer must have been something other than aggression, else presumably it would have wreaked more havoc than it elected to do. Maybe that business was simple observation, with the side effects unknown to the saucerians. Or maybe they just didn't care.

Finally, in going over some of Fawcett's accounts of contact in the sky, we are led to wonder how many reports involving mysterious plane crashes which left no survivors (or those which were hushed up) were of necessity not available to the compiler. Maybe such reports, were they available, could shed some light on the perplexing question of why plane accidents definitely occur in cycles.

But what of people on the ground who have complained of being variously frightened, irradiated, sickened, or even burned by saucers? Fawcett includes a bundle of these, too.

UFO HOSTILITIES AND THE EVIL ALIEN AGENDA

KEEP LOOKING UP

Although Fawcett's reports definitely show that people have been injured and killed as a result of encounters with UFOs, we don't believe that a concrete case for overt hostility has been built up. As for near-collisions with airliners, our own planes have near-misses almost daily and sometimes crash into each other, though one would like to believe the saucer people have better radar systems.

Close ground encounters seem to follow a pattern in that these contacts often cause redness of the skin, burns and other irritations, somewhat like, but not necessarily identical to radiation burns. In very rare cases have terrestrials complained of being "rayed" or otherwise attacked; instead it seems the injuries have been caused unintentionally by saucerians who either didn't know or didn't care whether such contacts could be damaging.

Some points are definitely proved by the Fawcett report, however. Whether or not intentional weapons, certain devices, either carried by the UFOs or a part of their propulsion or communication systems, appear to be powerful indeed, and far more effective than any weapon we have invented. If the saucers wished to attack on a large scale, there seems to be very little we could do about it.

Just as the existence of saucers has been proved by an overwhelming number of sightings by reliable observers, we believe this report offers indisputable proof that they also have the technological capability of hostile acts, and possibly the conquering of the earth should they so deign. We believe, however, their moral culpability for such acts is yet to be proven.

Some of these reports lead us to a conclusion different from those advanced by both the "hostile saucers" and the "spacemen are here to help us" schools. Maybe the saucerians are here on some strange business we haven't been able to figure out, and, as far as humans go, they just don't give a damn about us. If we get in their way, too bad; if we stay out of their way, well and good.

But whether or not they have any hostile design (more than 10 years of mass appearances have to the contrary tended to indicate that they have not), their capability of such behooves us, I believe, to keep our eyes continually on the heavens.

ADDED NOTES BY TIM BECKLEY

George passed away peacefully on Sunday, January 20, 2013. I had continued corresponding with him over the years while editing several newsstand periodicals such as UFO UNIVERSE and UFO FILES. In the Premier Edition of the latter publication, we published an extensive twelve page interview with GF. "UFOs From World War II to the New Millennium" was as in-depth as you could possibly go into the psychology of an individual so deeply rooted in investigating the UFO phenomena. I believe the interview was done at one of the annual MUFON conferences, so when a couple of years ago George and I ran into each other under the most bizarre of circumstances I assumed – incorrectly – that he was on his way to attend and possibly speak at one of the group's high profile symposiums.

The MUFON conference was being held in Denver and I was on my way there with a stopover in Saint Louis. With an hour or so to kill I was wondering around the airport with nothing much to do when I decided to stop into a nearby

cantina for a cold one. There, sitting at one of the tables near the bar, was someone I thought I recognized as George D. Fawcett.

It was, indeed, Mr. Fawcett, and I was delighted to meet up with him before going on to Denver, where no doubt we would have the opportunity to carry on more fully any conversation. .

"So I see you're on your way to MUFON. What are you speaking about this year?" I asked.

George just kind of looked oddly at me – "I'm not on my way to MUFON. I'm on my way to Roswell (NM) to make arrangements to donate my 25 file cabinets full of photos and clippings and reports to the UFO Museum there as part of my legacy."

We both had a hardy chuckle over the fact that we had "randomly" bumped into each other totally by "chance" (no way!) at Saint Louis' Lambaart International Airport before flying away to our desired destinations.

Never saw George after that. But anyone who has followed my research knows I place a great significance in the correlations between UFOs and synchronicities. Those who have their curiosity aroused should pick up from Amazon a copy of my relevant work, "The Matrix Control System of Philip K. Dick and The Paranormal Synchronicities of Timothy Green Beckley."

Those who want to peruse Fawcett's research can Google him by name or you can listen in on a conversation we had on our show "Exploring the Bizarre" that is archived on "Mr. UFO's Secret Files," our YouTube channel. The interview is with a close friend of Fawcett's. Kent Senter was co-director, along with GF, of the NC state chapter of MUFON which they started together and later encompassed The Center for UFO Research NC. Look for the episode titled "UFO Abduction and MIB in ole Dixie."

And, as George would have us do – watch out for those hostile UFOs. Man, they could be coming to get you!

UPDATE: PILOT DISAPPEARS IN UK – WERE UFOS INVOLVED?

A pilot in the United States Air Force, Captain William Schaffner, died a mysterious death while flying over the North Sea in Europe on September 8th, 1970. Stationed at the Royal Air Force base Binbrook in Lincolnshire, England, he was on duty when an unusual object was picked up on radar that evening. Schaffner decided to fly out in his Lightning XS894 plane and investigate.

Equipped with two air-to-air missiles, his aircraft took off from the Air Force base at 10:06 PM with the call sign Foxtrot 94. The object he was following was flying parallel to the coast and was being tracked by radar operators on the ground.

As he drew nearer to the object, Schaffner exclaimed: "Wait a second, it's turning… coming straight for me… am taking evasive action…" Shortly thereafter, all communication ceased. Captain William Schaffner was never heard from again.

MYSTERIOUS CIRCUMSTANCES

In the days following the incident, there was absolutely no sign of the pilot or the aircraft – it was as if they had been wiped from the face of the Earth.

That is, until three weeks later. The aircraft was then finally located by Navy divers and ultimately recovered from the seabed. It was brought back to RAF Binbrook for investigation. The canopy of the plane was closed but there was no sign of Schaffner. The aircraft itself had suffered very little damage.

FURTHER SIGHTINGS

The area in question had been a UFO hot spot throughout 1970-71, with sightings occurring all along the Lincolnshire coast, primarily around the town of Cleethorpes. On one particular day, a large UFO, estimated to be in excess of 180 feet, was seen hovering for hours at another nearby Air Force base.

To this day, nobody is sure what happened to Schaffner or his aircraft. In light of him being in pursuit of an unknown object, we can only surmise that some element relating to the chase caused his untimely death at just 29 years old.

Whether the craft itself was of extraterrestrial origin is up for debate, but the signs certainly point towards it. Considering the abundance of UFO sightings in the area at the time, along with some fairly detailed reports regarding the size and speed of various unknown objects in the skies at the time, the evidence seems to be in favor of it.

It is of course also possible that the UFO in question was intelligently and maliciously controlled, killing or abducting Schaffner as he approached it.

There have been some more mundane explanations for Schaffner's disappearance in the years since, but what we have read doesn't sound terribly convincing. We must also bear in mind that when dealing with a mysterious military death, there is often some type of cover-up involved. Either way, his case remains one of the more compelling, but also disturbing incidents we have come across.

UFO ON COLLISION COURSE IN SPAIN

November 11, 1979 - Manises UFO incident. A flight with 109 passengers heading to Las Palmas had to make an emergency landing because of a UFO. Pilot Francisco Javier Lerdo de Tejada noticed an object on a collision course with Flight JK-297. He changed altitude to avoid the lights coming head on with the plane, but the object followed about half a kilometer behind. The pilot decided to make an emergency landing at the Manises Airport. The object did not follow, but radar picked up the object and determined that it had a diameter of 200 meters.

Reports of this nature in the 1940s and 50s were abundant. Without a doubt pilots were ordered not to talk about such experiences. From time to time some reports did filter through – like the UFO that closed in on a jetfighter over Tehran in 1976 when the Shah was still in power. In the Cold War years there was the case of the American pilot Lieutenant Milton Torres, on duty over the UK, whose plane

'OBJECT' ALARMS HOCKLEY AREA; SOLON ASKS PENTAGON FOR REPORT

Levelland 'Flaming Thing' Brings World Knocking At City's Door

By BILL WILKERSON
Avalanche Staff Writer

LEVELLAND, Nov. 3 — While a curious world pounded at the door seeking information, Levelland residents today were pondering the strange, fiery "thing" seen by numerous people Saturday night as it flashed through the skies and landed on nearby highways.

By late today seven persons said they saw the object at close range, many others reported seeing it from a distance and reports still were coming in.

The object was described by most observers as being a 200-foot long, egg-shaped ball of fire that moved at great speed.

It was reportedly sighted during a 2½-hour period at points 4 to 20 miles from Levelland to the east, north and west.

Officer A. J. Fowler of the Levelland police department, who received most of the telephone reports, quickly alerted area officers when the first report was received at 10:30 p.m. Saturday.

Fowler called Lubbock police and Reese AFB was notified in the belief that the low-flying object might have been an airplane in distress.

For an hour, the area was quiet as patrolmen searched for the "thing." Meanwhile, RAFB reported none of its planes were in the air and CAA officials at Lubbock said they knew of no planes in the vicinity.

Then four reports were received in the next 90 minutes and by 1:30 a.m. today dozens of officers were combing Hockley County for a trace of the "fireball."

Spotted By Officers

The last report sent officers to an area about 4 miles northwest of Levelland where several patrolmen spotted the object from a distance.

Hockley County Sheriff Weir Clem said he saw a streak of fire flash across the highway in front of his car as he was driving south toward Levelland.

Highway Patrolmen Lee Hargrove and Floyd Gavin of Littlefield, who also were driving south several miles behind Clem, said they saw the flash of fire.

Fowler said he questioned the units he sent out to the area, and each man reported seeing the object from a distance at least once.

The series of sightings literally "swamped" the Hockley sheriff's office and the police station here with telephone calls today.

"All I've done all day is answer my telephone," Clem said.

See FIERY THING Page 18

SHERIFF WEIR CLEM
Saw Streak Of Fire

PEDRO SAUCEDO
Truck Lights Went Out

RONALD MARTIN
Too Astounded To Move

seemed to have been targeted by a UFO. Torres was ordered to shoot down the unidentified craft, which as always engaged in effective evasive maneuvers, leaving many questions behind.

UFO HOSTILITIES AND THE EVIL ALIEN AGENDA

Chapter 04

PATTERNS OF HORROR
By Brad Steiger

EDITOR'S NOTE: I first met Brad in 1967 on the floor of Manhattan's Commodore Hotel at the largest indoor UFO conference ever held, with ten thousand attending. The event was organized by the late James W. Moseley, the editor of "Saucer News" (then the number one selling UFO magazine in America). Steiger had just published his earliest work on UFOs. "Strangers in the Sky" was a pulp potboiler with some of the most exciting – and sensationalist – UFO reports of recent vintage. It was at the crest of a UFO heat wave and the public was hankering for information.

You could hear the enthusiasm in his voice as he eased into the various conversations going on around him on the convention floor. He seemed congenial and cool headed, and so there was no problem with Steiger hobnobbing among the delegates. But I must say some of us were rather surprised by the books he turned out next: "Flying Saucer Invasion: Target Earth," "Flying Saucer Menace" and "Flying Saucers Are Hostile" were not in keeping with the "friendlier" aspects of the Ultra-Terrestrials/Space Brother phenomena, whose adherents drowned themselves in peace and love with the various contactee books by Adamski, Menger, Fry and Bethrum. Steiger dug in deep in his pioneering works to terrorize us with tales of UFOlogy's most nerve-wrenching encounters.

Though these case studies may be a "bit older," they will be fresh and new to the majority of those who have never had to deal with the contention that UFOs could not only be hazardous to our health but doom life on earth. – TGB.

****

During the many, many years that UFOs have been actively researched, many widely divergent theories concerning their goals and objectives have been advanced. UFO actions have been interpreted as brotherly, benign, and protective; as indifferent, aloof, and superior; and as inquisitive, aggressively curious, and occasionally militant.

Several flight patterns have been noted, including the preponderance of sightings made when Mars is closest to Earth and the fact that certain UFOs follow the "straight lines" of Earth longitude and almost seem to be "mapping" the globe.

The majority of saucer sighters have reported eight basic shapes – the domed, the cigar, the half-globe, the crescent, the fireball, the disc, and the rare Saturn and gyroscope types. During a saucer "flap" one particular shape may be observed above military installations, atomic energy plants, and along coastal defense lines. Another shape may devote an excessive amount of time to electric power lines, power sub-stations, water-pumping stations, reservoirs, lakes, and other bodies of water.

In some cases, a pattern is set by the number of objects flying in a formation. Certain UFO formations constantly change from rectangular to diamond-shape. Others fly in circular formations. Still other UFO squadrons seem to describe geometric designs resembling characters in the Greek alphabet.

UFOs have always exhibited a frank interest in the military aspects of life on Earth. Since the 1940s, they have been seen buzzing Air Force planes inflight, swooping down on naval vessels in mid-ocean, hovering above air bases and rocket installations. Even the most secret projects seem to have been ferreted out. In the past five years it has become increasingly apparent that the strangers from the skies remain curious about military potentials, atomic armaments, the space race, power and water supplies, and methods of raising crops.

Certain saucer cultists, who have been expecting space brethren to bring along some pie from the sky, continue to deliver saucer-inspired sermons on the theme that the UFOs come to bring starry salvation to a troubled world. The self-appointed ministers who preach this extraterrestrial brand of evangelism ignore the fact that not all "saucers" can be considered friendly. Many give evidence of hostile actions.

There is a wealth of well-documented evidence that UFOs have been responsible for murders, assaults, burnings with direct-ray focus, radiation sickness, kidnappings, pursuits of automobiles, attacks on homes, disruptions of power sources, paralysis, mysterious cremations, and destruction of aircraft.

LOCK AND LOAD THE INTERPLANETARY WAY

The UFO crew members seem to have become bored with simple observational techniques or, perhaps, they have been made bold by our obvious weakness and inferiority. Dozens of reputable eyewitnesses claim to have seen alien personnel loading their space vehicles with specimens from Earth, including animals, soil and rocks, water, and struggling human beings.

The sightings of saucer occupants by responsible witnesses often send the investigator into one of the most frustrating aspects of the UFO field – the contactee aspect. At a contactee convention in Nevada, a young man named Hanno Mayberry told of meeting a seven-foot space girl who had ebony black skin, a high forehead, and two feline eyes of luminous green.

Reverend Hal Wilcox of Los Angeles opened his presentation with a mystic chant accompanied by a salaam-like gesture – he put his fingertips to his forehead and his thumbs to his upper lip. The chant, he advised those assembled before him, could be obtained in a thirty-six page booklet he was offering for sale. The booklet contained nuggets of essential wisdom which the "reverend" claimed to have gained in an Oriental monastery and on a trip through outer space to a planet named Celo.

The Nevada State Journal of July 11, 1966, listed other topics for discussion at the convention. Among them were: "From Earth To Alpha Centauri," "I Was Teleported Into A Flying Saucer," "I Was An Agent For The Extraterrestrials," "Message From Triangulum," "The Hollow Earth Mystery," and "Christmas On Jupiter."

One should not be too hasty in making general statements about the contactees. It would be both uncharitable and inaccurate to label all the contactee stories as tales told by idiots, full of space and fury, signifying nothing. There may be something ominous and insidious beneath the surface of these reports of beautiful, brotherly space people and their otherworldly utopias.

Many of those who claim to have encountered aliens have recalled that, immediately upon making the initial contact, a bright flash was directed at them. John Reeves of Brooksville, Florida, for example, thought that the alien had taken his picture with a flash camera of some advanced design. After the contact experience, there is almost always a marked change in the contactee's ability to remember details. There seems, in most cases, to have been a loss of time from the contactee's memory. Sometimes there have been certain lasting effects on the contactee that have prevented him from ever living a normal life. After his communication with benign brothers from the skies, the contactee seems to have a totally new concept of life, love, and prosperity, subtly introduced to him by the aliens.

The contactee has been told of fabulous utopias where no one works, where money is not necessary. Gabriel Green, one of those "appointed" to go forth and tell about such moneyless systems, has written: "space men say the Shangri-La systems on their planets combine the best merits of capitalism and socialism." Green's contact described the economy as "a super credit-card system, where everyone has a credit card and there is no limit to what anyone can charge..."

If the aliens have really come to share such an easy existence with us, why have they left men with no evidence of contact, other than radiation sickness, first-degree burns, or, at best, memory lapses? Why have the UFOs burned children with direct-ray focus, destroyed remote villages, and attacked aircraft and land vehicles?

The latest trend in UFO reports has taken the form of what one might call "the great American car chase." Dozens of reports have been concerned with UFOs diving over highways and buzzing terrified motorists. In some cases, saucers have pursued automobiles to the point where motorists have run off the road and crashed.

Another area which has received intensified UFO activity is our reservoirs. The Wanaque Reservoir in Wanaque, New Jersey, has been visited several times and a photographer has even taken pictures of a UFO beaming light down into the water. .

Why are they so interested in our reservoirs? Consider this rather unsettling theory: If the time has come for invasion, what would be the easiest manner by which an alien race could conquer us? By force? No, because, as they have learned, we are not that weak, not since we have developed missiles with atomic warheads. Although our weaponry is vastly inferior to theirs, we could still release a great deal of nuclear energy before we went under.

The increased radiation in the atmosphere might possibly affect the aliens as well as ourselves. They can hardly see a hollow victory that would award them only a burned-out planet as a prize. It would be much simpler and infinitely less messy to beam down hypnotic drugs into our drinking water and infiltrate people by degrees. If such an invasion plan were put into effect, there would be no resistance, no atomic bombs, and no destructive radiation. Some morning we would wake up and just find them in control.

We should not, however, exclude the possibility that there is more than one source for the UFOs. In addition to the aggressive and hostile them, there have been numerous reports of UFOs whose actions must be interpreted as solicitous of man and whose only purpose seems to be observation of an alien culture.

It may be that, on one side, we have them, and on the other, the benign observers. By one we may be regarded as chattel; by the other, we may be deemed worthy of patience and tolerance. An uneasy question immediately presents itself: In the event of attack by the exploiters, would the friendly observers come to our aid?

But, ask the skeptical, where is proof that unidentified flying objects exist at all? And − if it could be established that these objects do exist − how could it be demonstrated that any of them are hostile?

My purpose here is to document evidence of a steady, pervasive, and increasingly aggressive invasion from outer space, and to inform the scientist and layman alike that it is time to consider an effective means of defense and an intelligent course of action.

"I want scientists everywhere to start looking at the evidence," declared the late Dr. James McDonald, a respected physicist. "The matter is urgent. It must not be delayed. The world better wake up to flying saucers before it is too late."

THE ALIEN DEATH MASK CAPER

In August, 1966, Inspector Jose Venancio Bittencourt of the Rio de Janeiro police was faced with what he termed "the most baffling mystery in my twenty-three years on the force."

On August 20, police had discovered the bodies of two electronics technicians, Miguel Jose Viana and Manuel Pereira de Cruz, one-thousand feet up a hillside, in the Rio suburb of Niteroi. The bodies had been found after a woman, who lived nearby, had reported to the authorities that she had seen a flying saucer land on the jungled slope of Morro do Vintem. The woman had been so insistent in her claim that police had been dispatched to placate her. The officers had not known what they might find, but they had not expected to find two corpses.

Both men had covered their faces with lead masks before they died. Several slips of paper were found near the bodies. Notations on several of the papers had been made in some strange kind of code.

One note, written in Portuguese, read: "At 4:30 P.M. we will take the capsules with an orange. After the effect is produced, protect half the face with lead masks. Wait for agreed signal."

A second note seemed to prescribe a regimen for the taking of the mysterious capsules: "Sunday, one capsule before meal; Monday, one capsule in the

Miguel José Viana | **Manoel Pereira da Cruz**

morning; Tuesday, one capsule before meal; Wednesday, one capsule before sleep."

Exhaustive laboratory tests were unable to determine the cause of death of either of the electronics technicians.

"There was no medical reason, within the ability of the state police lab, to detect for the deaths," said Inspector Bittencourt. "Our lab men have ruled out the possibility of poison, violence or asphyxiation."

After the case had been reported in the newspapers, Mrs. Gracindo de Souza, wife of a member of the local stock exchange, told police that she and her daughter had been driving down Alameda Sao Boaventura when they had seen a UFO hovering over the clearing where the bodies were later discovered. Mrs. de Souza and her daughter had made their sighting on August 17 – the day which medical examiners had established to have been the date of the deaths.

A watchman, Raulino de Matos, saw the technicians arrive at the mountain in a jeep with two other men. When the four started to climb the hillside, de Matos had paid no further attention.

The dead technicians were found lying side by side, their arms at their sides. There were no signs of struggle. Investigators did find blood nearby, but laboratory tests established that it had not come from either of the victims.

The masks that covered the men's faces were the kind commonly used in electronics to protect the eyes from burns; it was later determined that the devices had been made in a shop in the town of Campos, where both men had lived.

Along with the two notes in Portuguese and the mysterious coded messages, investigators found a number of electrical charts, simple mathematical formulas, a lady's handkerchief, a raincoat, sunglasses, and a toothbrush.

After an extensive month-long investigation, the Brazilian police admitted its failure to have solved the "lead mask murders."

"Station zero," Inspector Bittencourt confessed in a candid summation.

Crude lead masks. Two dead men in their unruffled Sunday suits, lying side by side. Strange undecipherable codes. Two notes prescribing the intake of unknown and unfound capsules. A trace of blood that did not come from either victim. Two unidentified men who may also have been victims, their bodies yet to be discovered.

Are these clues to a murder mystery which has baffled the most determined investigators? Or are they additional manifestations of a malignant, yet intelligent, course of action which, when viewed in a certain perspective, will indicate an ever-increasing pattern of horror of world-wide significance?

To be direct: What of the UFOs that had been sighted hovering above and landing on the hillside on the day the technicians were killed and the day their bodies were discovered?

Consider the enigmatic notes that spoke of the "agreed signal." Ponder the capsules which were prescribed, and the lead masks which were to be worn "after the effect is produced." Most of all, weigh thoughtfully the fact that exhaustive police crime lab tests were unable to determine any cause of death for either of the men.

Had the two electronics technicians kept a rendezvous with the occupants of a UFO and found, to their ultimate terror, that the aliens were not the benign space brethren they had been led to expect?

Had Viana and de Cruz discovered that some extraterrestrials have come not to issue pronouncements of universal peace, but to conquer?

Had these two young men learned too late that the crew members of some alien craft, far

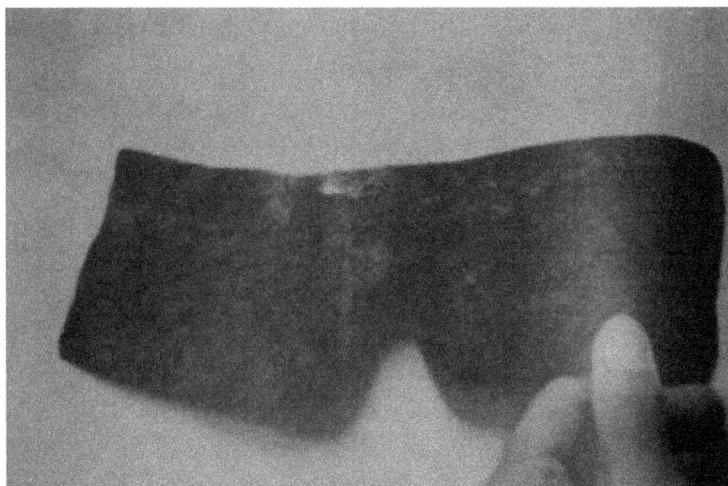

UFO HOSTILITIES AND THE EVIL ALIEN AGENDA

from being indifferent to earthlings and shunning contact with us, are decidedly aggressive and regard Homo sapiens as man might regard cattle? If the relationship of these aliens to us can be defined in such terms, these strangers from the skies would feel no compunction about shocking us with a "prod" if we approached their craft too closely, in paralyzing our heart muscles if we proved too troublesome, or in kidnapping us aboard ship to poke us or stick us or even to cut us open for biological investigation.

In the case of the two Brazilian technicians, there is no clear evidence that saucers were actually involved in the bizarre murders. After all, the fact that UFOs were reported hovering in the same area may have been only a coincidence.

But it is not necessary to rely on evidence that may be only circumstantial. There are other, explicit physical data that all too graphically demonstrate the thesis that "someone up there may not like us."

On June 2, 1964, eight-year-old Charles Keith Davis of Hobbs, New Mexico, was standing just outside the door of the DeLuxe Laundry while his grandmother, Mrs. Frank Smith, was gathering clothes from a washer.

Charles pointed to the sky and seemed about to say something, when, according to his grandmother, "there was a whooshing sound and a blackish ball of fire covered Charlie."

Mrs. Smith, who was standing approximately three feet from the youngster when the incident occurred (at about 4:00 P.M.), said her grandson was "covered with black, his hair standing on end and burning. Charles was just as black as he could be . . . his hair was standing up on top of his head. I grabbed him and tried to smother out his hair, which was on fire." Mrs. Smith's efforts to extinguish the black flames were supplemented by laundry employees, and the badly burned boy was rushed to Lea General Hospital. Listed in good condition after having been treated for burns, Charles told the police and doctors that he had been burned by "a fire that came out of the sky."

A MAGNETIC FORCE SPELLED DOOM

While driving on a lonely stretch of road about one hundred miles from Melbourne, Australia, in April, 1966, Ronald Sullivan was startled to observe that his headlights were suddenly being pulled to the right, as though drawn by some strange magnet.

"I braked as hard as I could and glanced over to the right," Sullivan was quoted in the Otago, New Zealand, Daily Times; "There, in the middle of the paddock, was a column of colored light about twenty-five feet high and shaped like an ice cream cone."

As Sullivan watched the illuminated "ice cream cone," it rose from the ground, noiselessly but at tremendous speed. After the object had risen, Sullivan's headlights returned to normal and refocused on the road.

UFO HOSTILITIES AND THE EVIL ALIEN AGENDA

At the same spot, three days later, a young man named Gary Taylor was killed when his car suddenly swerved into a tree at 10:00 P.M. A motorist traveling about half a mile behind Taylor told police that he could see no reason for the auto to have swerved. "It was as if something pulled the car into the tree," he told the authorities.

Learning of the alleged accident, Sullivan telephoned a reporter on the Mayborough Advertiser and arranged to meet him at the spot on Good Friday. The two men found a saucer-shaped depression about fifty yards from the road.

"It was right where I saw the column of light," Sullivan remarked. "When we were leaving the paddock, I saw the tree where the car had crashed and it gave me an eerie feeling. 1 can't help wondering if the boy died as a result of the thing I saw."

SEARED TO DEATH

In June, 1954, an eleven-year-old African boy was wondering about "the strange thing" that attacked an entire village!

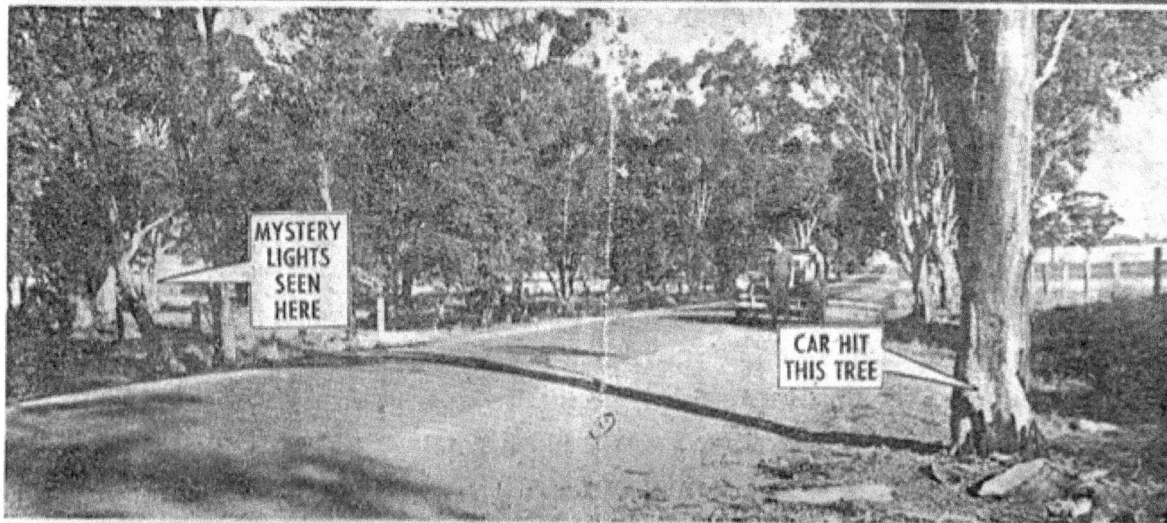

Bourkes Flat 4 April 1966

The Sun, Tuesday, April 12, 1966— Page 3

THEY CALL IT THE DEATH STRIP

MYSTERY LIGHTS SEEN HERE

CAR HIT THIS TREE

THE STRAIGHT stretch of the Dunolly-St. Arnaud road, showing the tree which Gary Taylor's car struck in Thursday's fatal accident.

BELOW: Mr. Sullivan examines the shallow depression in a ploughed field beside the road.

Did mystery light

UFO HOSTILITIES AND THE EVIL ALIEN AGENDA

For several nights Laili Thindu and his companions, while tending flocks of goats and sheep and sleeping outdoors on the outskirts of the village of Kirimukuyu, watched strange lights on Mt. Kenya. At first, the boys theorized that mountain climbers were working on an ascent of the mountain by night and carrying huge torches. When the lights soared into space, the boys' theories began to take on a more supernatural coloration.

One night, as Lalli Thindu lay in his crude hut, he could hear the drums of a nearby village throbbing joyously at a tribal marriage ceremony. He became perplexed when the odd lights left Mt. Kenya and hovered over the village, beaming down bright rays. The drums were suddenly silenced.

The next morning, Lalli Thindu learned that all the dancers, all the children, all the livestock – the entire population of the village – had been seared to death by terrible burning streams of light from glowing objects. It was not until Lalli Thindu ventured to Nairobi that he was able to tell his story to someone who recognized the tale for what it really was: the annihilation of an African village by a UFO.

This pattern of horror is by no means a particularly new one, although things have been stepped up a bit since World War II when the more ingenious of our own species began to find searing, burning power in nuclear energy.

In the April 30, 1964, issue of the Watford and West-Herts Post (Ireland), Lawrence Bradley wrote the editor about a UFO incident that he had witnessed in 1922.

". . . the place was County Donegal in the Irish Republic. A civil war was raging at the time and the army that I belonged to was fighting scattered rearguard actions, mostly in the mountains. One evening, tired and dispirited, I lay down at the entrance to an old cave. In the fading twilight I noticed that practically all the bushes and grass that grew around this entrance were scorched and burnt. The only occupants of the cave were sick and wounded men who were unable to walk. The six able-bodied soldiers who were looking after them told me a strange story which, at the time, seemed farfetched and unconvincing."

The soldiers told Bradley that they had been awakened by a strange whirring noise outside the cave early that morning. Stumbling to their feet and readying their weapons, the soldiers squinted into the pre-dawn darkness and decided that they were hearing the approach of an enemy armored car. They immediately opened fire.

". . . the object retaliated by firing jets of flame at the cave. The defenders had to withdraw in face of the fierce heat. All the undergrowth was now ablaze and smoke was billowing into the cave so that it was a case of facing the flame throwers or suffocating to death. .

"When they ran out, they saw the flame-throwing object ascending into the sky. It was clearly visible in the first light of dawn – circular in shape and bright in appearance, as if made of aluminum. I daresay some of the men who saw this strange phenomenon are alive today and can vouch for this story."

Bradley is convinced that UFOs are no "laughing matter" and submitted his story as "another bit of evidence in support of the theory that flying objects do exist."

Bradley has also presented us with another uncomfortable bit of evidence that these flying objects may be extremely hostile. The Irish soldiers may have provoked the saucer's flames by firing upon the object first, but what had Charles Davis done to bring about the ball of flame that engulfed him? Had the crew members of some UFO been so offended by the tribal drums of an African marriage ritual that they destroyed the ill-fated Kenya village?

On April 7, 1938, three innocent people may have been involved in some perverse galactic experiment in cremation.

At 1:14 P.M., P. F. Phillips, second mate on the tramp steamer S.S. Ulrich, noticed that the ship was yawing badly as it sailed off the coast of Ireland. Phillips' fears about what might have been wrong in the wheelhouse could not have included the actual horror that awaited him. When Phillips pushed open the door to the wheelhouse, he found that helmsman John Greeley had been transformed into a human cinder.

Except for the heavy, acrid stench of burning flesh that permeated the wheelhouse and the inescapable evidence of Greeley's hideously charred body, there was no other sign of fire in the compartment. The deck, the ship's wheel, even the dead man's shoes were unmarked by flame. Medical examiners later declared that Greeley had been literally "fried from the inside out."

A few hundred miles from the baffling cremation aboard the S.S. Ulrich, police officers of Upton-by-Chester, England, were stunned when, during the course of their investigation of a runaway truck, they opened the cab to discover the incinerated body of George Turner, the driver.

UNEXPLAINED HUMAN COMBUSTION

The temperatures necessary for human cremation are very high, yet the investigating officers were amazed to find no evidence of fire within the truck. The seat cushions bore not the slightest burn, the windows had not cracked, buckled, or melted as they would have in an ordinary fire of such intense heat. The clock on the dashboard had stopped at 2:14 P.M.

At 3:14 P.M., in Nijmegan, Holland, a young man identified as William Ten Bruik was burned beyond recognition while seated in his Volkswagen. The automobile itself bore no evidence of fire. The exposed gas tank had not burst into flame. The car's upholstery was not even smudged.

While these mysterious cremations are sufficiently enigmatic in themselves,

UFO HOSTILITIES AND THE EVIL ALIEN AGENDA

let us consider an additional element that further compounds the mystery. John Greeley, helmsman of the S.S. Ulrich, cremated at 1:14 P.M. off the coast of Ireland; George Turner, incinerated in his truck in Upton-by-Chester, England, at 2:14 P.M.; William Ten Bruik, burned beyond recognition in his automobile at Nijmegan, Holland, at 3:14 P.M. – even though separated by several hundred miles, these men were cremated at precisely the same moment! The hour interval between the incinerations was due only to the difference in time zones.

Although cases involving the spontaneous combustion of the human body have been noted before, it defies probability that three such bizarre occurrences could have taken place at exactly the same moment, in such widely separated geographical areas, without someone or something providing a catalytic agent that induced such ghastly self-immolation.

Is it only in imaginative science-fiction stories that awesome weapons are placed in the hands of an alien race? Consider the story of James Flynn, 45, a rancher of East Fort Myers, Florida.

On April 12, 1965, Flynn started on a combination hunting and camping trip in the Everglades. Taking along his dogs and his swamp buggy loaded with provisions, the rancher looked forward to the diversion of a few days on his own.

While he was rounding up his dogs at 1:00 A.M. on Monday, April 15, Flynn noticed a huge light hovering above the cypress about a mile away. When the object began to lower itself to the ground, the rancher drove his swamp buggy closer in order to investigate.

Through his binoculars, Flynn could see a large, cone-shaped object which he later estimated to have been between thirty and thirty-two feet tall and about sixty-four feet wide. The object had three rows of two-foot square windows that reflected a dull yellow light. The area under the object gave off a reddish glow, and there was ample radiant illumination from the mysterious craft for Flynn to examine its surface. Flynn later remarked that the object had definitely been metallic. In the half hour during which he studied the UFO, he was able to determine that the craft had been put together from squares of material approximately four feet by four feet. Flynn was even able to see the rivets that held the metallic squares in place.

Unafraid of the unidentified aircraft, Flynn started his swamp buggy and began to travel the quarter of a mile separating him from the strange visitor to the Everglades. A dog that traveled with Flynn on the buggy did its best to veto its master's plans, and began trying to tear its way out of the cage in which it was riding.

When Flynn was within a few yards of the craft, he switched off his swamp buggy and doused his lights. Without hesitation, the rancher walked to the edge of the UFO's red circle of light. Then, in what he assumed to be a universal gesture of goodwill, he raised one arm and waved. He waited thirty seconds and then began to advance.

Flynn had walked about six feet when a beam of light shot out from under the bottom of a row of windows and struck him on the forehead. He instantly lost consciousness.

When the rancher awakened, he was alarmed to discover that he was blind in his right eye and had only partial vision in his left. Dimly, Flynn could see a symmetrical circle of scorched ground where the cone-shaped object had been hovering. A number of cypress trees had been burned at their tops. If Flynn's loss of vision were not sufficient proof of the UFO's presence, the terrain bore obvious evidence of the extraordinary visitor to the swamp.

It was not until Flynn walked into the office of Dr. Paul Brown that he faced the startling realization that he had been unconscious for twenty-four hours. Dr. Brown was most emphatic that the day was Wednesday, April 17, not Tuesday, April 16, as Flynn had believed. The rancher shook his head in bewilderment. He had lost an entire day.

Dr. Brown was more concerned about James Flynn's loss of vision. Due to hemorrhaging in the anterior chamber of the eye, Flynn's right eye had the general appearance of a bright red marble. His forehead and the area around his eyes were inflamed and swollen. The man was almost totally blind.

James Flynn's story of sighting a UFO in the Everglades and being struck down by a ray from the interior of the craft made the national wire services; because of the man's solid reputation, the incident was also given considerable attention by serious investigators. Upon release from Lee Memorial Hospital, Flynn accompanied researchers to the site of his observation. The physical evidence of the scorched cypress and the burned circle of grass was still there for all to see. Although seared tree tops and singed grass may not constitute conclusive proof that a UFO stopped to rest above that particular spot, James Flynn is left with an ever-present reminder that one should approach UFOs with utmost caution. The results of Flynn's checkup, after he was released from the hospital, indicated that he still had cloudy vision in the right eye and a depressed spot of about one centimeter in diameter in the skull area above the same eye.

On October 2, 1956, a night watchman in Trenton, New Jersey, managed to collect medical compensation for disabilities caused by the close approach of a UFO while he was patrolling his firm's property.

Harry Sturdevant had been a night watchman for over twenty years and had proved himself capable of handling any situation that might present itself. But no man could have been prepared for the object that appeared first "as a red light in the sky," then shot down at him.

"It was about sixty to one hundred feet in diameter, shaped like a cigar. It had no wings and no fins. I heard no propulsion from it except a hissing sound like escaping steam.

"It gave me the greatest shock of my life. There was a smell like sulfur or brimstone, but it was different. I don't know what it was really except it was very nauseating and it made me very sick. I lost my sense of taste and smell; my throat would not swallow properly. My stomach felt worse than the time I was overcome with mustard gas while fighting with the allies in France in World War I.

"I collapsed in pain and lay there on the ground for half an hour before I was able to drive."

Sturdevant finally made his way back to the trailer that served as his home while he was patrolling the partially constructed thoroughfare his employers, the Herbert Elkins construction firm, were building. Once inside the trailer, Sturdevant picked up the telephone and notified police headquarters that he had just been buzzed by a flying saucer.

* * * * *

EDITOR'S NOTE: This case is reminiscent of the famous "Scoutmaster Incident" of 1953. One wonders how many miles apart the incidents may have taken place.

* * * * *

A CASE FOR WORKER'S MEDICAL COMPENSATION

Six weeks later, a New Jersey workmen's compensation referee decided in Sturdevant's favor and decreed that the night watchman should be paid for medical expenses incurred in the treatment of his temporary loss of hearing and sense of smell. For the first time in history, a UFO was officially held responsible for injury to an earthman.

Ten-year-old Sharon Stull of Albuquerque, New Mexico, was just as disabled as Harry Sturdevant and a UFO was just as responsible as in the watchman's case. But the painfully burned girl was unable to claim compensation from anyone. Her case did, however, provoke an official response. Police Chief A. B. Martinez issued an unprecedented warning to Albuquerque residents, asking them to stay away from mysterious objects. The law-enforcement official declined to state his views on what the UFOs might be, but he strongly emphasized his warning that "they should be treated with respect and caution."

Little Sharon Stull's doctor stated that he had found membrane inflammation of both eyes and first-degree bums under the eyes and on the nose. The physician had asked the girl to draw a picture of the strange thing which had burned her with the painful ray emanations. Sharon, and some playmates who had witnessed the incident on April 28, 1964, told authorities that the object had been shaped like an egg and was slightly smaller than an airplane. They had seen no windows of any kind.

UFO HOSTILITIES AND THE EVIL ALIEN AGENDA

A strange ray that burns and blinds is frightening enough, but consider an unknown force that can rip apart airliners as if they were made of balsa wood.

Braniff Airways' Flight 542, a Lockheed Electra turboprop airliner, took off from Houston, Texas, en route to New York at 10:37 P.M., September 29, 1959. Flight 542 carried twenty-eight passengers and six crew members. The flight plan called for the airliner to travel over the Buffalo, Texas, area, sixty-eight miles southeast of Waco. The countryside was quiet and the sky was partly cloudy as Flight 542 neared Buffalo at an altitude of fifteen thousand feet and an air speed of two hundred seventy-five knots.

About the time that Flight 542 was approaching Buffalo, Major R.O. Braswell was flying a C-47 at six thousand five hundred feet between Shreveport, Louisiana, and Lufkin, Texas, which is east-northeast of Buffalo. It was there that the major saw the "thing."

"It was colored like a large red fire and looked like an atomic cloud. It was a massive thing, about five degrees above my plane. The base was at an altitude of twelve thousand to fifteen thousand feet. The top was at about sixteen thousand feet."

Billie Guyton of Centerville, Texas, was observing the aerial phenomenon from the ground at the same time that Major Braswell was flying below it. Guyton reported that he saw an object emerge from the glow. Major Braswell declined comment on this observation.

Jackie J. Cox, a thirty-nine-year-old schoolteacher, was driving toward Buffalo when he heard the "noise of a plane which sounded louder than normal." Then he saw a "bright light in the sky that spread to cover the entire sky, as if phosphorous or magnesium were burning."

W. S. Webb of Buffalo had just gone to bed when he happened to glance out the window and see "a ball of fire that looked like a star, shoot through the sky. Then I heard a noise like something falling from the sky. It was a whooshing, shrill sound."

Farmer Richard White and his wife had just turned off the television set and gone to bed. White had decided to sleep on the breezeway that night in an effort to beat the summer heat. As he lay gazing at the summer stars, he noticed a brilliant light high in the southern sky.

"The whole sky seemed to be lit up by a huge fireball passing overhead, only to be followed by a tremendous explosion, so violent it seemed to shake the ground itself."

Seconds later, the night was filled with shrill whistles as debris fell through the still summer air. The giant $2,300,000 airliner had been torn into thousands of fragments.

Officials investigating the wreckage said that an airplane crash usually leaves twenty or thirty major pieces on the ground. Such was not the case with Flight 542. Braniff vice-president of operations R. V. Carleton told newsmen, "I've investigated lots of crashes but I've never seen one where the plane was so thoroughly demolished, the wreckage so widely scattered and the people so horribly mangled. And there was nothing among the wreckage which indicated a fire or bomb aboard the plane."

The bodies of the thirty-four victims aboard Flight 542 had been so widely scattered and torn into so many fragments that recovering them was a slow, grim task.

Chunks of bloody pulp were barely recognizable as the remains of human beings. Some parts of bodies were plastered so firmly to tree limbs that workers had to use great effort to remove them.

Investigation and the resultant inquiry produced four important facts: 1) There had been no fire or explosion aboard the airliner while it had been in flight or after it had crashed. 2) Scorch marks found on glass window ports, the rear of the fuselage, and across the parting of the tail showed signs of having been exposed to tremendous exterior heat. 3) The force that caused the airliner to disintegrate had not come from within the plane. 4) Moments after the fireball had disengaged itself from the red cloud, every dog within miles of the Buffalo area started howling.

For the past two decades, certain as yet unexplained phenomena have been associated with the observation of UFOs: mysterious, glowing clouds; fireballs; shrill, aerial whistling sounds; the strange behavior of animals in the area. All these phenomena were in evidence at the scene of the crash of Flight 542.

If there had been no explosion or fire aboard the airliner, if there was no discovered cause for the mid-air disaster, can we afford to ignore the testimony of the residents of Buffalo, Texas? They saw an unidentified flying object launch a fireball moments before the airliner crashed.

According to eyewitnesses at the scenes of other air disasters, hostile UFOs use many methods in swatting down an occasional aircraft.

METHODS USED TO SWAT DOWN OUR AIRCRAFT

On the night of June 10, 1960, Edward Henry Tindale, radar observer in charge of the Mackay, Australia, meteorological office, told investigators that he had plotted an unidentified, stationary craft twice on the night in which an airliner mysteriously crashed. Both "plots" of the UFO were made in the area where the main wreckage of the airliner was found.

Tindale testified that he had plotted the UFO at 10:30 P.M. and again at 11:30 P.M. One need not belabor the point that there is no known Earth aircraft that can remain stationary in the sky for an hour. The Brisbane, Australia, board of inquiry stated that their aircraft experts could not explain the cause of the crash.

UFO HOSTILITIES AND THE EVIL ALIEN AGENDA

On February 19, 1961, the pilot and co-pilot of the private airliner used by the then vice-president Lyndon B. Johnson were killed when the plane crashed on a brushy hillside sixty miles west of Austin.

The plane had left Austin Municipal Airport at 7:08 P.M. for the landing strip at the Johnson ranch. At 7:16 P.M., when the airliner was twenty-five miles away, ranch personnel received a report advising them to prepare to receive the plane. Fourteen minutes later, the pilot radioed to upbraid the ranch for not turning on the landing strip lights. "We've got them on," the ranch radioed back.

"Well, I sure can't see them," the pilot complained. "I would have landed several minutes ago, but I have been circling up here trying to find your field. What is your ceiling?"

The pilot was told that the ceiling was three hundred feet and was again assured that the lights had been turned on in preparation for the landing of the airliner.

"I repeat that I cannot see them," the pilot said. "We're not going to attempt a landing on a field we can't see. We're going to return to the Austin airport."

That was the last radio contact received at the LBJ ranch. But at that same moment, radar at Bergstrom Air Force Base was picking up a UFO in the area. Within a few minutes, the vice-president's private airliner had crashed.

Some investigators of UFO activity have theorized that alien spacecraft might be bringing down earthly aircraft that has ventured into the saucer's force field. This field, a depth and frequency controlled radial zone, could affect a plane flying into it by throwing off the plane's direction finder, compass, radio, and other sensitive instruments. This same field might cause the interior of the aircraft to heat up, catch fire, or have a sudden power failure. As a result of increasingly powerful vibrations, as the plane ventures deeper into the force field, the aircraft may perform freakishly or spin completely out of control. The crew might be completely incapacitated. If the aircraft should penetrate deep enough into the force field, the airplane and its occupants could be pulled to shreds.

This kind of force may have been responsible for the crash of an Air Force C-118 transport plane near Orting, Washington, at 8:19 P.M. on April 1, 1959.

At McChord Air Force Base, Colonel Robert E. Booth, commander of the 1705th Air Transport Group to which the C-118 was attached, told reporters that a radioed report from the pilot indicated that there had been a mid-air collision. About an hour and fifteen minutes after taking off from McChord on a routine training flight, the pilot had radioed: "We have hit something, or something has hit us!"

Colonel Booth said that the pilot then called "mayday," the international distress signal, and radioed that he was returning to the base.

Then came the shout: "This is it!"

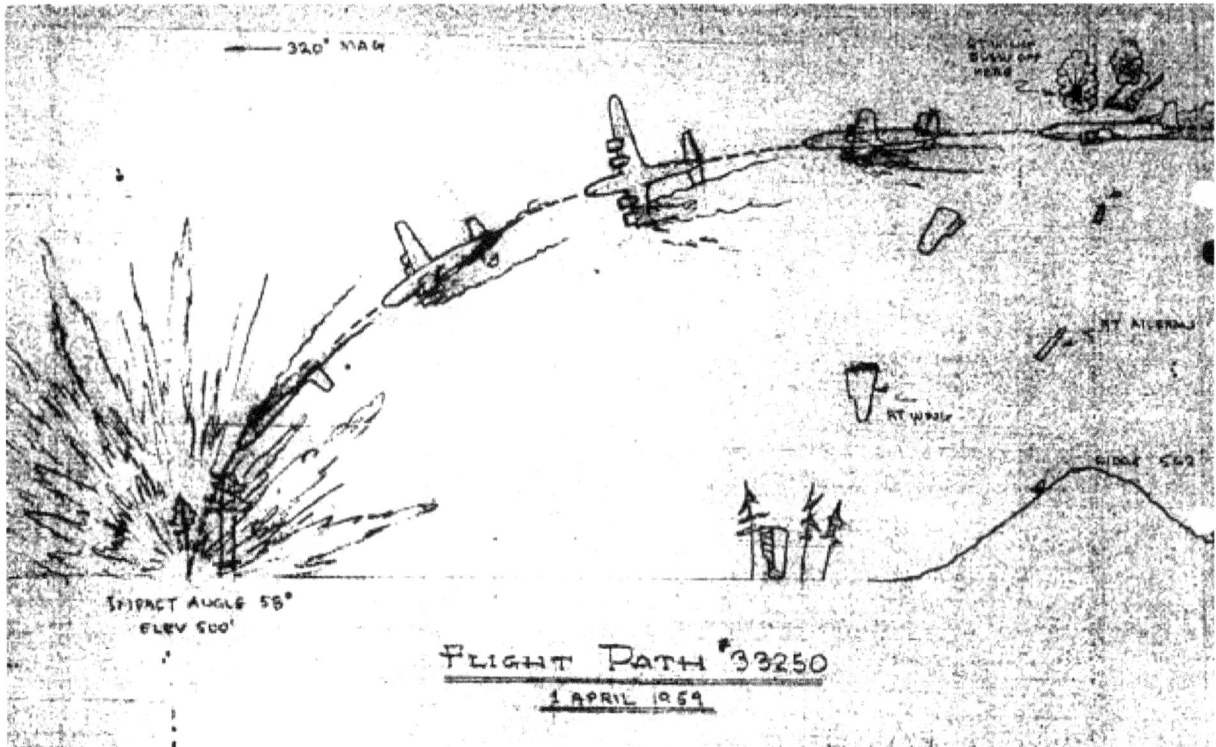

FLIGHT PATH #33250
1 APRIL 1959

There was no further word from the stricken plane, which was torn to shreds in a tremendous explosion when it crashed in the mountainous Rhodes Lake area. The two officers and two enlisted men aboard were killed.

Then the Air Force erected its "iron curtain." Barricades were set up around the area of the crash and civilians, newsmen and police officers were refused permission to survey the scene of the disaster.

"Do you admit there was a plane crash?" a reporter asked the investigating Air Force officer.

"I have no comment," the officer answered.

Because of the peculiar behavior of the Air Force during the course of this investigation, and because there had been reports of strange aerial phenomena both before and after the crash of the C-118, Robert Gribble, then director of the Aerial Phenomena Research Group (Gribble was later director of the National Investigation Commission on Aerial Phenomena of Seattle), began to suspect that UFO activity may have been a cause of this crash, or at least associated with it. Several witnesses to the disaster were interviewed and a strange and frightening case began to emerge.

Although Gribble had been prevented from inspecting the wreckage, he was able to talk to witnesses who had been on the scene. The four-engine transport plane was described as having been "shredded beyond belief, smashed to bits." Two bodies had been found buried deep in the ground by the impact. This report agreed with statements to A.P.R.G. – Aerial Phenomena Research Group – investigators by residents of the area that they had seen the plane "slammed to the ground, straight down" while in a horizontal position, as if it had been thrown by some unseen force. Only three bodies had been found. The fourth was still missing.

A.P.R.O. Director, Robert Gribble

The A.P.R.G. investigators' reconstruction of the evening of the crash determined that at 7:00 P.M., a series of mysterious aerial explosions had shaken the north Seattle area. At 7:20 P.M., the entire greater Seattle area was shaken by an aerial explosion. Damage was reported.

At 7:45 P.M., Mr. and Mrs. Sam Snyder of Graham, Washington, which is near the crash area, said they and a guest witnessed a brilliant glow through the timber in the direction of the crash scene. The glow was followed by another flash.

During the early evening hours, several area residents reported seeing mysterious lighted objects in the sky. The activity of the UFOs covered an area extending from twenty miles north of the crash scene to about eight miles southeast of it. Unidentified objects were also reported in the sky in the Mt. Rainier area. These sightings were confirmed by the Orting, Washington, Chief of Police and by the Public Information Officer at McChord Air Force Base. The P.I.O. attributed the lights to flares dropped during a parachute jump exercise at Fort Lewis, which is located next to McChord. But the News Chief at Fort Lewis emphatically denied that any such exercise had been in progress on the night of the crash.

Several witnesses told Gribble and his crew of investigators that none of the four engines of the C-118 had been running as the plane passed over their area. These same witnesses told A.P.R.G. that two glowing, parachute-like objects had been following the transport. They also noted that part of the tail assembly on the C-118 had been missing as the plane passed overhead.

Mr. and Mrs. Bill Jones of Orting spoke of having seen three or four parachute-like objects in the air as the plane passed near their home. The Air Force had no comment regarding these objects.

At 10:00 P.M., nearly two hours after the crash, another series of mysterious explosions rattled windows in the Seattle area.

At about 10:00 A.M. the morning after the crash, Gribble received an anonymous telephone call informing him that "radar at McChord Air Force Base picked up UFOs prior to the crash." In subsequent conversations with the P.I.O. at McChord, Gribble was unable to obtain either an admission or a denial that base radar had plotted the UFOs.

On Saturday, April 4, three days after the crash, a group of A.P.R.G. investigators drove to the Orting-Sumner area to investigate the reports of UFOs and their role in the destruction of the C-118. Fred Emard, Orting's Chief of Police, had told Gribble that he would be glad to furnish A.P.R.G. with information regarding the sightings of mysterious lighted objects, but that the UFO researchers would have to come to Orting so the Chief of Police could "see who he was talking to."

When the A.P.R.G. group arrived in Orting, they found Chief Emard in conference with an Air Force colonel. The streets of Orting almost seemed to be patrolled by Air Force personnel. There was an atmosphere not unlike martial law about the little Washington town.

Gribble and his researchers waited for their interview at Emard's home. After about fifteen minutes the Chief of Police arrived and told them that he had nothing to say to A.P.R.G. The UFO researchers were advised that any statement Emard had for their group would be made through McChord Air Force Base.

"The Chief and other officials of Orting had been silenced," Gribble reported. "Since the crash, the Air Force has been working frantically, and silently, trying desperately to silence anyone and everyone who witnessed the crash or who has any information pertaining to the crash. Residents of the Sumner-Orting area have been advised not to discuss the accident with anyone. They have been bluffed into submission. Airport tower operators at some local airports have been 'forced' by Air Force officers to sign affidavits which compel them to remain silent about the crash. No sign, no job.

"One gentleman, who was one of the first to arrive at the crash site, talked frankly and freely of what he had seen and heard. Then he was silenced. His experience with the crash was bad enough, but his experience with censorship left him 'shook up,' but good.

UFO HOSTILITIES AND THE EVIL ALIEN AGENDA

WHAT WAS THE MILITARY ATTEMPTING TO HIDE?

"The citizens of the Sumner-Orting area are silent and scared. They know something is wrong but they can't put their fingers on it. What is the Air Force trying to hide? Did the Air Force radar see a UFO hit the C-118? Were there eye-witnesses to the 'contact' which the pilot reported? Is this the reason for all the secrecy?"

The authors do not at this point wish to entangle the reader in a wearisome restatement of the various charges that the Air Force has been conducting a "conspiracy of silence." Such allegations are abundant in the written and spoken testimonies on the UFO enigma. To the authors, it seems that in certain cases the Air Force has conducted itself in a manner strongly indicative of overt censorship of information. In other cases, the official statements of the Air Force's investigating officers have been quoted out of context and military procedure misinterpreted as a lack of respect for the individuals involved. At times, "censorship" may have been inadvertent. At other times, the Air Force may have assumed various guises and roles in order to prevent the release of certain information when they deemed that its publication would not have been in the best interest of the American people.

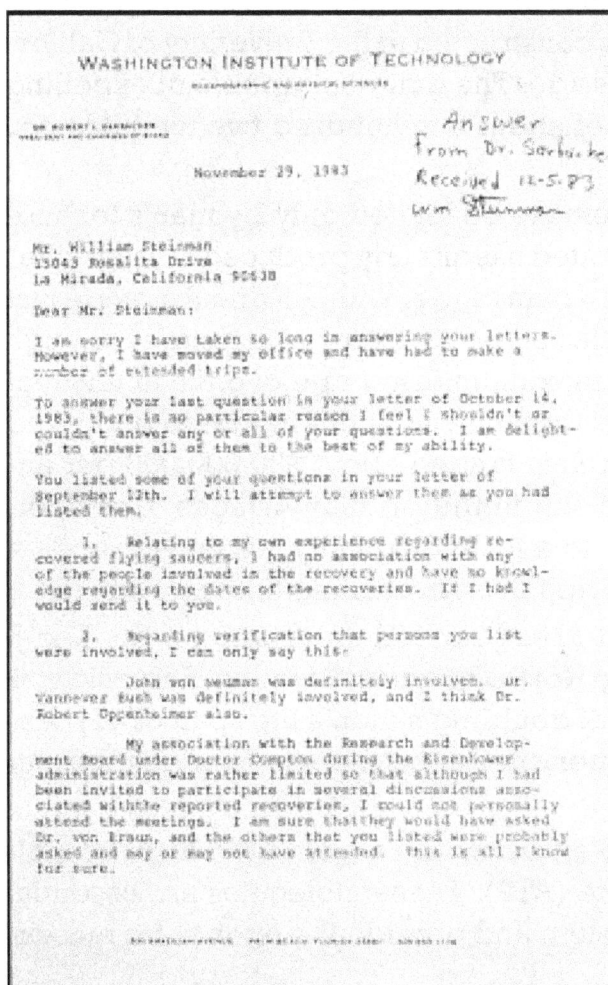

We are not primarily concerned about which government agency or educational institution does the investigating, evaluating, and analysis of UFO data. What we are concerned about is that after the final assessments have been made and the scientists and military men have convened, a course of action shall have been plotted to establish a defense for the public, for those who stand to lose the most in a possible war of the worlds.

Energy beams, paralyzing rays, force fields, and death rays should not seem incredibly fantastic in the world of today. A mechanism capable of cooking a human body inside its unscorched clothing can be found in most modern hospitals. It is called a diathermy machine and is constructed of a short-wave radio generator arranged so that the output of energy is absorbed by a human body.

UFO HOSTILITIES AND THE EVIL ALIEN AGENDA

The majority of us are familiar with the microwave oven which, unlike standard ovens, cooks food from the inside out in a matter of seconds. If the microwave oven should be set at too high a temperature, it is possible to completely char the interior of a thick steak while leaving the outside raw.

Exposure to intense radar waves can "fry" a man's internal organs and still leave his skin visibly untouched. In World War II, the development of high-frequency waves for radar encouraged some scientists to actually seek to perfect a "death ray" for military use. Recently, physicians have begun to warn of the dangers of prolonged contact with such high-intensity microwaves as those employed in the Distant Early Warning radar chains. A medical team at Johns Hopkins University has released findings which indicate an association between mongoloid children and fathers who work around powerful radar sets.

Controlled use of lightning balls, or fireballs, is being considered as a military weapon by both the United States and Russia. It has been demonstrated in the laboratories at the University of California and the Bendix Research Laboratories that an electrodeless discharge should, under suitable conditions, form a spherical plasmoid. Powerful microwaves can be focused into a confined space, and the artificial lightning ball has a diameter equal to about one quarter of the wave length of the microwave field. A plasma gun was constructed at the University of California Radiation Laboratory nearly ten years ago. The gun was capable of expelling doughnut-shaped plasmoids at a speed of about one hundred twenty miles per second.

The potential of the laser beam seems to be limited only by man's technological knowledge. Maser Optics Incorporated has already produced a "laser rifle" for the U.S. Army. The rifle is able to blind a man or to set him on fire at a range just under one mile. The twenty-five pound rifle carries a battery which stores enough energy for ten thousand flashes. It has a maximum firing rate of once in ten seconds. Government sponsored research in radiation weapons has been budgeted at $2,000,000 per year. Total laser research is funded about $20,000,000 per annum. Admittedly, our various "rays" are still in the primitive stages. Consider, however, the weapon capacity available to a technologically advanced race—a race capable of breaching the stars, perhaps even breaching time and space.

In February, 1963, the Associated Press carried an article on Dr. Leo J. Baranski, an experimental psychologist at North American Aviation. Baranski was working on a ray that would be capable of doubling a man's life span or of paralyzing or slaying entire populations, depending upon the use to which the ray would be put.

Baranski's research had led him to an advanced study of the organic molecules of the acid adenosine-triphosphate (ATP). These molecules are essential to the function of the body's muscular system and provide the energy for muscular contraction. Baranski theorized that if one could find the microwave frequency of ATP molecules and control the release of energy by artificial means, one might be able to direct the muscular system into acts of superhuman strength. On the other hand, Baranski recognized the fact that such a manipulation might bring about muscular paralysis or death.

The scientist was quoted as saying: "Using masers [microwave amplifying devices] we could extend the range of effectiveness to almost any distance on earth and in space. A weapon that incorporates ATP's critical microwave releasing frequency could produce effects ranging from completely reversible traumatization [temporary paralysis] to virtually instant lethality.

"Such a weapon is within the present state of the art; even down to the size of a hand gun, since all the necessary components are already on the market."

Once such a molecular manipulator was developed to a high level of sophistication, the ray could instantly charge a muscular system with power and enable a man to accomplish heretofore unimagined feats of strength. Negatively, the ray could flood the muscles with too much energy and cause them to "lock" against one another. In this case, the man would become literally "musclebound," completely paralyzed. In still other applications of the ray, certain muscle groups, such as the cardiovascular system, could be singled out and paralyzed in order to bring about instantaneous death.

Again, we need not belabor the point that a technologically superior race, which is capable of interplanetary or intergalactic travel, may certainly have arrived at a practical application of Baranski's theory. Reports of paralyses suffered after close approaches by UFOs are many, as are reports of mysterious cases of accelerated growth.

WARMINISTER'S 'THE THING'

In the strange phenomena that have been plaguing Warminster, England, for so long, we can find almost the entire range of activity possible with microwave stimulation of ATP and subsequent energy release. (A more detailed account of this eerie case is cited as: "The Paralyzing Force That Stalked An English Village" in Brad Steiger's *Strangers From The Skies.*)

Mrs. Madge Bye, 38, was suddenly aware of a "crackling" sound above her head as she walked to church on Christmas morning, 1965. She felt herself hurled against a church yard wall and "pinned there as though by invisible fingers of sound." Mrs. Bye was then frozen into a temporary state of shock.

The weird crackling sound paralyzed a child's dog and set a nine-year-old girl's limbs to jerking about when she attempted to carry the dog indoors.

The Reverend P. Graham Phillips spoke for many of the villagers when he told newsmen that he and his entire family had often observed a "brightly glowing, cigar-shaped object" hovering over the village of Warminster. David C.

UFO HOSTILITIES AND THE EVIL ALIEN AGENDA

Holton, a botanist, geologist, and biologist, told of seeing a flock of pigeons suddenly killed in flight for no apparent reason. He examined the birds as soon as they struck the ground and found that almost instant rigor mortis had set in.

"There can be only one explanation which is at all feasible," Holton told a representative of United Press International. "All the evidence points that way. This is neither natural nor supernatural. It is extraterrestrial, I assure you."

Mrs. Dora Horlock was quick to second Holton's extraterrestrial thesis. It seemed to her the only logical explanation for the twelve-foot thistle that sprouted virtually overnight outside her cottage.

Several reports of giant plants sprouting up after extensive UFO activity in a particular area have come to the author's attention, but it is not only plants that seem to have experienced dramatic growth spurts.

There is a most bizarre postscript to the story of ten-year-old Sharon Stull of Albuquerque, New Mexico. Previously, we described how the girl was burned by a UFO on April 28, 1964. Those facial burns may have far greater implications than the physicians at Bataan Memorial Hospital realized when they released Sharon. For in the four weeks that followed, Sharon Stull grew five-and-one-half inches and gained twenty-five pounds.

"A month ago she was just a child who liked to play with dolls," Mrs. Stull said. "Now she is suddenly mature and grown-up, cooks meals by herself, cleans house, and takes care of the younger children."

Mrs. Stull told reporters that Sharon had recovered from the eye burns, although she needed to wear dark glasses in the sunlight and could read only a few paragraphs before her eyes began to smart.

"I'm so confused I don't know what to believe," Mrs. Stull said of her daughter's sudden growth and nearly complete change of personality and habit.

"I just feel funny," was Sharon's reply to queries about the state of her health.

When Sharon Stull saw the strange object in the sky, she had been four feet, eight inches tall and weighed eighty-five pounds. Four weeks later, Sharon stood nearly five-feet-two and weighed one-hundred-ten pounds. The ten-year-old girl had quickly outgrown all her clothing and rapidly stretched out of even her new dresses and shoes.

"I know she definitely saw something in the sky but I don't know what," Mrs. Stull said. "It has been a nightmare for us ever since. I wish I had kept her inside that day."

Chapter 05

RADIOACTIVE! - THE MOONBEAM SYNDROME

There have been physical trace cases where UFOs have singed or left an impression in the soil and when a Geiger counter was passed over the area it showed the landing place to be "hot"! The degree of radioactivity varies from case to case, but we can say with a degree of certainty that most UFOs normally do not leave you healthier than before you encountered them (not always true, as some healings have transpired, but this is not the topic for discussion here).

Researcher Micah Hanks calls it the "Moonbeam Syndrome," and while such cases are not unknown, thankfully they do seem to be rare.

One witness might have been trying to cause a degree of hysteria when he claimed a domed UFO was omitting an unknown substance over a residential neighborhood. The tabloid the story appeared in tried to create a buzz when it posted the sensationalistic headline on December 22, 2016: "SHOCK CLAIM: Alien UFO 'Sprayed Radioactive Green Chemicals' Over Homes." Apparently, a video of the incident was sent to the Mutual UFO Network for analysis but we know of no posting offering any conclusions.

Beyond ghastly, I would say. It's like being on the front lines in Syria and coming under a poison gas attack. But led by an alien foe instead of an earthly madman.

It's hard to say what category one should place the death of Max Spiers into. It sounds like something the Russians are accused of doing to spies who fled to the UK.

Spiers was deeply into the political side of UFOs - cover-ups, Disclosure. He was attending a conspiracy conference in Poland where he fell ill and vomited two liters of some unknown black fluid. An associate called a physician, who was unable to help revive the researcher. The Polish authorities said Max died due to "natural causes," but, upon having his body returned to his native UK, the coroner there said his death was "unexplained."

His mother was quoted as having said: "I think Max had been digging in some dark places and somebody wanted him dead." One source is insistent that

UFO HOSTILITIES AND THE EVIL ALIEN AGENDA

Spiers' interest in the "weird" goes back to his childhood. In a diary entry written by Max, aged 5, he recounted seeing "lights outside of the window and strange men wearing 'v' shapes around their necks." Max had been interested in conspiracy theories from a young age and he would often stare outside of the window expecting to see UFOs. Some think that he had watched a program called Monkey Magic as a child and imagined himself to be Tath_gata, "a character with the Buddha's power."

I am going to put this down as a dose of radiation or something far worse. Though I do believe that if you are exposed to a lethal dose of radioactivity you are not going to puke black goo and certainly not find yourself keeling over unexpectedly. Nevertheless, I guess a few Russians would have something different to say — IF THEY WERE STILL ALIVE!

A HORRIFIC DEATH FROM A UFO

Harry F. Koch, Publicity and Research Director of Universal Research Society of America, who made this sighting and reported it, was burned by radiation, and weeks later, died mysteriously of a supposed heart attack which had none of the usual symptoms of heart disease. Did radiation kill him?

It was 8 P.M., on April 3, 1966. I looked at my barometer that hangs between two windows that face almost west. After I turned off the room lights, I looked out the window to see if the stars were out because the barometer was in the stormy area. As I looked out the window I noticed what looked like a high moving orange-red ball of light in the northwestern sky. I stood and watched it to be sure it was moving, thinking it could be a star or something else.

Weather: A clear cold night; bright moonlight; many bright, twinkling stars; a very calm night. The object appeared to be moving eastward along the northern azimuth of 50 degrees. I watched the object two or three minutes; still doubtful, I decided to go outdoors. Standing near the west wall of my house, I couldn't see anything moving as I did in the house. I moved about three feet away from the house. If the object of orange-red was still moving, the limbs of the tall trees would have blocked my view.

Moon position: To my left, still low at about 40 degrees azimuth the moon was bright and almost full. Looking at my watch, it was 8:10 P.M. Still thinking, maybe the object I saw from inside the house was a satellite or a high flying jet plane; but I could not hear any jet sounds, the night was quiet, limbs of the trees very still as I peered through them toward the north and northwest.

When I first sighted the object in the house, I didn't see any pulsating lights, as the law requires all planes to have. But I didn't see the orange red object again. I don't know if this has any connections. Suddenly, as I stood near the house, I caught various colored bright flickering lights out of the corner of my eye. I looked

70

up toward the eaves of the two story house. I noticed a ball-shaped object about the size of a baseball as the bright flickering lights came into view. It came across the house rooftop from an angle; I figured at about a 25 degrees angle from where I saw it and moving away from me.

But it had to come from about 70 degrees azimuth, otherwise it would have hit the house. The object must have just cleared the roof peak because of the angle. It also missed my TV antenna and the tree limbs that hang over the front part of the house roof.

Action of the object: The object appeared to be breaking up into quarter-size flake pieces, each flaming fiery red, orange, yellow, whitish and blue-green - at least those were all the colors I remember. These colored flakes trailed behind and within a tail, each piece rolling and tumbling, which caused the flickering affect. These flakes seemed to flow behind the main dark ball within pencil-thin streaks of various colored flames. And the colored flakes kept within the tail area, which was as wide as the dark ball object. The tail extended away from the main ball about a foot or a little more. The flakes also appeared to be dropping, or they could have been disintegrating or revolving around within the tail. All the flakes appeared to be separated, causing its weird fiery light or fire to flicker. It is hard to explain (see drawn chart of object). The nose of the bail (front end) seemed to be solid, it was a fiery orange-red, the side and back appeared black. There were no sounds, odors, vapors or heat. Nothing seemed to fall toward the ground, such as ashes, dust etc. I say this because as it passed directly over my upturned face, I would have felt something on my face, head and hands. It all appeared to be in a vacuum.

Reasoning: At first, I thought the object was a comet or a meteor because I have witnessed them many times out in the western desert country. But I never saw them so close to the ground and moving so slow. A few I did see hit the desert floor, but it was just a matter of seconds as they streaked through the sky and plunged nearly straight down into the sand.

Course of object: As the object moved near a fence line of a hedge that was 200 feet from me, (I measured the distance) and as it came about 10 feet above the hedge, all the colored fiery lights or flames just slowly faded away.

Speed of object: From the time the bright object left the eaves of the house, and passed over my upturned face, till the flares slowly disappeared near the hedge or over them, I would say it took 8 seconds to travel the 200 feet.

Notation: When the fiery colored object stopped, I tried to line the object up with a highway street light that was about one and a half blocks away and at an angle from me. I was hoping I would see something of the object ball drop toward the ground. But I didn't see anything go up or down. Watching this object, I felt that if it dropped into the dry hedge or weeds it could have started a fire because it was still the size of a baseball. Or if anyone at that time would have walked across the lawn and got hit by it, it would have or could have burned the person. With all the weird fiery appearances, it still didn't look like the fire flames as we know it. The colored lights or flares were very fiery bright, yet it had the appearance of seeing it through a slightly frosted glass, like the glass one views x-ray or slide films with.

Radiation test: The following day, after I sighted that strange ball of fiery light, (on the afternoon of April 4, 1966) my scalp and the back of my ears broke out with a reddish rash. It was visible and itched. This discomfort continued until April 8, 1966, when all the rash and itching stopped. I had my sister, Lorraine A. Koch look at my scalp. After close inspection, she assured me the visible rash was all gone. She being a former Beauty Technician, having graduated and passed the State of Indiana license board as a Beauty Operator Technician, stated she didn't know what the rash was, nor what caused the rash. Without washing my hair or using any medication, the rash disappeared and the itching stopped. All I could think of was that, as I stood under that ball of fiery light passing over my upturned face and head, (although I didn't sense any feelings of any particles of dust dropping from the object), it could maybe have been the rays from the lighted object. It was an oversight on my part that we didn't take a radiation test of my hair and head when Orvil Hartle and I took the radiation test of the flight line of the object early the following morning (April 4, 1966).

The tests proved negative in all areas but one place, and that was under a tree limb that was 25 feet above the ground where the object passed under about 12 feet above the ground on its downward flight. On the ground under that tree limb the Geiger counter recorded 37 counts per minute, all other areas was on the average of 25 counts per minute, which was normal. I also remembered, as the object was under that tree limb, the colored lights appeared to flare and fizzle without sound, as if the object was beginning to start a fission or to explode. This could have been the reason for the increase in radiation under the tree limb.

April 11th, 1966. TO WHOM IT MAY CONCERN: On April 8th, 1966, I viewed a negative for Mr. Orvil Hartle. My opinion of the negative is of a normal outside exposure of an unidentified object on a sheet of 127 size film and in no way has the negative been altered or tampered with, to the best of my knowledge.

App. 60' tree 24' from next tree

App. 50' tree 21' from next tree

App. 40 to 50' tree 29' from next tree

App. 50' tree - 40' from corner

Stop sign

20' phone pole

Oberreich Street

Mail Box

Cable

App. 25' tree 21' from corner

TV Ant.

Area object faded away

I stood here, object passed 15' above me

200'

Object flight path

Hedge Line

Tree limbs that extend out over lawn 35' from tree trunk and 15' above the lawn. Object passed under it.

N

Sidewalk

Basserman Street

Highway 2

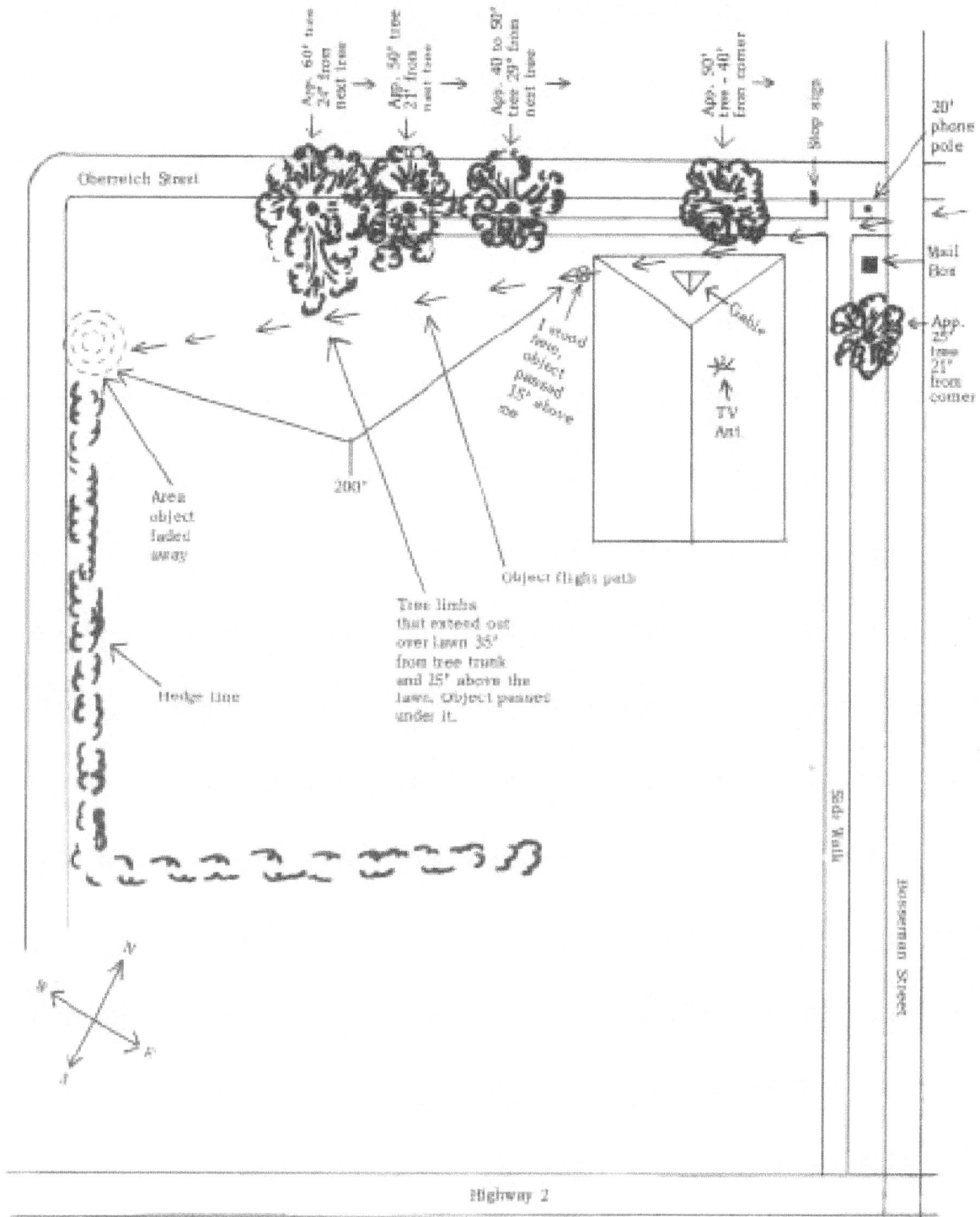

The area and flight path of the object

12

From nose of ball to end of tail about 1 ft. long

Various colored pencil line fiery streaks

Fiery, flickering, flaming colored quarter size light

Size of a baseball

Point object faded away

About 75 ur u/m

15'

10' Above the ground

Hedge fence

Where I stood

200'

In sight for 8 seconds

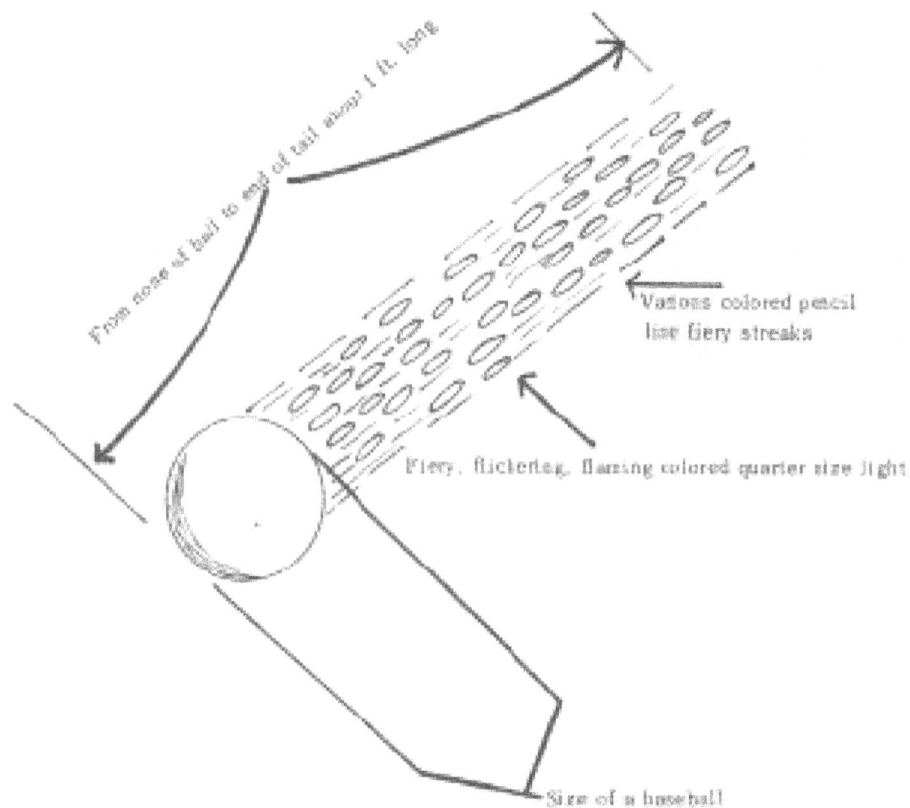

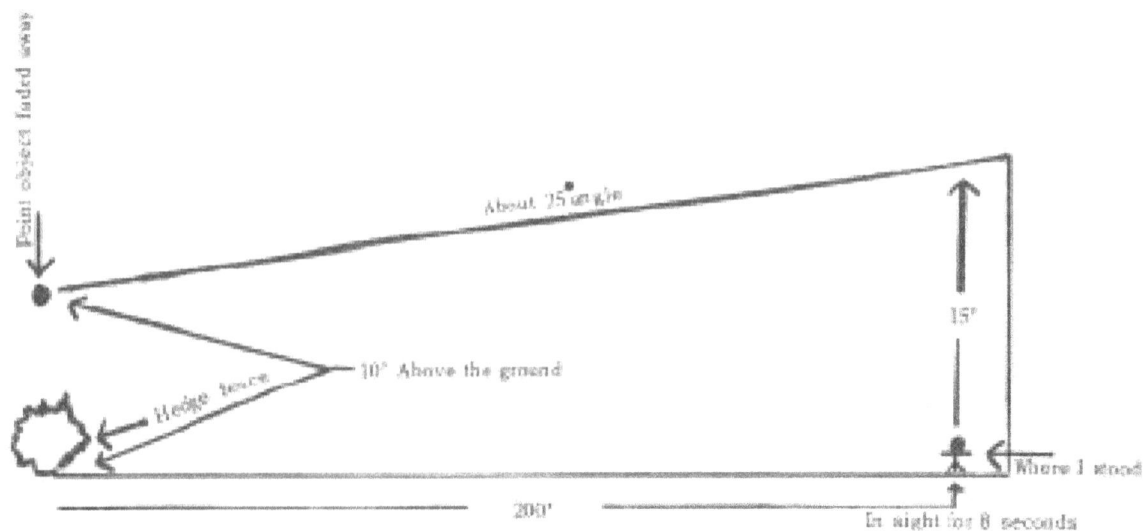

The angle and what the object looked like.

Chapter 06

A SERIOUS CASE OF RADIATION BURNS
By Sean Casteel and Tim Beckley

It's the best known case of its kind!

It's very well documented!

The incident has never been satisfactory explained!

It proves that not all UFOs have our "best interests" at heart!

This work contains numerous cases where UFOs have come a-zapping, seriously burning, scalding, giving out lethal doses of radiation, and in a few cases incinerating human flesh.

But none stands out in the annals of UFOlogy for its real-world authenticity as does what has become known as the "Cash-Landrum Case." It has become without a doubt the best example of a close encounter of the second kind.

The best abbreviated account was penned by long time UFO investigator/author Jerome (Jerry) Clark in his "The UFO Book: Encyclopedia of the Extraterrestrial":

"On the evening of December 29, 1980, near Huffman, Texas, three occupants of a 1980 Oldsmobile Cutlass observed a remarkable sight. Betty Cash, Vickie Landrum, and Vickie's seven-year-old grandson, Colby, on their way home to Dayton, were driving through the southern tip of the east Texas piney woods when they noticed a large light above the trees some distance ahead.

"The light was briefly lost to view," Clark continued, "but they saw it again when after rounding a curb they found themselves on a straight stretch of two-lane road on Highway FM 1485. This time it was approaching them, floating above the road at less than treetop height and belching flames from its bottom. Cash and the Landrums were only about 130 feet from the object."

Landrum told Cash to stop the car, fearing they would get burned if they got too close. However, as a born again Christian, she quickly reinterpreted the object as a sign of the Second Coming of Jesus. She even told Colby that the light was Jesus and would do them no harm.

UFO HOSTILITIES AND THE EVIL ALIEN AGENDA

The Cash-Landrum incident as depicted by concept artist, Carol Ann Rodriguez

Meanwhile, it was impossible to simply drive away because the road was narrow and the shoulders soggy. Betty Cash, who was driving, knew that the car would get stuck if she tried to turn around. Since there was no other traffic on the lonesome stretch of road, she stepped out of the car, as did Vickie Landrum. Vickie returned to the car as soon as she heard Colby screaming in fear.

"The object, intensely bright and a dull metallic color," writes Clark, "was shaped like a huge upright diamond, about the size of the Dayton water tower, with its top and bottom cut off so that they were flat rather than pointed. Small blue lights ringed the center and periodically, over the next few minutes, flames shot out of the bottom, flaring outward, creating the effect of a large cone. Every time the fire dissipated, the UFO floated a few feet downwards toward the road. But when the flames blasted out again, the object rose about the same distance. The witnesses said the heat was strong enough to make the car's metal body painful to touch."

The object then moved to a point higher in the sky. As it ascended over the treetops, the witnesses claimed that a group of helicopters approached the object and surrounded it in a tight formation. Cash and Landrum counted 23 helicopters and later identified some of them as tandem-rotor CH-47 Chinooks, routinely used by military forces worldwide.

The UFO and the helicopters were then "lost to view," according to Clark, and the three resumed their journey. The whole incident had taken 20 minutes.

But what followed for Cash and the Landrums after their return home was unexpected sickness and misery.

Cash began to suffer a headache and nausea that would not go away, and large knots formed on her neck and scalp. Soon they became blisters. Also, her skin was reddening and her eyes swelled. She threw up repeatedly and experienced severe diarrhea. The Landrums were experiencing similar symptoms, though not as intensely.

Cash's friends soon feared she was near death because she had lapsed into near-unconsciousness. She was taken to the emergency room at Parkway Hospital on January 3, 1981. She could not walk and had lost large patches of skin and clumps of hair. The Landrums improved slightly, though the sores on their skin and the damage to their eyes persisted. Vickie suffered periodic sickness over the next few years and Colby had problems with chronic illness, sores and hair loss.

"A radiologist who reviewed the victims' medical records for the Mutual UFO Network," writes Clark, "concluded 'We have strong evidence that these patients have suffered damage secondary to ionizing radiation. It is also possible that there was an infrared or ultraviolet component as well.'"

Cash and Landrum undertook a long and frustrating campaign to get compensation for their injuries from the government but were unable to prove that either the diamond-shaped craft or the military helicopters were the property of any governmental or military entity.

WEEKLY WORLD **40¢**

NEWS

Amazing psychic predicts the big winners of this year's Academy Awards
CENTERFOLD

NEWS EXCLUSIVE

3 SURVIVE UFO ATTACK

They tell of nightmare encounter with spacecraft

PAGE 19

UFO HOSTILITIES AND THE EVIL ALIEN AGENDA

Cash died at the age of 71 on December 29, 1998, eighteen years to the day after her close encounter.

Landrum died September 12, 2007, seven days before her 84th birthday.

HARMFUL EFFECTS ON WITNESSES BY UFO

Here is a more detailed checklist of the seriously negative effects the incident had on the witnesses.

MEDICAL EFFECTS

Colby's face was sunburned. He had problems with his eyes and that problem remains even now, but not to an extreme.

Both Vickie and Colby felt strange when they arrived home. She used three bottles of baby oil on Colby's and her own face over the next several days. She and Colby had diarrhea for several days (at least three days). Colby's was so bad she had to clean him up like a baby. They both had bad stomachaches.

The hair on the right side of Vickie's head came out, but it later grew back. She said her scalp felt like it was asleep.

Vickie's fingernails, left hand only, were damaged. That hand was on top of the car.

Colby had nightmares for two or three weeks. He would wet the bed because he was afraid to get up at night. His school grades fell off for a week or so; but he is back to normal now – A's.

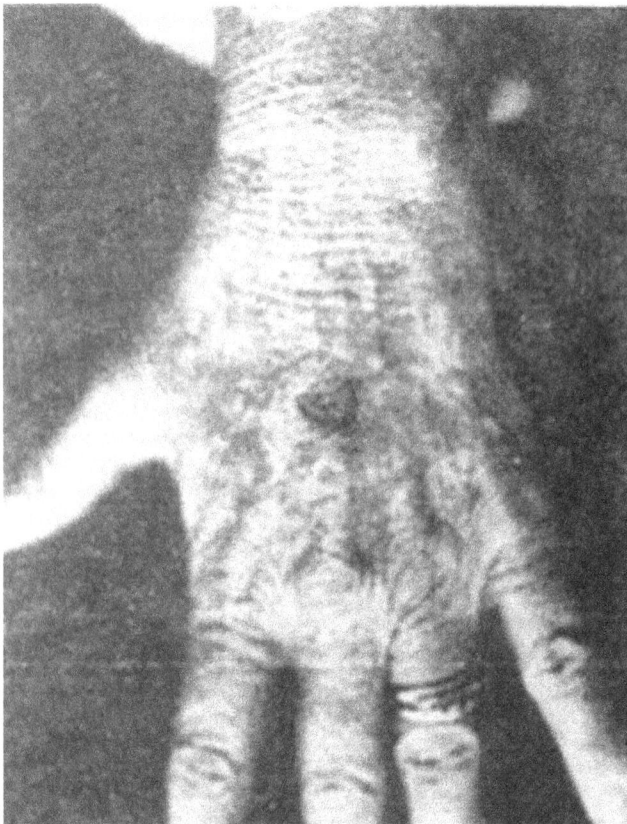

Vickie thought Betty Cash was going to die. Betty was extremely ill until Vickie insisted she go to the hospital on the fourth day. Betty had a swollen neck, large nodules (blisters) on her scalp, face and right eyebrow. She couldn't eat, and that is unusual for Betty. Vickie explained that Betty is usually a big eater.

For example: Betty and Vickie had a big lunch on December 29, and that evening they stopped at New Caney to eat again. Vickie just had coffee while Betty had sausage, eggs and extras. After the incident Betty could not eat. Vickie would make her some soup or other meal – she would eat a bite or so and get deathly ill and vomit. She was unable to eat from the time of the incident until she entered

78

the hospital. She too, had diarrhea very badly.

When Betty was admitted to the hospital at the emergency room by Dr. Shenoy, the nurses asked if she was a burn patient. After two weeks she was released. She was home 13 days, of which she spent seven or so in bed all the time. Because she remained ill, could not eat properly, and seemed to be getting worse, she returned to the hospital.

Vickie had a problem with her eyes – the right one being the worst. They were both very inflamed and the right one drained constantly. Before the incident, Vickie wore eyeglasses for reading only. After the incident, she had glasses she must wear all the time, prescribed by a Dr. Chandler, an eye doctor from Liberty, Texas.

HELICOPTERS

Vickie and Betty only saw the object during the close encounter – no helicopters. As it rose and flew away, there were more than 20 helicopters, although Vickie admitted that they moved around a lot and a few might have gotten counted more than once. Even so, she is positive that there were 10–12 or more – no doubt about it.

Colby says he saw helicopters all during the event, even when the object was low over the road. He is quite sure of that.

Vickie says there were two kinds of helicopters involved – maybe more. One kind was large and smooth running with a very large rotor on top. Another had two rotors on top, but one was above the other slightly.

The three witnesses drove on after the object rose up and the fire stopped coming out of the bottom. At three different vantage points, they observed the object gaining in altitude and distance with the helicopters in pursuit.

Vickie truly feels that this was not anything unnatural. She believes the U.S. government was transporting and escorting something dangerous through the area. (Her son mentioned a similar incident near the lake about six months earlier in which a fiery object landed and burned 300 square feet of grass).

Colby said the helicopters appeared to be trying to find out what the object was. He counted 18 - 20 helicopters and stopped counting.

SUGGESTED READING

CASH-LANDRUM UFO INCIDENT by John Schussler

THE UFO BOOK: ENCYCLOPEDIA OF THE EXTRATERRESTRIAL, by Jerome Clark

Chapter 07

ETs - DO THEY HAVE BLOOD ON THEIR 'HANDS'?
By Timothy Green Beckley

The material in this chapter has been edited and assembled by editor/ publisher Timothy Green Beckley based upon the actual words of Alvin Moore as put forth in his *"Diary of a CIA Operative"* published in a very limited edition by Tim's Inner Light - Global Communications.

WHO WAS COMMANDER ALVIN E. MOORE?

Of all investigators of UFOs, the late Commander Moore has one of the most impressive backgrounds. His skill in engineering and other sciences, his expertise in history, law and intelligence work uniquely qualified him to evaluate the mass of data about the alien presence, which he defined as Skymen, or extraterrestrials. Educated at the U.S. Navy Academy, plus the American University, George Washington School of Law and Louisiana State University. Moore held both B.S. and MA degrees. As part of his extensive contribution to his country, Moore specialized in aeronautical engineering and was the patent engineer and attorney for the Wernher von Braun team of space scientists. He also served as a US. Patent Office Examiner specializing in aeronautics and propulsion; an assistant nautical scientist with the Navy Oceanographic Office; a CIA intelligence officer and an American Vice Consul. He was granted more than 50 U.S. patents on his own inventions, mostly on aircraft, marine craft and automobiles. He also authored numerous technical and historical articles and books.

UFO HOSTILITIES AND THE EVIL ALIEN AGENDA

* * * * *

STRANGE CONCLUSIONS REACHED BY COMMANDER MOORE

Aliens known as "Skymen" have been coming to Earth and exploiting it for many years. Some of them have homes in caverns on the moon, Mars and its satellite Phoebus, Jupiter, as well as the Asteroids. Many more originate much nearer to the Earth's surface, from "Sky Islands," or even from within the hollows of our planet, and possibly underwater hangers! Sky chemicals and electrostatic gravity-like force of the alien sky islands and skycraft have caused legions of accidents. Skymen have kidnapped a multitude of people and have long extracted blood from animals and men, as well as committed mysterious murders!

During his life, Commander Alvin E, Moore (of Naval Intelligence and with close ties to the CIA) spent nearly a quarter of a century researching the possible association of mysterious events attributed to UFOs and extraterrestrials. His conclusions are utterly shocking and certainly deserve further investigation despite attempts by so-called "serious UFOlogists" to sweep this aspect of the phenomenon under the cosmic rug.

There was a report on a New Orleans television news program the evening of 10 September 1977 of a woman who had been found dead with "punctures in her abdomen." And the next morning's New Orleans Times-Picayune gave a few more details of her strange death. She was an "unidentified black woman in her early twenties found dead in bushes of a hedge. Her blouse had been pulled up over her face and her jeans had been unzipped. Police stated she had apparently died of a penetrating wound to the vagina, although they were unsure what kind of instrument had been used to inflict the fatal wounds."

I note that the plural word "wounds" was used - tying in with the plurality of "punctures" in the abdomen referred to in the television report; and I wonder if one of those wounds was made by an instrument that penetrated the womb of the young woman. As apparently was done in the case of Betty Hill, and might have been done in the case of Pat Price. Also blood might have been extracted via one or more of the "punctures." The police would easily recognize a slit made by a knife or dagger. Was the puncturing instrument round or triangular?

On the night of May 10th, 1951, police found a young woman screaming on a street corner in Manila, crying out that she was being bitten by something, and seemed to be fighting something in the air. She was eighteen years old and her name was Clarita Villanueva. The police arrested her. She said she was being bitten by something that looked like a man with big bulging eyes, wearing a black cape, and able to float in the air.

After she was locked up in a cell, she began screaming that the thing was coming at her through the prison bars. A policeman unlocked the door and brought her out, still screaming. And then he was astounded to see punctures that looked like teeth marks being formed on her arms and shoulders. The next morning in the presence of policemen and Medical Examiner Dr. Mariana Lara, the screaming girl was again attacked by a being or object, invisible to the men. For five minutes the punctures appeared on her arms, back of the shoulders, and back of the neck. Then she fainted.

Reporters and the Mayor of Manila Lascon came. The Mayor and Medical Examiner went with her in a car to the prison hospital. On the way she began screaming again, and the horrified men saw what looked like teeth punctures appear on both sides of her throat, on a finger; and while the Mayor held one of her hands it was deeply punctured by an invisible wound-making instrument.

They saw that her arms and hands were badly swollen from previous punctures. Obviously, this was a case of blood extraction by a floating being or object - probably object - that could be made invisible.

Concerning invisibility of alien sky objects: In 1953, Associate Editor John DuBarry of True Magazine showed me a photograph that a woman had taken of her rose bush. There was a little object in the air near the plant that looked like a swastika. The woman who sent the picture to True had not seen the swastika, but the camera had pictured it. If it had whirled fast enough - with the speed of a bullet in flight - she would not have seen it. This ancient-earth symbol, the swastika, was one of the characters in the strange writing on the footprint casts of the skyman that George Adamski reportedly saw in 1952.

A number of persons who have been captured by alien skymen have reported their extraction of "blood samples." This taking of blood probably involves more than mere samples.

Some earlier examples of mysterious, blood extraction wounds: In June and July 1899, in the eastern United States, an outbreak in many cities of a large number of so-called "bug bites" - for instance, eleven persons in New York were said to be "bitten" on 8 July; and children in Iowa, New Jersey and New York were "bitten" and one died; and on 10 July on a woman in Chicago: "the marks of two small incisors could be seen." About 16 September 1910, in Portugal, a child was found dead. Her body was bloodless. The last man seen with her was arrested. Somehow, the police got a confession from him; he said he was a vampire. On 7 May 1909, on Broadway, New York City, at least six persons reported being jabbed by something like a hat pin. In March 1901, at Cambridge, England, Lavinia Farrar was mysteriously killed. Her body, fully dressed, was found on her kitchen floor. There were no signs of robbery. Near her, on the floor, there were some "drops" of blood and a blood-stained knife. Policemen could see no connection between the knife and her death. But in the coroner's examination, her clothing was removed, and on the innermost garment the examiners saw a very slight blood stain. They removed this and found that the woman had been "stabbed to the heart." But the wound and her body were almost entirely bloodless! Her dress was not punctured and had no fastenings opposite the wound. Obviously, it had been replaced after the stabbing. What had happened to the blood? In my opinion, a skyman of the evil type had killed her and taken her blood.

On 19 July 1975, Charles D. Savelie, 21 years old, was in his truck on North Beach Road, Bay St. Louis, Mississippi, when, according to his later report, something bit his neck, puncturing an artery. "Savelie jammed a finger into the wound to halt the bleeding," and drove to a hospital. After an operation his condition, in spite of blood loss, was satisfactory. Deputy Sheriff Lethon Garriga said apparently he was stabbed. Probably he was - by an invisible blood-extraction device.

Our maid, Mrs. Leona Toomey, cleaned a room where a man of Waveland, Mississippi, allegedly shot himself and died. She said she was puzzled by the very small amount of blood on the carpet where he fell.

An extract from an entry in my UFO journal, dated 29 February 1972, at Waveland, Mississippi: On Saturday night, February 19th, as we were returning at about 11 PM from playing Canasta, Laura received one of the strange, slightly bloody, small cuts she gets from time to time. She was unaware of the cut until we got home. On the 21st Laura again was mysteriously cut - in a small cut, with blood, on one of her fingers. (In an entry I also referred to several of her mysterious occurrences on the 18th to 21st.)

On the morning of 31 December 1930, 19-year-old Beulah Limerick of Washington, D. C. was found dead on a mattress in her bedroom. No blood was seen on or near the body; and a hospital intern reported her to have died of natural causes. But the undertaker who was preparing the body for burial found that a lock of her hair had been compactly placed in a hole in the back of her head. A news item of about fourteen years later concerning this strange murder stated that the hole marked the path of a small-caliber bullet into her brain.

But I wonder if a bullet actually was found. For: a small hole was discovered in one corner of her mattress, and a small blood stain was by it; and the body obviously was bloodless. For: when it was found, no blood was seen; the intern saw none; and apparently no hair was matted with blood near the hole in her head. The instrument that neatly punctured her skull probably went through the corner of her mattress and into the back of her head, as she slept - probably paralyzed by an alien skyman's power. The police made a long, intensive investigation; but no one guilty of the murder ever was found. I refer back to my account of the three mysterious 1953 wounds of the dog "Molly", which looked like .22-caliber bullet holes, but no bullet was found. According to my memory, I saw only a little blood around those holes.

On 24 March 1955, at Green Meadows, Maryland, the rather happy family of Mr. and Mrs. Cecil Harp ate supper. The children, 15-year-old Ken William, his

UFO HOSTILITIES AND THE EVIL ALIEN AGENDA

14-year-old sister Lois and their younger brothers Charles and Robert, were in high spirits. After supper, Charles went out to buy ice cream and Ken went upstairs to his room. About twenty minutes later, Lois went to Ken's room to call him to come down for ice cream. She found him on the bed unconscious. An emergency team arrived and tried to revive him, but he was pronounced dead. One of the volunteers noticed a small hole in his sweater just over his chest, with a corresponding hole in his chest. No bullet was found then or later.

Ken's father had given him a .22 caliber rifle with instructions not to use it until he wanted to go hunting. It was stored securely in a nearby closet. No one downstairs had heard a shot.

The sweater and Ken's body had no blood on them or it would have been noticed by the emergency volunteers. What happened to the blood? Was it sucked out by a skyman's blood-extraction apparatus, including a small, round, body-puncturing instrument?

On 26 September 1954, a mystery arose at Alexandria, Virginia, as to who or what killed 25-year-old Mrs. Grace Smith, a beautiful blonde dancing instructor. She was found dead in her locked, combined home and studio; and her 2-1/2 year-old son, Sergoff, was also in the locked upstairs bedroom, where his mother's body lay on the floor. Obviously, the woman had been killed, for on her bed there was a blood-stained sheet, and there were other blood stains in the room. But how had she been killed? There was no visible mark of violence on her body. Was this a mysterious blood-extraction death?

About 9 P.M., 6 August 1954, at Bonner Springs, Kansas, 44-year-old Mrs. Eva A. Wagenknecht, lay down on her bed, removing her clothes. In another room her husband Henry, son Henry, Jr. and two others were playing cards. The card game ended about 10:30 P. M., and Henry, Sr., turned on the television for a while. But Henry, Jr. said that soon after he went to bed he heard a noise like the house was vibrating. Thinking a storm had come up, he went to the window to check outside, but seeing no sign of stormy weather, returned to bed.

According to Henry, Sr.'s account (and he said he was willing to take a lie detector test in support of it): He quit watching television about midnight and went into the kitchen to see about some plum preserves cooking there. Uncertain that they were done, he went toward the bedroom, calling to Eva to come and taste them. Unanswered, he went into the bedroom; and he found his wife gone. Horrified, he saw blood spots where she had lain.

Policemen were summoned, and a little after midnight Marshals Alva Kerby and Elbert Woolf found Eva Wagenknecht's body in weeds, nude, face upward. An 18-inch, two-by-four piece of lumber was found, but it was about fifty feet from the corpse. In a nearby culvert Eva's blouse had been stuffed. Her slacks were found in the house. The blood spots apparently were few - on the bed, an adjacent curtain, and below the window. A window screen had been moved. Why was the dead or nearly dead body taken outside? Did murderous skymen quickly remove it, before extracting blood, in view of what happened in the later-described mysterious murder of Marilyn Shepard near Cleveland, during the month before this August, 1954?

The following circumstances indicate that this murder of Eva Wagenknecht might have been by an alien skyman or skyman-controlled device: (1) the vibratory noise and apparent brief motion of the house; (2) the fact that there was no sign at all of struggle in the bedroom; (3) the fact that the husband in a neighboring room heard no noise of the killing and removal of the body, apparently through the window; (4) the stripping of the body, possibly for blood extraction; (5) the fact that Eva, mother of seven children, apparently was not raped, and doubtless had no ruthless enemies; (6) the face-upward position of the body - a position convenient for extraction of blood, probably left unchanged in a hasty getaway.

In the morning of 8 September 1974, at Kennett, Missouri, the body of Mrs. Lee Ann Garrison was found in her home. She had been beaten, presumably shot, and was dead. On that morning and the night before, her husband was with seven persons on Pickwick Lake in northeastern Mississippi, including Ronald Windsor, prosecuting attorney of Alcorn County, Mississippi. Nevertheless, Missouri authorities tried to extradite Dr. Garrison from Mississippi to be tried for the murder. Dr. Garrison opposed the extradition; and the testimony at his hearing by Melvin Duckworth, Missouri Highway Patrol investigator, not only further established that the physician was not guilty, but indicates that a collector of various types of blood killed Lee Ann Garrison, for: (1) a surgical glove was found near her home with type-O blood on it; (2) type-A blood was in a footprint at the base of the stairs of the ill-fated home; (3) type-B blood was on the murdered woman's leg; and (4) apparently blood on her body was type-AB. She had type-A blood; and the only type-A blood found was in the footprint. Four types of blood at the murder scene! If a skyman had killed her and was taking her blood, the blood-holding device must have been leaking.

I have long been suspicious of some of the reportedly accidental deaths as a result of mishandling firearms. I have numerous reports and clippings concerning several deaths of this type during that hot, dangerous 1954 and early part of 1955. Early October 1954 was particularly hot around Maryland, Washington, D. C. and further westward. The hottest September 30th in Washington since 1861 had occurred, and on October 5th, a writer at the Washington Post stated:

"Since Friday we have had a sense of being constantly in a Turkish bath." And on the 6th, the Post included an Associated Press item that during the first three days of West Virginia's hunting season seven men had been shot to death or had succumbed to fatal illnesses while hunting.

Charles Fort

UFO HOSTILITIES AND THE EVIL ALIEN AGENDA

Early in the morning of 1 January, 1955, Michael A. Kendall of Bethesda, Maryland, thought he heard a prowler in or outside of his home. He got out of bed, loaded his .22 rifle and started across the room. Something caused the rifle to discharge and the bullet went through the flesh of his right shoulder. Police stated the rifle did not have a trigger guard. With such a rifle, Mister Kendall must have been very careful. But he was shot. This, of course, might have merely been an accident, but what of the prowler?

On the same New Year's Day in the same town, 17-year-old Douglas Shae Hutton left his family downstairs and went up to his room, which contained his .22 rifle and a number of National Rifle Association trophies he had won in competition. No one in the family heard a shot, but they later found Douglas dead, shot between the eyes.

Twenty days before these bizarre events, on a Sunday night, 12th December, 1954, in her apartment at Arlington Towers, across the river from Washington, D. C., 23-year-old Sallie Wood of the National Security Agency, was wrapping Christmas presents. In the room was a 12 gauge shotgun with the price tag still on it, which she had bought a day or two earlier as a present. Her door was locked and chained and the windows were locked.

On Monday, she and her fiancé, Herbert Gallegly, were to select an engagement ring. Instead, he and apartment officials broke into her room and found her shot to death. She was lying on her side and her shoulder had a gaping wound, apparently from a shotgun blast. The new gun had the butt on the floor leaning against and on the far side of a table.

How could she have been shot by a gun whose barrel was pointed toward the ceiling? An intensive investigation finally resulted in a freakish report supposing that she loaded the gun, placed it on the table, got back down on the floor to wrap more presents, and jarred the table, dislodging the gun which then tumbled off the table and discharged, then flung itself back against the wall on the far side of the table where it was found. This, clearly, was not possible.

The following events cause me to suspect alien skymen influence in Sallie Wood's murder. 1) The neighbors in the apartment across the hall heard no shot (and shotguns make very loud noises when they are discharged). 2) Her apartment was on a high upper floor and thoroughly locked and chained. 3) The new apartment complex was at the riverside. 4) There was a "blackened indentation" on the adjacent door. (Blackened by sky chemicals?)

During an evening of March, 1929, Isidor Fink, owner and operator of the Fifth Avenue Laundry, New York City, was ironing clothes. Numerous robberies had occurred in the neighborhood, so Isidor had bolted his doors and locked the windows. Only the transom was unlocked. Outside the laundry, a woman heard him screaming and other noises, but heard no shots. She called the police.

UFO HOSTILITIES AND THE EVIL ALIEN AGENDA

Isidor Fink was found dead on the floor with wounds that appeared to be bullet holes in his chest and in his left wrist. There were powder burns on the left wrist. Later, New York Police Commissioner Mulrooney said the killing, in a locked room, was an "insoluble mystery."

About 8 November 1954, Peter Pivaroff of Los Angeles began having severe pain in his heart. After about a day of the pain, he went to a hospital, thinking he had a heart attack. X-ray pictures were taken, but he died before he could see them. They showed a darning needle that had been inserted between the fourth and fifth ribs and into his heart. But this wasn't his only puncture. There was another, between his seventh and eighth ribs. The darning needle was identified as one his daughter Diana had borrowed, which had disappeared.

On 4 December 1913, Mrs. Wesley Graff, sitting in a theater box, felt a thing "scratching her hand," and pain there like a wasp sting. She rose to her feet. There was a man near her, the only person, she thought, who could have harmed her, so she accused him, and fainted. Policemen came. On the floor they found a pricking instrument - a darning needle. But the marks they saw on Mrs. Graff apparently were not made by the needle. Was this darning needle put on the floor as a cover-up for blood-extraction? Was the darning needle of Peter Pivaroff's daughter also such a cover-up?

Late in the night of 2 February 1913, the body of Maud Frances Davies was found on tracks of the London Underground Railway. Her head had been cut off by train wheels. At the inquest someone said that she probably committed suicide. But a Dr. Townsend (apparently the coroner) testified that, while living, she had been punctured more than a dozen times by a hat-pin-like instrument, and that in one of the thrusts it had penetrated to the heart. I say: Probably, while rendered immobile, blood had been extracted from Miss Davies. And then, in cover-up, probably a sky-man or skymen, carefully placed her body on the rails with her head positioned to be run over by the next train.

In January, 1975, nine men, mostly middle-aged drifters, were killed in the Los Angeles area in similar mysterious murders. The murderer came to be called the "Skid Row Slasher," because, with a surgeon's or skilled butcher's type of precision, and in one powerful cutting-instrument stroke, he slashed the victim's throat all the way to the spinal cord, nearly cutting off the head, in each of the nine murders of that month. Someone theorized that the murderer was a very powerful man who could hold an attacked man while he neatly nearly beheaded him. If the victim was conscious and in possession of his full strength, such a slashing would not be possible. But he could be made immobile, temporarily paralyzed, as so many victims of skymen have been, and his blood extracted, and then his head nearly cut off as a cover-up.

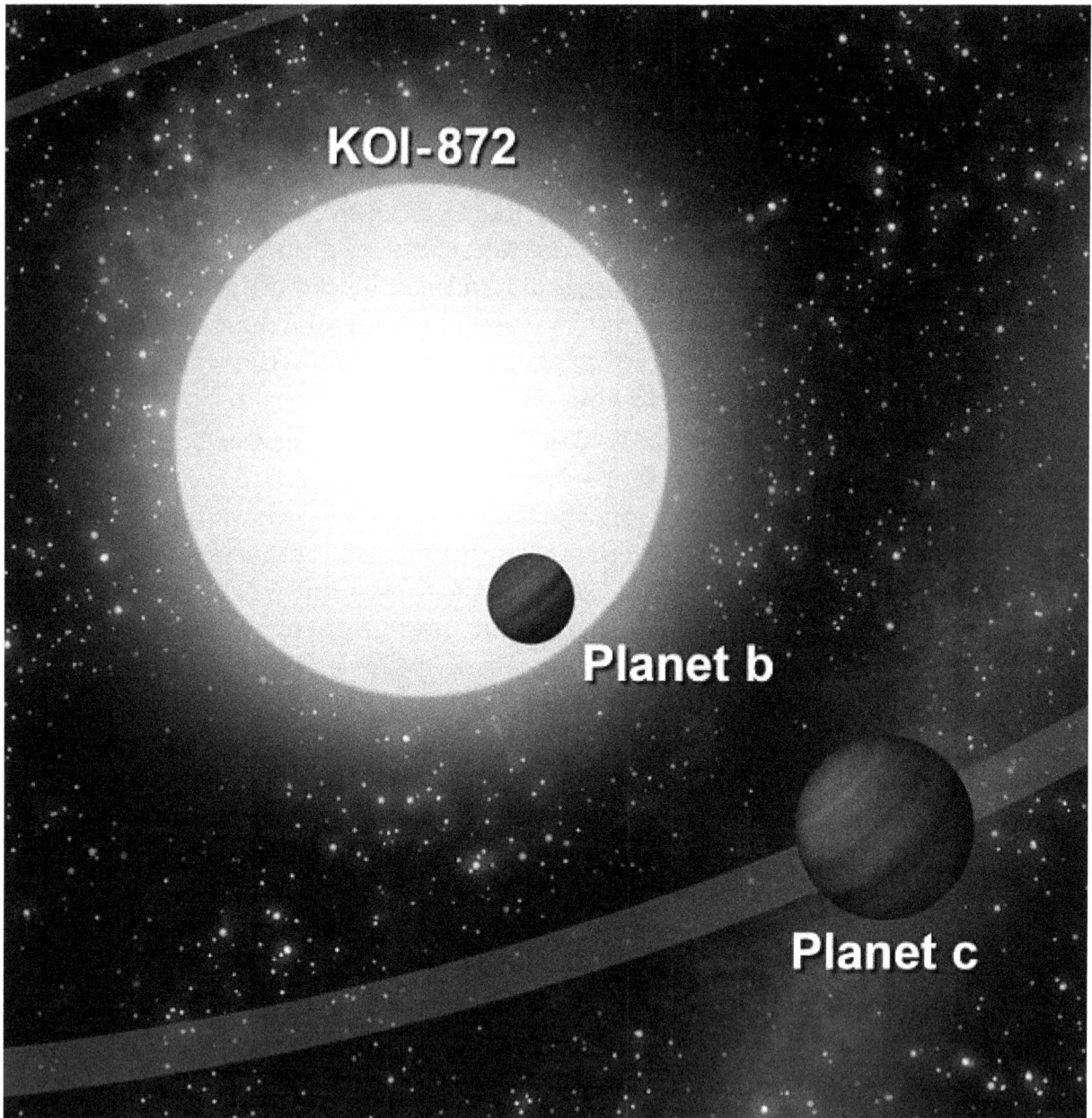

Alvin Moore's "Invisible Worlds."

Wartime UFOs: Combat & Military Atrocities

Chapter 08

THEY MARCHED INTO OBLIVION

By Timothy Green Beckley and Arthur Crockett

Nobody knows what happened to the full regiment of Chinese soldiers. When the Japanese tanks arrived in Nanking, they found no one to fight. Other armies have disappeared, too, leaving not a flake of skin or strand of hair as witness they had ever existed on Earth. It was as if they had all been lifted from the surface of the planet.

Disasters take many forms, and nearly all involve great loss of life and destruction of property. The earth trembles and a thousand houses fall in on their occupants, crushing them. A sleek airship explodes in the air and kills more than a score of people. A hurricane dashes a city to rubble, leaving thousands dead or homeless.

We are all familiar with the kind of devastation such calamities can manifest on the innocent. But there is another kind of disaster which may have escaped the notice of most people.

With these disasters, there is no residue of a fiery holocaust, no hospitals jammed with broken bodies, no screaming headlines to announce the numbers of the dying and the dead.

The disasters to which we refer are silent and insidious, yet over the years they have claimed millions of lives, and not only have so many innocents perished, but their bodies have somehow – through some mysterious means – been removed from the face of the earth as well.

They are the strange instances in which large numbers of military men have literally marched into oblivion.

One of the first cases on record of a massive disappearance is also one of the saddest. The so-called "military men" were children – fifty thousand of them – who had marched off for the pathetic Children's Crusade. They had been inspired by a teenager in France, Stephen of Cloyes, who heard voices telling him to round up a child army to fight the Muslims in the Holy Land. A favorite explanation is that they were taken by slavers, but that theory has never been proven.

UFO HOSTILITIES AND THE EVIL ALIEN AGENDA

Members of the crew of the Eurydice

In more recent times, whole regiments have vanished. Military planes have flown into peculiar-looking clouds and never come out. Submarines with full crews have sailed into a limbo from which there is no return.

One amazing example occurred on December 10, 1939. After the Japanese seized the city of Nanking, Chinese Colonel Li Fu Sien set up a delaying action south of the city. Some 2,988 men were involved, but two hours after the front had been established, radio communication with the regiment came to a sudden halt. An investigation was conducted and aides found cooking fires still burning and the soldiers' guns neatly stacked. But there was no sign of the men.

The immediate assumption was that the entire regiment had deserted, but it was clear that nearly 3,000 soldiers could not have crossed open country without being detected by the Japanese. To this day the matter remains unsolved. Later, Japanese records were checked to see if the Chinese soldiers had been taken as prisoners. They had not. Not a single man of the missing regiment has been seen or heard from since that day. It is a puzzle that may forever go unsolved.

A similar incident occurred in the early days of World War One. During August, 1915, the Turks and the Allies were fighting on the Galipoli Peninsula. On the 28th, the fighting on Hill 60, Sulva Bay, Anzac, was in its final stages. According to eye-witnesses, the morning was cloudless except for six or eight odd clouds that remained motionless over Hill 60.

UFO HOSTILITIES AND THE EVIL ALIEN AGENDA

All were shaped exactly alike, were light grey and looked like huge loaves of bread. Despite a five-mile-an-hour breeze, the clouds remained fixed above the hill. Beneath them, at ground level, was still another cloud. Observers reported that it looked almost solid. It was about 800 feet long, 220 feet high and 200 feet wide. This huge cloud rested on a sunken road known as the Kaiaick Dere.

The witnesses of this odd cloud formation were twenty-two men of the Number Three Section of the First Division Field Company, NZEF. They were in trenches on Rhododendron Spur and five hundred feet higher than the road on which the cloud had settled. They had a perfect view, and during that hot morning they were to witness something far more terrifying than the strange cloud formation – they watched in horror and astonishment as an entire regiment of their comrades vanished into thin air!

The twenty-two men watched the 1/4 Norfolk Regiment march up the sunken road toward Hill 60, apparently on its way to reinforce the troops already on the hill. The observers were in position to see the road on both sides of the cloud, and they could clearly see the men march into the cloud, but did not see them emerge on the other side.

A British Regiment in those days comprised anywhere from eight hundred men to as many as four thousand. Although it is not known how many men were in the regiment, it was claimed that it took more than an hour for the last soldier to enter the cloud.

UFO HOSTILITIES AND THE EVIL ALIEN AGENDA

At that time the cloud was seen to lift from the ground and join the others, which were still hovering above. They all then moved against the wind and were observed by ground troops for nearly an hour before they completely disappeared.

Officially, the 1/4 Norfolk Regiment was listed as missing, or possibly wiped out. When Turkey surrendered in 1918 the British demanded the return of the missing men, assuming that they had been captured. The Turks, however, insisted they did not have the men as prisoners of war, nor did they even know the regiment ever existed. A complete recounting of the event can be found in official histories of the Galipoli Campaign.

Sudden and unexplained disappearances of military men are not as rare as one might imagine. In nearly every conflict one can find at least one instance in which soldiers, sailors, airmen or marines have simply vanished from the face of the earth.

Families expect their soldiers to be injured or even killed, but they have great difficulty believing their loved ones have simply vanished without a trace. How terrible it must have been for the families of the lost British platoon in the Sudan in 1898, in circumstances similar to the loss of the 1/4 Norfolk at Galipoli. In the Sudan case, search parties found footprints in the sand and followed them, thinking that eventually they would catch up with the missing platoon. But that did not occur. Instead, the searchers found that the tracks ended suddenly. There were no tracks to the sides or further ahead, and it was apparent the platoon had not backtracked. The entire armed cadre seemed to have been lifted right off the face of the planet.

Thousands of children were slaughtered or vanished during the Crusades.

UFO HOSTILITIES AND THE EVIL ALIEN AGENDA

The very same scenario was repeated in the area of the Khyber Pass when a company of men, possibly British engineers, were in combat at the northwest frontier of India and Afghanistan. After the company was reported missing, search parties were dispatched. Again, footprints were found and, again, they ended abruptly. None of the company was ever found. What happened? Was it possible that an entire company of men was abducted? No one has the answer. And these episodes are by no means isolated events.

On July 24, 1924, British Flight Lieutenant W. T. Day and Pilot Officer D. R. Stewart took off in a single engine plane for a reconnaissance flight over the desert of Mesopotamia where the Arabs were fighting with the British there as observers.

The estimated flying time was four hours. They did not return. The plane was found the following day. It had landed perfectly. There was no damage. The fuel tank had plenty of fuel to complete the flight and the engine was started and the plane was flown back to the base.

The riddle was only compounded when searchers found boot prints in the hot sand. The men had apparently jumped from the plane and walked away for about forty feet before their boot prints simply stopped. It appeared that the two men had been standing side by side when they vanished. Their tracks did not continue on in any direction. A massive search was conducted for many miles around the site of the landing but no trace of the men was ever found.

It would appear that no arm of the military is exempt from the powerful unknown forces which apparently come into play during the course of war. In 1939, just before the beginning of World War Two, four submarines belonging to America, Great Britain, France and Japan, vanished within a four-month period.

During the week of January 21, 1968, a French submarine and an Israeli submarine disappeared in the Mediterranean. They were hundreds of miles apart at the time. Both sailed into the limbo of the lost at almost the same time. Sophisticated electronic equipment failed to locate any large metal objects during extensive searches. Four more submarines have vanished since that time. They belonged to America, France and England.

On Wednesday, March 4, 1970, again in the Mediterranean, the French submarine Eurydice was swallowed up and forever lost. Repeated searches found nothing. The list of strange disappearances of military men continues almost endlessly. More than fifty pilots and men of the United States Air Force have vanished recently with no clue as to their whereabouts. Officials are now pondering the possibility that many of the men missing in action in Vietnam and Korea and World War Two are not lying in the jungle or in unmarked graves near farming villages, or languishing in prison camps, but may have been snatched off the earth by some unseen and unknowable force.

UFO HOSTILITIES AND THE EVIL ALIEN AGENDA

But why the military? And why, as in many cases, have these weird occurrences taken place during the height of battlefield madness? Why is the disaster of bloody combat compounded by the mystery of massive kidnappings?

There is evidence that during the times of great emotional upheaval, something happens which breaks the laws of normalcy. Call it a combined psychic energy, mass hallucinations, mass hypnosis or a crack in the continuity of time itself. No matter what tag we place on it, the fact remains that unexplainable things do occur, particularly in times of war.

Vast commotions occurred during the opening days of World War One concerning phantom armies and visions in the skies, all happening while real, live soldiers were marching off into oblivion.

Savants of these matters theorize that missing members may someday return as legions of phantom armies, the supposition being that they have entered a time zone unknown to mortals and, like unhappy ghosts, continue fighting for eternity. The theory holds that on the battlefield a great amount of energy is expended in a small area and that, in that space, men suffer much anguish, both mental and physical, enduring heights of euphoria during victory and the depths of despair during defeat.

With so many conflicting emotions concentrated in one location, it is believed that the area is haunted by the dead and dying, and the phantom armies

Entire regiments of soldiers vanished without a trace during fierce battles.

battle on blindly in a nether world beyond our understanding.

The Battle of Marathon occurred in 492 B. C. For several years following the victory of the Greeks over the Persians, many people visiting the site could hear the clash of swords and the screams of the wounded and dying men. Some claimed they could smell the blood that had been shed there. It was reported that those who were gifted enough to see the actual battle died soon after. Legends recount episodes of warriors disappearing into great cloud formations.

Another episode concerns the days of late August, 1915, when the Allies' hopes of victory were dashed. They were in retreat, weary and dying in vast numbers. Lance Corporal Headly-Johns of the Lancashire Fusiliers described an eerie sight to Harold Begbie, a London newsman. Johns' statement, in part, reads:

"I was in my battalion in the retreat from Mons on or about August 28th. The weather was very hot and clear, and between eight and nine o'clock in the evening, I was standing with nine other men on duty. Captain Leaton suddenly came up to me in a state of great anxiety and asked if any of us had seen anything astonishing.

"Later Captain Leaton returned and leading a few us some yards away, gestured toward the sky. I could observe quite clearly in mid-air a strange light distinctly outlined which was not a reflection of the moon, nor were there any clouds in the vicinity.

"The light became brighter and I could see quite distinctly three shapes; one in the center having what appeared to be spread wings. The other two were not as large but quite plainly distinct from the one in the center. They appeared to have a long, loose hanging garment of gold tint, and they were above the German lines and facing us.

"We stood watching them for about three-quarters of an hour. All of the men with me saw them, and other men came up from other groups who had also observed the same things. I am not a believer in such things, but I have not the slightest doubt but that we really did see is what I now tell you."

Corroboration comes from a nurse, Miss Phyllis Campbell, who attended a wounded R. F. A. man. He told her, "We all saw it. First there was a yellowish mist sort of rising before the Germans as they came to the top of the hill. I just gave up. No use fighting the whole German race.

"The next minute comes this funny cloud of light, and when it clears off there's a tall man with yellow hair and golden armor on a white horse, holding his sword up. Before you could take a breath, the Germans scattered and we were hot after them."

Miss Campbell heard the same story from dozens of other wounded soldiers who had seen the vision on August 28th. A month later a letter came from a friend, also a nurse, in the Potsdam Hospital in Germany. The woman wrote:

"There has been much comment here because a certain regiment which

UFO HOSTILITIES AND THE EVIL ALIEN AGENDA

Major Colman Von Keciczky (right, seen here with German rocket pioneer Hermann Oberth) stated that hundreds of American solders reported missing in Vietnam might actually been "skynapped" by UFOs.

had been ordered to take a small section of the front failed to carry out command, declaring that it was impossible!

"When they went forward, they were powerless, their horses turned sharply and fled Nothing could stop them. They saw at the same time strange shapes in the skies, and lower down a huge man on a white horse. It was like going full speed ahead and being suddenly pulled up before a precipice. That is the way they talk here. Is there anything to it, Phyllis?"

Also on August 28th, 1915, Lieutenant Colonel F. E. Sheldon saw something which prompted him to report it in a letter to the London Evening News on September 14, 1915. The colonel was riding ahead of a weary column with two other British officers. There was a bright moon. Sheldon was amazed to see two long lines of horsemen on either side of his column.

Thinking he was "seeing things," he did not mention it to anyone, but a short time later the other two officers asked if he had seen the strange white cavalry riding parallel with them. They were so convinced that they had seen the troops that one of them took a party out to investigate. They found no one.

Sheldon told the newsman: "I am absolutely convinced that I saw those horsemen, and I feel sure that they did not exist only in my imagination. I do not attempt to explain the mystery; I only state the facts."

The really strange part is that all the 1915 visions occurred on the same day that the 1/4 Norfolk Regiment vanished at Galipoli. Were the incidents related? Who can tell?

While we do not possess sufficient data on mass disappearances of military forces during World War Two, some of which have never been officially documented, according to retired Major Colman S. von Keviczky, "hundreds of American soldiers were reported missing in action during the Vietnam conflict in the demilitarized zone, after strange blips were observed on radar screens, blips that did not appear to be from enemy aircraft."

UFO HOSTILITIES AND THE EVIL ALIEN AGENDA

Von Keviczky suggests that the GIs were actually "skynapped" by UFOs, and that vehicles from other worlds might have been responsible for the disappearance of entire armies in the past.

Oddly enough, although World War Two may have been light on disappearances, other astonishing oddities did occur which are worthy of note and appear to be connected. The 51st Highlanders, for example, had an experience worth recording during the disastrous days of June, 1940. At that time, the Allies were being pushed toward Dunkirk by the advancing Germans.

The Scottish Highlanders spent 48 hours in a wooded area outside of Dunkirk, but refused to stay another night. They were convinced the thicket they were in was haunted. Most felt a "presence" and were frightened by it.

Researchers at the Dunkirk library discovered after the war that during the summer of 1415, sometime after the Battle of Agincourt, English and French soldiers had fought in that same wooded area. It was also learned that at the same time, a contingent of British soldiers had vanished from the French countryside without a trace.

Although the following story has little to do with disappearing armies or phantom armies, it is woven of the same fabric and deserves mention if only to point out the fact that in troubled times of war strange things happen to people under duress.

It was April, 1944. The Normandy invasion was still two months in the future. The British had endured five long years of blood, sweat and tears, and were bone weary. Worse, wild rumors of German superweapons (V-1s and V-2s) were circulating though the lines.

On April 27th, hundreds of people in Ipswich allegedly saw a vision of a cross in the sky during an air raid alert. On May 7th, the Chicago Tribune Press Service sent the following dispatch:

"...numerous residents of eastern England stoutly maintain that the sign of a cross was seen in the sky for fifteen minutes. Those who have given detailed descriptions include a naval commander, a carpenter, housewives and others. The consensus of the statements is that the vision gradually drew clearer until it was most distinct. The local pastor is investigating."

William Graham, an engineer, said "I saw the sign of the cross actually begin to form. There was no mistake that it was a crucifix."

Pastor Rev. Harold Godfrey Green, vicar of St. Nicholas Church, personally interviewed 2,000 people. His conclusion was, "There was scarcely any variation in the accounts. I have verified the fact of the vision quite definitely. I am satisfied myself, beyond doubt, of the authenticity of the vision.

"There were clouds in the sky which drifted by while the vision remained stationary."

UFO HOSTILITIES AND THE EVIL ALIEN AGENDA

Disappearing armies, spectral battles and visions. Nearly all represent disaster to men caught up in the strange phenomena. Where did they go? Who elects them to be the special casualties of war or natural calamities? No one will venture a valid answer, but author John Keel may have offered a solution in his book "Our Haunted Planet": "Perhaps the planet Earth is nothing more than a farm. We, unfortunately, are the crop."

Chapter 09

PRE-FIRST WORLD WAR SCARES AND SIGHTINGS
By Nigel Watson
Extracted from 'UFOs of the First World War' by Nigel Watson
(The History Press, 2015).

There were numerous "phantom airship" scares in the lead up to the First World War in New Zealand, Australia and Great Britain, which can be mainly attributed to "war nerves" and the threat posed by the German Zeppelin fleet.

Yet, there were also sightings in Germany or its allied territories. At the end of 1912 two aircraft were seen to come from Austria and fly over Kamenetz-Prodosk, Russia. It was feared that the Austrians were using aircraft to spy on Russian territory, including Russian Poland.

Russian aircraft with powerful searchlights flew over military sites in Jaroslaw (Galicia), Austria, on 18 January and the following nights, leading the Austro-Hungarian authorities to allow them to be shot at. One Russian aircraft crashed after it was shot at by sentries at Jaroslaw; when the body of the dead pilot was recovered he was identified as a member of the Russian General Staff. Another aircraft was accused of flying over Jassy, Romania, on 30 January 1913. It switched off its searchlight after being fired at by soldiers. The next night at Lembery, Austria, an aircraft with a powerful light was shot at.

A mayor of a town near Plock, Rus-

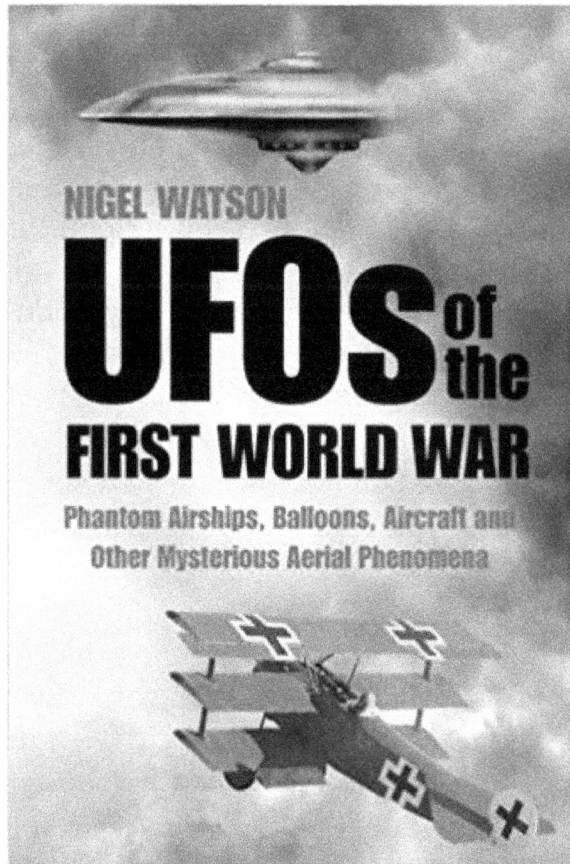

sian Poland, was even abducted by two Austrian pilots and flown a good distance before being released. Each side of the border accused the other but it is noteworthy that few aircraft at that time were capable of carrying powerful searchlights.

On 24 February 1913 a large 'German' airship was seen in Poperinghe, Belgium. The Belgium-France border was said to be the haunting ground of nightly aerial phantoms equipped with searchlights. In the Netherlands six 'German' dirigibles were seen passing over Noord-Brabent. At the end of February both countries feared German espionage from the air, land and sea. Also, on the 24 February at 8 p.m. an airship visited Dunaburg (Dwinsk), Russia. And an Austrian balloon followed by an aero-plane came to explore Kelets, Russian Poland, but they were driven away by gunfire.

The German press had made a lot fun of the 1909 and 1913 British airship scares, but the laughing soon stopped when the tables were reversed. On 4 March 1913 a 'Russian' airship was seen at Tarnowitz, Prussia. A burning airship was even seen to crash on the Caputh side of Lake Schwielow, Germany. A search was made for the wreckage, but not surprisingly nothing was discovered to confirm this sighting.

On the Eastern frontier regions of France there were many rumor of German aircraft flying around at night. To detect them cars were fitted with searchlights to probe the night sky. When a real aircraft was forced to land, at the beginning of April, near Luneville, a crowd came out and it wasn't very welcoming!

SPY STORIES

Britain also suffered many spy scares and rumors, some of which ran in combination with the phantom airship sightings. On 13 July 1908, *The Times* (London) stated that the Secretary of War was to be asked about a rumored Staff ride made through England by a foreign power. The same paper also asked whether the Chief Constables of the eastern counties knew of any foreign spy activity in England. According to The Observer, German officers were said to be active off the southeast coast of England, and similar stories were spread by other sections of the press.

In 1909 this trend was repeated, *The Illustrated London News* even went as far as to publish a map of the United Kingdom showing the fifty-four invasions, which had taken place since 1066.

On 19 May 1909, Sir J.E. Barlow asked in the House of Commons if the Secretary for War knew anything about the 66,000 trained German soldiers in England, or about the 50,300 stands of Mauser rifles and 7,500,000 Mauser cartridges stored in a cellar within a quarter mile of Charing Cross. Mr. Haldane said that this story was ludicrous, and it lowered our reputation for common-sense abroad.

Mr. Haldane might not have wanted to say anything, but the *Daily Telegraph*

did reveal that these arms were probably the 500,000 arms stored by the Society of Miniature Rifle Clubs, in a sub-basement of Lloyds Bank in The Strand.

Another story said that in the garrison town of Colchester the police had been receiving many reports of foreigners who were seen to be noting the whereabouts of crossroads and buildings throughout the neighborhood. The police themselves noted several incidents of a similar nature in May 1909.

Throughout Europe there were parallel worries about foreign intrusion. A Brussels ministerial source claimed that large numbers of German soldiers were maneuvering on the south-eastern border. In Verviers, Belgium, a number of postal officials were said to be searching the countryside on the orders of their superiors, looking for strategic and topographical information. A German newspaper in Triers advised that the town be fortified to withstand a French attack via neutral Luxembourg, as French officers had been known to have motored through the Grand Duchy with strategic aims. In reply, Luxembourg claimed that Germans had not only motored through their land, but surveyed it from the safety of airships. This controversy served to force the Belgium Clerical anti-militarists to amend their objections to a pending Army Bill.

These airship and spy scares seem to support the view that foreign enemies were attempting to reconnoiter military locations. Conversely one could argue that people living in 'strategic' areas would be more sensitive to rumors related to spying activities.

UFO HOSTILITIES AND THE EVIL ALIEN AGENDA

The ambiguity and bizarre nature of the airship sightings, the fear of alien invasion, the existence of foreign spies and inventors, allied with secret inventors and government investigations, in the 1900s, parallels the modern-day UFO phenomenon, which embraces stories of strange encounters, the fear of alien (extraterrestrial) invasion, men in black, and secret government projects.

Whether these airship phantoms were the product of mass delusion or aliens cloaking themselves in the futuristic technology of that era, we can conclude that politicians and journalists were equally guilty of enhancing the public fear of invasion, in order to secure more funds for military invasion. The generation of rumors of war were turned into actuality by a process of self-suggestion. In this state of mind the European powers marched inexorably towards the most bloody and destructive war in the history of mankind.

> I was on duty in Cromwell Road and was coming out of Cobden Street into that thoroughfare when I heard what I took to be a motorcar, which I judged was some 400 yards distant. It was 5:15 [a.m.] and still quite dark. I walked along Cromwell Road, expecting to see the lights of an approaching car, but none appeared. Still I could hear the steady buzz of a high-powered engine, and suddenly it struck me that the sound was coming, not along the surface of the road, but from above! I looked up, and my eye was at once attracted by a powerful light, which I should judge to be some 1200 feet above the earth. Outlined against the stars was a dark body . . . somewhat oblong and narrow in shape . . . about a couple of yards long. . . . It was going at a tremendous pace, and as I watched, the rattle of the engines grew gradually fainter and fainter, and it disappeared into the northwest. Altogether I should say I saw it for about three minutes.

AERIAL PHANTOMS START THE WAR

German newspapers understandably made a big fuss about French aircraft bombing the neighborhood of Nuremberg on the 2 August 1914. Yet, there was no substance to these reports and they were simply used to justify giving an ultimatum to Belgium and to declare war on France the following day.

The German Ambassador in Paris, Baron Wilhelm Eduard Schoen, delivered the declaration of war, which stated: "French acts of 'organized hostility' and of air attacks on Nuremberg and Karlsruhe and of the violation of Belgian neutrality by French aviators flying over Belgium territory...' were some of the key factors.

Germany had already declared war on Russia on 1 August, and within two days its troops had moved through Luxembourg and were pouring into Belgium, which led Britain to declare war on Germany on 4 August 1914.

Thus, phantom aircraft helped set the tragic horrors of the First World War grinding into motion.

References

1. Bullard, Thomas, *The Airship File* (Bloomington, Indiana: Privately published, 1982), pp.293-296, p.298, p.300 and pp.304-306; Bullard, Thomas, 'Newly Discovered "Airship" Waves Over Poland', *Flying Saucer Review*, Vol. 29, No. 3, March 1984, pp.12-14; Watson, Nigel, 'Airships and Invaders', *Magonia*, No. 3, Spring 1980; 'At the border of Galicia (1913)', at:

souvenirsdenosgreniers.unblog.fr/2013/05/29/a-la-frontiere-de-galicie-1913/)

2. Playne, Caroline E., *The Pre-War Mind in Britain* (London: Allen & Unwin, 1928), pp.118-119.

3. *Illustrated London News*, 27 March 1909.

4. *Irish Times*, 19 May 1909; *Sheffield Daily Telegraph*, 19 May 1909; *Grimsby News*, 28 May 1909.

5. *Bath Chronicle*, 27 May 1909.

6. *Irish News*, 20 May 1909; Ibid., 21 May 1909.

7. Tuchman, Barbara, *August 1914* (London & Basingstoke: The Macmillan Press Ltd., 1980), p.114 and p.126.

About Nigel Watson

Nigel Watson has researched and investigated historical and contemporary reports of UFO sightings since the 1970s. He is the author of "Portraits of Alien Encounters" (VALIS, 1990), "Phantom Aerial Flaps and Waves" (VALIS, 1990), "Supernatural Spielberg" (with Darren Slade, VALIS, 1992), editor/writer of "The Scareship Mystery: A Survey of Phantom Airship Scares, 1909 – 1918" (DOMRA, 2000), "The UFO Investigations Manual" (Haynes, 2013), and "UFOs of the First World War" (The History Press, 2015).For the UneXplained Rapid Reads e-book series he wrote: "UFOs: The Nazi Connection," "Spontaneous Human Combustion," "UFO Government Secrets," "The Great UFO Cover-Up" and "Ghostships of the Skies" (all 2015).He has also written for numerous books, publications and websites, including Magonia, Paranormal Magazine, Fortean Times, Wired, Flipside, How It Works, All About Space, Fate, Strange Magazine, Beyond, History Today, Aquila, Alien Worlds, UniLad, The Unexplained, Flying Saucer Review, UFO Magazine India and UFO Magazine (USA). In the 1980s, he gained a BA degree in Psychology (Open University) and a BA (Hons) degree in Film and Literature (University of Warwick).

Chapter 10

HOSTILE ALIENS, FOO FIGHTERS AND UFOS DURING WORLD WAR II

As reported by Jerome Clark in "The UFO Book," during World War II there were frequent sightings of "unconventional aerial phenomena" reported in both the European and Pacific theaters of war. The phenomena were named the "foo fighters," after a phrase repeatedly used by the newspaper cartoon character Smokey Stover, who was fond of saying "Where there's foo, there's fire."

The foo fighters are usually thought of as amorphous nocturnal lights, but later research has established that they represented a broad range of aerial anomalies. Two years later, after Kenneth Arnold had his sighting in 1947, phenomena identical to the foo fighters would be called "flying saucers" and, subsequently, unidentified flying objects.

"Perhaps the earliest known sighting," Clark writes, "is from September 1941. The witnesses, two sailors onboard the S.S. Pulaski, a Polish vessel converted into a British troop ship, were in the Indian Ocean in the early morning hours of a clear, starry night. One of them, Mar Doroba, spotted a 'strange globe glowing with greenish light, about half the size as the full moon as it appears to us.' He alerted an English gunner, and the two watched the object as it followed them over the next hour."

Another report, from England in 1943, was provided by Barbara Whiteley. At the time, she was stationed at an RAF base in Compton Bassett, Wiltshire.

"Up in the sky, not very high – I would think as high as a church tower maybe – over the camp was this long, cigar-shaped thing. I just couldn't believe my eyes. There were five or six lights, like searchlights, but the lights were coming from the thing onto the ground. The thing was like a barrage-balloon but more narrow at the ends and quite long. I couldn't say anything to anybody that night, and then the next morning I thought, well, I must be imagining things because we didn't hear about UFOs in those days."

THE NEW YORK TIMES, THURSDAY, DECEMBER 14, 1944.

Floating Mystery Ball Is New Nazi Air Weapon

SUPREME HEADQUARTERS, Allied Expeditionary Force, Dec. 13—A new German weapon has made its appearance on the western air front, it was disclosed today.

Airmen of the American Air Force report that they are encountering silver colored spheres in the air over German territory. The spheres are encountered either singly or in clusters. Sometimes they are semi-translucent.

SUPREME HEADQUARTERS, Dec. 13 (Reuter)—The Germans have produced a "secret" weapon in keeping with the Christmas season.

The new device, apparently an air defense weapon, resembles the huge glass balls that adorn Christmas trees.

There was no information available as to what holds them up like stars in the sky, what is in them, or what their purpose is supposed to be.

From 1944 through the end of the war, there were numerous sightings of all manner of unusual flying objects. For example, a Canadian soldier in Antwerp, Belgium, in September 1944, observed a "glowing globe traveling from the direction of the frontline toward Antwerp. It seemed to be about a meter in diameter and looked as though it was of cloudy glass with a light inside. It gave off a soft white glow. Its altitude seemed to be about 13 meters, speed about 50 km/h, and there was no sound of any sort. It was obviously powered and controlled. It was followed by another which in turn was followed by others, five in all."

According to a New York Times article from January 1945, while flying over the Rhine Valley in Germany, pilot Lt. D. Meiers reported, "A foo fighter picked me up recently at 700 feet and chased me 20 miles down the Rhine Valley. I turned to starboard, and two balls of fire turned with me. I turned to the port side, and they turned with me. We were going 260 miles an hour and the balls were keeping right up with us."

UFO HOSTILITIES AND THE EVIL ALIEN AGENDA

Clark writes further that, "Little is known about official investigation of the phenomenon, but apparently some effort was made to understand it."

According to what Clark calls a "cryptic reference" in the minutes of the CIA-sponsored Robertson Panel meeting, convened in January 1953:

"Instances of 'foo fighters' were cited. These were unexplained phenomena sighted by aircraft pilots during World War II wherein 'balls of light' would fly near or with the aircraft and maneuver rapidly. They were believed to be electrostatic (similar to St. Elmo's fire) or electromagnetic phenomena or possibly light reflections from ice crystals in the air, but their exact cause or nature was never defined. If the term 'flying saucers' had been popular in 1943-1945, these objects would have been so labeled."

The testimony of the witnesses quoted here all supports the idea that the foo fighters were under intelligent control and could fly alongside pilots on both sides of the conflict with apparent ease. In the many decades since World War II, the UFOs have demonstrated a keen interest in our nuclear missiles and have even overflown fields of battle as the combat raged.

Our ability to make war on each other is something the aliens study closely, it seems, whether for good or evil. Are they warmongers of a kind? Have they come to gloat over their handiwork, to celebrate their skill in pitting us against enemies both foreign and domestic?

Chapter 11

MUSSOLINI AND THE MILANO UFO CRASH

William Brophy was invited by Italy's primer UFO investigator Dr. Roberto Pinotti to the San Marino Republic UFO Symposium as a speaker to talk about the 1933 UFO crash near Milano, Italy and the 1942 Battle of Los Angeles But there was to be much icing on the cake as he revealed for the first time Mussolini's interest in the subject and the dictator's attempt to get to the bottom of a mystery that must have been new to him.

On June 13, 1933, a bell-shaped UFO crashed near Magenta, Italy just west of Milano.

The occupants were tall blond Nordics with oriental-like features on their light blue eyes. The Italians called the UFO "la Campania"; the Germans called it "Die Locked." Benito Mussolini, the Prime Minister of Italy, informed Pope Pius 11 of the crash and placed G. Marconi in charge of the special RS-33 Study Group, which served as a model for our later MJ-12 group. The Japanese told the Italians and Germans that those tall blonde-haired people were in their legends; this led to the Axis Alliance.

NEW DOCUMENTS "WILL REVOLUTIONIZE UFOLOGY!"

The discovery of some documents from the fascist era of Italy opens up new horizons for research on UFOs, and compels students of this subject to amend the official history of the flying saucers. Under "Il Duce," Mussolini, there was a secret government commission of inquiry, known as the RS-33 cabinet, which studied the UFOs. Then a second series of documents, in three different packages, posted in March-April 1996, were sent by "MR. X" to the top Italian ufologist Dr. Roberto Pinotti, Director of the Italian UFO research journal NOTIZIARIO UFO. The sighting of a flying cigar over Venice and Mestre on the morning of August 17, 1936, was reported showing a "cigar" – described as "torpedo-shaped" – along with two spheres beside it, one of which resembled the planet Saturn, being pursued by a fighter aircraft. It was an *original document* – not a copy – about a UFO sighting LONG BEFORE THE SUBJECT OF UFOS EXISTED! (More than ten years before Kenneth Arnold. G.C.)

Three telegrams from the Milan office of the Stefani Agency (the Fascist

UFO HOSTILITIES AND THE EVIL ALIEN AGENDA

A.N.S.A.), which gave instructions, on the order of Dictator Duce Mussolini, to recuperate a flying saucer which had landed on June 13, 1933. There was also a Senatorial letter describing in detail the strategy to be followed after the craft had been recovered; i.e., censorship of the newspapers; arrest of the eyewitnesses; elaboration of a series of conventional explanations for the UFO (i.e. *sonde* balloons, meteors, perihelia) to be fed to the public via the Brera Astronomical Observatory at Milan.

It was discovered that the RS-33 CABINET had been a top-secret study commission, created in the bosom of La Sapienza University in Rome following upon the recuperation [sic] of the UFO. Headed by the physicist Guglielmo Marconi (known for his belief in Martians). After Italy's accession to the Berlin-Rome Axis, by arrangement with Hitler, this material was all subsequently to be passed to the Nazis (who did, in actual fact, as a result, attempt a few years later to construct revolutionary discoidal aircraft called the V-7.)

In the first place, it has come to light that ever since the end of the Thirties a German aircraft designer named NORDUNG had been attempting to build a "solar" flying disc to be sent out into space – subsequently re-baptized as the "flying wheel," and so it is no wonder if the Italian Fascists, finding themselves confronted with a UFO, should have opted for the explanation that it was a secret weapon! The "Moretti" trail led me to the Varese region. My hypothesis is in fact that the saucer recovered in 1933 had been hidden in one of the nearest and most discrete hangars in that region – namely the hangars of the aeronautical establishments of the Siai Marchetti at Vergiate in the Varese region of Ticino which at that date were under the control of General Italo Balbo who became a member of the RS/33 CABINET). The "Varese trail" had been suggested to me through several clues: namely that the messages about recovery of the UFO came from the nearby Telegraphic Office of Milan: that in those very days Blackshirts were suddenly dispatched to that region.

UFO HOSTILITIES AND THE EVIL ALIEN AGENDA

The Varese newspaper, the Cronaca Prealpina, of June 20, 1933, gave the first report, emphasizing that forms of life on Mars were in contact with men of Earth – the existence of agreements between Mussolini and Hitler for the study of alien technology, agreements that had been made in 1938; these documents were: an Agency Stefani message from Florence containing an interview with the Fuhrer Hitler when he was visiting Italy. Marconi claimed he could develop a death ray, which, so Mussolini said, had for one day, by remote control, halted all automobile and aircraft engines as far away as Ostia – a weapon which, by virtue of 1938 agreements between Mussolini and Hitler, would evidently reappear in the following year at Essen, Germany, causing a "total blackout" of the city according to the account given by the American ufologist Leonard Stringfield in his book UFO SIEGE (1977.) It is likely the Fascists and the Nazis managed to achieve "back- engineering" from the UFO of 1933. The Fascist UFO Files nevertheless has a very special importance. For it demonstrates the existence of a governmental commission on UFOs before the official birth of UFOlogy itself

If, at the end of the 1930s, the Italians were beginning to hear of German experiments with disc-shaped aircraft (which we do know existed and which certainly were test-flown), then is it surprising that many Italian scientists may well have suspected that any weird craft seen flying over Italy in the years from 1933 to 1936 might well also have been German? And then later, during WWII, they might logically have suspected that they were seeing "secret weapons of the enemy Allies"?

"Mussolini" himself was a keen flier, and held a pilot's license. Who else

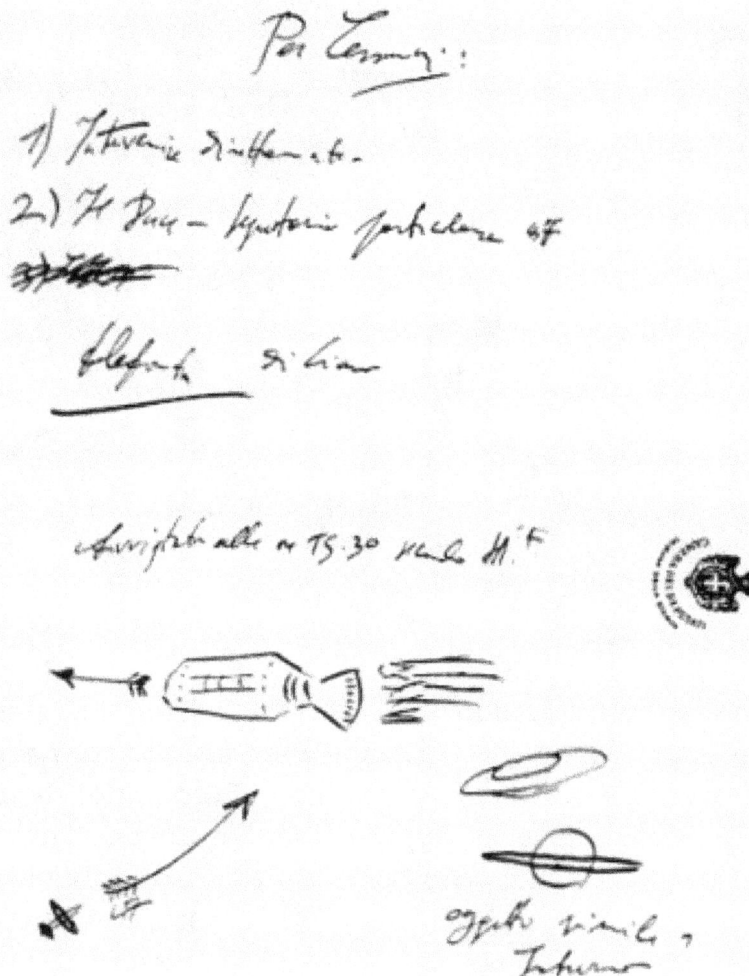

among the leaders in WWII did that? Well, it seems that as Germany tightened her grip on Italy, the Germans were able to command the Italians to hand all the alien craft's evidence over to them! Remember the remarkable German success in building at least two types of "saucer-shaped" multi-jet aircraft. Roberto Pinotti, probably the best of Italy's UFO researchers, confirms that it is 100% certain that the "Fascist UFO Files" are completely genuine.

It is believed the Italians gave the Germans access to the alien craft and they were able to back engineer the craft to develop the so called Foo fighters often reported by allied bombing aircrews. The key was the advanced propulsion system.

The famous wartime UK Prime Minister Winston Churchill has always been a man of mystery and controversy and that continues to this day. Numerous histories and biographies have been produced about him that portray him in all kinds of different lights. Fifty-two years after he died new evidence is still appearing, and the latest revelations are the most remarkable yet. The old statesman had many hobbies and interests, and one of those was popular science. He was interested in the idea of extraterrestrial life and wrote an article about it in 1939 that has only just been discovered. He intended it for the newspapers, but it was never published. He understood the concept of the "Goldilocks zone", the narrow gap between the minimum and maximum distance from a star that a planet must be in order for water to exist in liquid form, an essential ingredient for life as we know it. Too close and the star's heat will turn any water to steam; too far away and it will only be found as solid ice. Churchill knew that life was possible on planets around other stars, but he thought only Venus held any prospects for it being found in the solar system along with Earth. We knew almost nothing about Venus in those days; since then we've found out from the Venera programme that it is probably the least likely place life could ever exist with its dense superheated atmosphere. It also seems that the Goldilocks zone is bigger than we first thought because there is liquid water on the moons of the outer planets and possibly life there too.

Winston Churchill also took the idea of UFO's seriously throughout his premiership. Recently released British government files show that there was a real "X-Files" style project to address the UFO issue that was active in the late 1950's and probably earlier. Churchill himself was briefed on the subject during World War II when RAF aircrews reported encounters with strange objects while flying on combat missions. This is probably connected with the stories of "foo-fighters" that were well known among aviators in the war. The Prime Minister was so alarmed by the phenomenon that he ordered the matter classified for at least fifty years. He was concerned that it might cause "mass panic" if released. Considering that the British public coped very calmly with the Blitz, U-boats and their sons dying in foreign fields, it indicates how gravely the government regarded the UFO situation.

Also in the newly released files are details about the Berwyn Mountains incident and the "spaceman" at the Blue Streak missile test in Australia in 1964. At the same moment on the other side of the world in Cumbria, England a second spaceman was captured in an amateur photo; see background links below. The files are at the National Archives and are free to download for a month, so grab them while you can.

Chapter 12

WARTIME JITTERS
THE BATTLE FOR LOS ANGELES

It was less than three months after the attack on Pearl Harbor and it's no wonder most Americans had war jitters. If you lived on the West Coast, chances are you would find yourself hyperventilating at the slightest provocation.

Thus you can imagination how on Feb 25, 1942, those living in Los Angeles and its suburban, "high end" communities along the coast were at their wits' end when they heard the sound of artillery fire and looked up into the night sky to find a huge unidentified object being fired upon by the military. Whatever the object might have been it was of unknown origin and the military was giving it its best shot in an attempt to bring it down. People were injured and several died as a result of this aerial bombardment. For miles around, shells were landing on rooftops and in people's backyards. It was a "take cover" situation.

It was every man, woman and child (and household pets) for themselves.

Here is a rundown of the evening's highly charged activities:

It is estimated that approximately one million residents saw what has become known as "the battle for Los Angeles."

Several radars detected an object about 120 miles west of Los Angeles

* Within minutes, anti-aircraft batteries went on high alert

* At approximately 2:20 am, the object was tracked on radar to within a few miles of the coast and a city blackout was ordered

* Shortly after 3:00 am, the object appeared right over the city and anti-aircraft batteries opened fire

* Approximately 1500 rounds were fired into the sky, not over the ocean but directly over the city

* Three citizens died from the shelling and three more from heart attacks attributed to the shelling

* A great deal of property damage was inflicted

UFO HOSTILITIES AND THE EVIL ALIEN AGENDA

* Dr. Bruce Maccabee, an expert in photographic analysis, believed the objects to be roughly 100 feet or more in diameter.

Below - courtesy Collective-Evolution.com - is one eyewitness's testimony from the night of this event. Scott Littleton is a Professor of Anthropology, Emeritus, Occidental College, Los Angeles, CA.

"The two of us stood side by side in front of the house, huddling together in the chill night air and staring up into the sky. The planes we'd heard were not in sight, but what captured our rapt attention was a silvery, lozenge-shaped "bug," as my mother later described it, that was clearly visible in the searchlight beams that pinpointed it. Although it was a clear, moonlit night, no other details could be discerned, despite the fact that, when we first saw it, the object was hanging motionless almost directly overhead. Its altitude is hard to estimate, especially after all these years, but I'd guess that it was somewhere between 4,000 and 8,000 feet. This may explain why we didn't see the orange glow reported by several eyewitnesses in Santa Monica and Culver City, where the object was apparently much lower. (One witness suggests that this glow may simply have been the reflection of shell bursts against the object's "silvery" body.)"

As might be expected, in order to squelch a possible panic by the public, within hours of the end of the air raid, Secretary of the Navy Frank Knox held a press conference, saying the entire incident was a false alarm due to anxiety and "war nerves." Knox's comments were followed by a nonsensical statement from General George C. Marshall that the incident might have been caused by commercial airplanes used in a psychological warfare campaign to generate panic so as to gauge the reaction of the public to a possible Axis invasion.

Some contemporary press outlets suspected a cover-up. An editorial in the Long Beach Independent wrote, "There is a mysterious reticence about the whole affair and it appears that some form of censorship is trying to halt discussion on the matter."

Representative Leland Ford of Santa Monica called for a Congressional investigation, saying, "...none of the explanations so far offered removed the episode from the category of 'complete mystification' ... this was either a practice raid, or a raid to throw a scare into 2,000,000 people, or a mistaken identity raid, or a raid to lay a political foundation to take away Southern California's war industries."

Thousands of rounds of anti-aircraft shells were fired at the mysterious object over Los Angeles but only a few Americans were killed, dozens wounded and a lot of property was damaged.

DEMOCRATS WIN IN NEW JERSEY, N.Y.

Los Angeles Times 9 A.M.
FINAL

MYSTERY AIR OBJECTS SEEN IN SKY OVER L.A.

Join Russia in 'Commonwealth of Sp
Khrushchev Proposes to U.S., Other

The object, apparently undamaged after hours of shelling, simply drifted away. To this day, no one can offer a real explanation of what the object was.

Searchlights and Anti-aircraft Guns Comb Sky During Alarm

CAPTAIN

THOMAS F. MANTELL, JR.

Born in Franklin, Ky. 30 June 1922. Graduated Male High School, Louisville. Joined Army Air Corps. 16 June 1942. Graduated Flight School, 30 June 1943. During WW II, Mantell assigned to 440th Troop Carrier Group, 95th Troop Carrier Squadron, 9th Air Force. Awarded Distinguished Flying Cross, and Air Medal w/3 OLCs for heroism. Following the war returned to Louisville. Joined newly organized Kentucky Air National Guard- assigned as Flight Leader, "C" Flight, 165th Fighter Squadron, Kentucky Air National Guard on 16 February 1947. 7 January 1948, while on training flight with three other P-51Ds (Mustang) Mantell was directed by flight tower at Godman Field to pursue an unidentified flying object. While in pursuit of object, died in plane crash near this site. The story of Mantell's death while chasing UFO made headlines across the country. Intense military investigation of incident became part of Project SIGN later BLUEBOOK, the military's investigations into UFOs. Much speculation and conjecture has been written about the incident. It is still uncertain what Mantell was pursuing at the time of the crash. Mantell is the first flight casualty of the Kentucky Air National Guard. Buried Zachary Taylor National Cemetery, Louisville, Ky.

Chapter 13

LOCK AND LOAD AND PREPARE TO FIRE
By Sean Casteel

The Mantell Incident is the stuff UFO legends are made of.

To hear the Air Force relate what happened, it was all a matter of misidentification. The case of an ordinary object being taken for something it was never intended to be - a craft from another planet. Yet, if you dig deeper - even after all these years - anyone can plainly see that something truly extraordinary happened on that January day in Kentucky.

The object was seen from the ground at first by private citizens.

It was observed by military personal from a well-equipped air base control tower.

Later on, it was chased by some of our most advanced aircraft. It was said to be tremendous in size and, after standing still and hovering for over 90 minutes, it could jet away easily, provoking concern from the experienced pilots in hot pursuit.

But let us travel back to the day in question. . .

Captain Thomas F. Mantell was an experienced pilot, with a flight history that included over 2100 hours in the air and being honored for his part in the Battle of Normandy during World War II.

In what became one of the most publicized of the early UFO incidents, Mantell died in the crash of his F-51 Mustang fighter after being sent in pursuit of an unidentified flying object.

Historian David Jacobs, who is also featured elsewhere in this book, says that the Mantell case marked a sharp shift in both governmental and public perceptions of UFOs. Before, the news media often treated UFO reports with a whimsical or lighthearted touch. But the fact that a person had died in an encounter with a UFO, according to Jacobs, meant that flying saucers might be not only extraterrestrial "but potentially hostile as well."

The story began on January 7, 1948, when the Godman Army Airfield at Fort

UFO HOSTILITIES AND THE EVIL ALIEN AGENDA

Members of the Kentucky Air National Guard
before one of their North American P-51 Mustangs.

Knox, Kentucky, received a report from the Kentucky Highway Patrol of an usual aerial object near Madisonville, Kentucky. Reports of a westbound circular object, 250-300 feet in diameter, were received from Owensboro and Irvington.

At about 1:45 P.M., Sergeant Quinton Blackwell saw an object from his position in the control tower at Fort Knox. Two other witnesses in the tower also reported a white object in the distance.

A report was also filed by base commander Colonel Guy Hix, who went on record about an object he described as "very white," and "about one fourth the size of the full moon. Through binoculars it appeared to have a red border at the bottom. It remained stationary, seemingly, for one and a half hours." Meanwhile, observers at Clinton County Army Air Field in Ohio described the object as resembling a "flaming red cone trailing a gaseous green mist." Still another report came in from Lockbourne Army Air Field in Ohio, where the object was seen to come very near the ground, staying down for about ten seconds, then climbing at high speed back to its original altitude of 10,000 feet, before leveling off and disappearing into the overcast at a speed greater than 500 mph in level flight.

As the observers discussed the strange sighting, four F-51s approached, led by Captain Mantell. Their routine mission was to return the aircraft, stranded a few days earlier at an air base in Georgia, to their rightful place at Standiford Air Field in northern Kentucky. Sergeant Blackwell asked Mantell and his companions whether they could get close to the object.

"One of the pilots, whose fuel was running low," writes Jerome Clark in "The UFO Book," "continued on to Standiford. Meanwhile, Mantell had spotted the object. He radioed the Godman tower that it was 'in sight above and ahead of me, and it appears to be moving at about half my speed or approximately 180 miles an hour.' He said, 'It appears to be a metallic object and it is of tremendous size.' He turned right abruptly and climbed sharply, without informing the other two aircraft of his intentions, and they scrambled to catch up with him."

When they reached 16,000 feet, the pilot on Mantell's right put on his oxygen mask. Already the thin air was hazardous, and neither Mantell nor the pilot on his left had brought oxygen masks with them. The two pilots followed Mantell up to 20,000 feet and were then over Bowling Green Kentucky, with no idea of what they were supposed to be looking for.

Mantell pointed it out to them, saying, "Look, there it is out there at 12 o'clock!" One of the other pilots would later tell investigators that "I was able to discern a bright-appearing object, very small, and so far away that I was unable

A crowd of stunned local onlookers gather
around Mantell's shredded Mustang.

to identify it as to its size, shape, color. Its position was slightly lower and to the left of the sun." He suggested that they level off, accelerate, and try to get under the object.

Mantell replied that he wanted to follow it up to 25,000 feet for ten minutes. If they got no closer to it after that, they would abandon the chase. Mantell then reported to Godman Tower that he was trying to get closer for a better look. At 22,500 feet, with oxygen running low, the two other pilots resumed the flight to Standiford. When they told Mantell what they intended to do, he did not respond. The last the pilots saw of him, he was "still climbing almost directly into the sun."

A minute or two later, William C. Mayes, a resident of rural Franklin, heard the noise of a plane entering a power dive. The plane exploded halfway between where it began the dive and the ground. Another resident, Carrie Phillips, heard the explosion and saw the plane come down in her front yard 750 feet from her house.

"When Franklin firemen dragged the body of the partially decapitated Mantell from the wreckage," writes Clark, "they noticed that his shattered wrist-watch was stopped at 3:18. The evening edition of the Louisville Courier read: 'F-51 and Capt. Mantell Destroyed Chasing Flying Saucer.' At 3:50 the object disappeared from the view of the observers at Godman tower."

The military's official response to the incident was to say the planet Venus was the mysterious object that Mantell died pursuing. While Venus was in the same position in the sky as the UFO reported at Godman and by Mantell, it was for all practical purposes invisible to observers. They would later blame the incident on a Skyhook balloon, which at the time was a secret Navy project.

British writer Harold T. Wilkins suggested that "some lethal ray of immense power and unknown type had been directed at Mantell and his plane by the entities, who may have wished to demonstrate to terrestrial military power the folly of any close approach." Comparable rumors and speculations persisted for many years afterward, according to Clark.

Sometime in the 1990s, Sgt. Quinton Blackwell was interviewed by the Fox television show "Sightings," along with Mantell's family, in an attempt to bring closure to the surviving Mantells regarding the young pilot's death. What Blackwell told "Sightings" differed greatly from his original 1948 testimony.

Blackwell had never spoken with anyone about the incident but was still haunted by his memories. His son convinced him to break the gag order imposed by the Air Force and tell the story of what really happened. The official story was still that Mantell had blacked out from lack of oxygen while mistakenly chasing an experimental balloon.

But Blackwell told a more interesting story.

He recalled that when he got to the observation tower, he was amused to see clearly, with field glasses, a flying saucer. He notified the base operation officer, who replied, upon seeing the object himself, "Boy, you've got me in trouble now. I have to call the base commander." Soon after, some of the generals came up to the tower for a closer look.

Blackwell also revealed that the real reason the two pilots on either side of Mantell broke off from the pursuit of the UFO was to return to home base to load their weapons with live ammo, and this on orders from Mantell himself. This command can only be given by a wingman commander when he can see a verifiable threat and permission from his superiors has to be granted.

Blackwell stated that Mantell described the object as a 200 feet across by 70 to 75 feet thick metallic object with observation windows on the top portion of it. The last thing Blackwell heard from Mantell's radio transmission was, "I will move closer to get a better look." The Air Force said that Mantell was hallucinating due to the lack of oxygen at the time of the sighting but Blackwell firmly believes that Mantell was fully in control of his aircraft. Blackwell had experience dealing with pilots with anoxia and Mantell showed no signs of that at that moment.

Blackwell also divulged the identity of three other colonels on the tower at the time of the event. Once they got the notice that Mantell's plane was down they quickly vanished from the tower, realizing that an investigation would be conducted. They did not want to be forced to write an official report on the incident given that this kind of saucer business was not exactly a "career enhancement."

Chapter 14

UFOS ATTACK BRAZILIAN MILITARY FORTRESS
By Dr. Olavo Fontes

Dr. Olavo Fontes, now considered a pioneer of Brazilian UFOlogy, was born on June 9, 1924 and was the son of Congressman Armando Fontes. Olavo's best received case studies involved the world famous incident involving Antonio Vilas Boas's seduction at the hands of a female "alien," and the explosion of a possible UFO over Ubatuba in 1957, which involved the strange heavenly fall of particles of ultra-pure magnesium that could not have been duplicated or manufactured at the time on this planet. Fontes was held in high regard by UFO researchers worldwide and was the South American representative for the prestigious Aerial Phenomena Research Organization, headed by Jim and Coral Lorenzen.

Dr. Olavo Fontes was an advocate of the "UFOs are hostile" theory.

Through his military and governmental contacts Dr Fontes said he was convinced that the governments around the world were in agreement not to reveal - due to its subversive nature - anything about the UFO problem. "Some military groups," he insisted "believe that such revelations would cause a huge shock, enough to cripple life in our countries for many years."

Furthermore, he became convinced that, "These visitors from outer space are dangerous when approached, and become hostile if attacked. We have already lost many planes trying to shoot them down. We have no defense against them as so far they easily outclass any hunter, and they have no chance against them. Guided missiles are also useless. They can fly even faster, and even maneuver around them as if they were toys. Or they can interfere with our electronic instruments and render them inoperable after launch, or they can blow them up before they arrive nearby. They caused the crash of military aircraft by stalling their engines by interference with their electronic systems."

Dismissing the contact claims of those who maintain they have met with extraterrestrials, Fontes persisted in saying that "they have so far expressed no desire in contacting us."

Which sets the tone of the late physician's investigation of a case in which several sentries were horribly burned by a UFO which swung down over a Brazilian military fortress. The case was considered highly controversial when first published and was neglected by most researchers seeking a more "comfortable" and collective "friendly" view of the Ultra-terrestrial appearance in our atmosphere and on the ground.

Dr Fontes' life was cut short by an aggressive cancer while he was still in his thirties. Some saw this as part of a conspiracy to prevent him from making further revelations of such a shocking nature - though it was never determined who - or what! - might have caused his early passing and silenced him.

Here is his complete report as presented in the APRO Bulletin, which has been out of print for decades. It presents a valuable contribution to the body of evidence that at least some UFOs have hostile intentions.

TGB

****

SHADOW OF THE UNKNOWN — UAO'S FRIENDS OR FOES?

There are such things as UAOs from other planets, and these things from space may have entities controlling them that mean to do us harm. The UAO situation has come to the point where it would be wise for mankind to start turning their eyes and thoughts toward outer space, because there is more danger lurking there than on the Earth itself.

We must be ready. You are not going to like this report. For one reason, it will make the existence of hostile UAOs self-evident. It will prove beyond any doubt that "they" are testing weapons against harmless civilian airplanes as well as against military fortifications and soldiers.

I am aware that such things will represent a hard blow for most civilian UFOlogists in every country, but not for the military - they already know. They cannot talk. I can, and I do not think I have the right to conceal the sinister angle of the UAO problem.

The incidents I am going to report are real, they really happened, so I cannot do anything except to get them published. In doing this, I am assuming a calculated risk, mostly a top-secret risk, because one of them is a military case.

The matter is too important to be considered under the perspective of my personal security. The "contacters" are going to be disgusted with this report, but that is not important. Far more important should be the effect of my cases on the group who thinks that UAOs are U. S. secret weapons.

Take Dr. Leon Davidson, for example: I would like to know how he will explain my cases in the light of his theory. He found a CIA "tie-up" in Adamski's case. However, I would like to know his explanation for UAO hostility against my country - a friendly nation, allied to the U. S. A. in World War II, and still linked to it by military treaties. There is no explanation possible unless my evidence is rejected without consideration.

I will present only three incidents, all of them occurring in the second half of 1957. The first case is a report of UAO hostility against an airliner. Some kind of weapon was used - possibly a microwave ionizer-i.e., the same device that has been tested again and again against grounded vehicles. The plane did not crash, and there was no physical evidence that the story was true. The second report

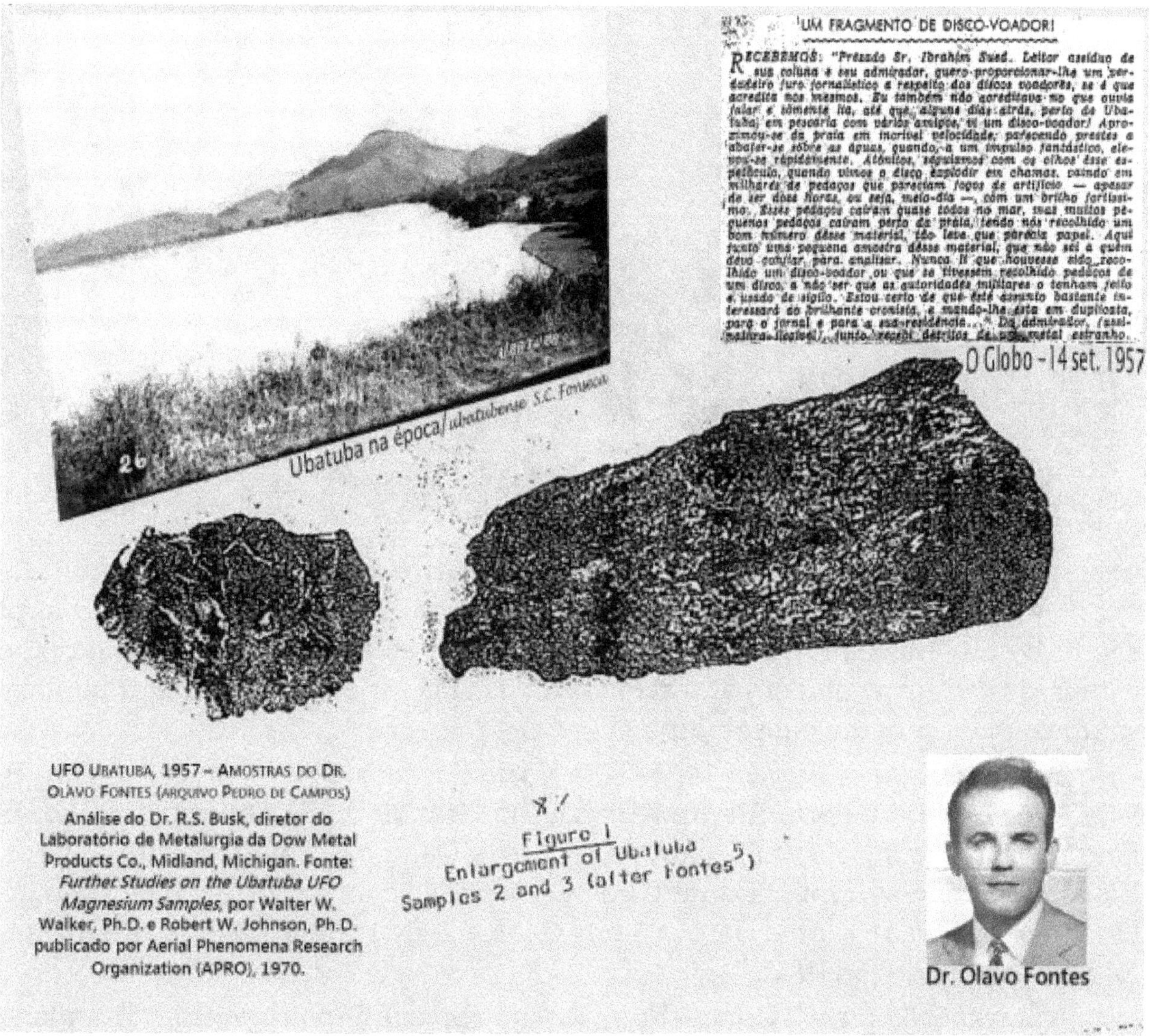

Ubatuba na época/ubatubense S.C. Fonseca

O Globo - 14 set. 1957

UM FRAGMENTO DE DISCO-VOADOR!

UFO UBATUBA, 1957 – AMOSTRAS DO DR. OLAVO FONTES (ARQUIVO PEDRO DE CAMPOS)

Análise do Dr. R.S. Busk, diretor do Laboratório de Metalurgia da Dow Metal Products Co., Midland, Michigan. Fonte: *Further Studies on the Ubatuba UFO Magnesium Samples*, por Walter W. Walker, Ph.D. e Robert W. Johnson, Ph.D. publicado por Aerial Phenomena Research Organization (APRO), 1970.

Figure 1
Enlargement of Ubatuba
Samples 2 and 3 (after Fontes⁵)

Dr. Olavo Fontes

describes what happened to another airliner when the same weapon was used against it. This plane did not crash either, but showed physical evidence that the UAO was not joking. The last report tells about a vicious attack by a saucer against two poor soldiers; two sentinels from the Itaipu Fortress.

At least two weapons were used: one against the sentinels, possibly an ul-

trasonic beam of some sort; another against the fortress itself, to paralyze its electrical system, probably the same tested against the airliners. This case was kept secret by the Brazilian Army, but I was able to get all the details from one of the witnesses - an Army officer - and also to confirm his report through other military sources, in spite of the censorship.

On the night of August 14, 1957, a Varig Airlines C-47 - the cargo ship P-VCC - took off from Porto Alegre Airport, Rio Grande do Sul, en route to Rio de Janeiro. At the controls was Commander Jorge Campos Araujo, a veteran pilot. His first officer was copilot Edgar Onofre Soares, also an experienced airline pilot. The plane was over the state of Santa Catarina, flying at 6300 feet, when copilot Soares spotted the UAO.

"It was 8:55 p.m.," Commander Araujo reported to the press. "The plane

UFO HOSTILITIES AND THE EVIL ALIEN AGENDA

OFFICE OF THE AIR ATTACHÉ
BRAZILIAN EMBASSY
Washington, D. C. 20008

At two o'clock on the morning of November 4, 1957, in Fort Itaipú, located in the vicinity of Santos, São Paulo, according to a Report by Dr. Olavo Fontés, two guards where on watch duty at the top of the highest lookout tower facing the ocean. Suddenly an orange color light appeared and quickly positioned itself at approximately 50 meters above the guards. With inoperable automatic weapons, they watched the starange glow and high pitched noise emerging from the strange object. Suddenly the two guards were hit by an intolerable heat wave, knocking one of the guards uncounscious. The other thought himself to be in flames, and screamed to alert others. The troops in the Fort quickly became alarmed due to a failure of all power systems, including lights, elevators and weapons. Three minutes later the power returned with many officers claiming to have seen the strange object rapidally climb towards the sky. The two guards sustained first and second degree burns over 10% of their bodies. The alarm system and the automatic electric clocks which were originally set for 5 am, went off at 2:30 of the same morning. After the incident the Fort was shut down and investigations by the Brazilian Secret Police and afterwards by North American Officials were conducted. The results were not made available.

---

had crossed over the town of Joinville just five minutes before. I was absorbed with the instruments' control panel when my attention was called by copilot Soares. He was pointing to a luminous object which was flying at the left side of the airliner. I began to watch it. It was not another plane, neither an astronomical body. I am absolutely sure. It was a strange craft. When I spotted it for the first time, it seemed to be placed far to the left of our aircraft.

"We were flying on a ten degrees course. There was no chance of any mistake. Though there was a thick layer of clouds below us, at 5,700 feet, all the sky above that layer was absolutely clear. We had a visibility of about 80 miles.

"Suddenly, in an unexpected maneuver with unbelievable speed (obviously supersonic), the mysterious craft was ahead of us and then it crossed to our right side, following a horizontal trajectory that made it pass just in front of the airliner, at the same level. After such a dangerous maneuver, the object apparently stopped in midair for a brief time, motionless. Then it abruptly went into a dive and was out of sight - lost into the cloudbank below."

Besides the commander and copilot, radio operator Rubens A. Tortilho and stewards Jose D. S. Machado and Afonso Schenini also saw the unknown object. They were called to the cockpit and came there still in time to watch the UAO.

There were some passengers aboard but none of them witnessed the sighting. One of them was an Army officer - he was called by the crew to be a witness. However, when he entered the pilot's cabin together with other passengers, it was too late; the UAO had already disappeared into the thick layer clouds 600 feet below the airliner.

Incidentally, the crew was sure it couldn't be seen from any of the passenger's windows for it had cut off the way in front of the plane.

All members of the crew declared that the object looked like one of the so-called "flying saucers."

"It was shaped like a saucer with a kind of cupola or dome on top of it," reported Commander Araujo. "The whole cupola glowed with an intense green light. The flattened base glowed with a less intense yellowish luminosity. No windows or portholes were visible on the object. As we didn't know its real size, we cannot estimate with accuracy the speed and distance of the mysterious flying object in regards to our aircraft. Its apparent diameter, however, was about 6 feet. The speed was incredible - obviously many times the speed of sound. I believe it was about six miles from us, but this was just an impression," concluded Commander Araujo.

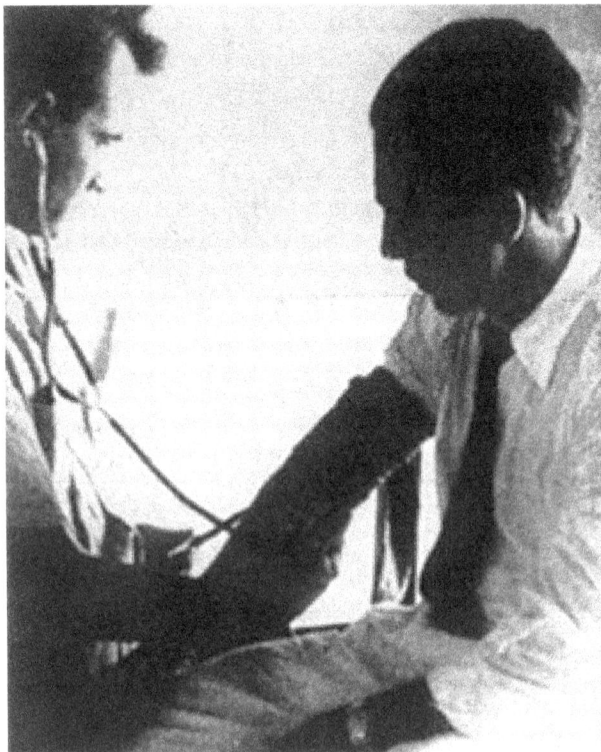

A rare photo of Dr Fontes examining Antonio Boas who maintains that he had sexual congress with a feline-like space being.

Soon after the sighting, Commander Araujo radioed a UAO report to Varig Airlines communications at Congonhas Airport, Sao Paulo City. Others heard about it and, a few days later, someone told a Sao Paulo newspaper about the sighting. The PP-VCC's crew was then interviewed and confirmed it. Commander Araujo's report hit the headlines all over the country on August

UFO HOSTILITIES AND THE EVIL ALIEN AGENDA

20th.

Despite this national publicity, the Brazilian Air Force refused to make any comment about the incident - not even to debunk it. Why? Because they had very strong reasons to "ignore" the incident. In fact, the story as published by the press was not complete. There was something more, something not told to the press - something "too hot to handle."

It would be unwise to apply any kind of pressure on the crew, for they might get angry and talk too much - so they were left alone. What was it?

Sometime after the sighting of the UAO the Varig airliner landed at Sao Paulo Airport. There Commander Araujo met a man who was a close friend of his, a former airliner pilot who still worked at the airport as traffic chief for another airline. This man is also a close friend of mine. He found them seated around a table, silent, and scared. He felt there was something wrong with them and asked about it. Commander Araujo reported the incident. At the end they were silent again, the whole crew.

My friend tried to break that uneasy silence with a joke. "I see that you are shocked about your uncanny experience, all of you. Don't be so worried about it. After all, it was only a saucer, not a ghost."

"It was more than you think, worse than you think," answered Commander Araujo. His hands were trembling, betraying deep emotion. "There is something more, but keep it confidential. When the object reached our right side and stopped for a brief time - just at that moment - the engines of the airliner began acting up, coughing and missing, and all lights inside the cabin dimmed and almost went dead. It seemed that the whole electric system of the plane was going to collapse. But a few seconds later the UAO dived abruptly into the clouds, and everything was normal again. However, we cannot forget those terrible seconds when we were suspended between life and death. I tell you, I saw my plane crashed and everybody killed. It was like Hell. Damn it, it is not pleasant to play the sitting duck."

It is not difficult to understand why no member in the crew had courage enough to report such a thing to the press. But they did not refuse to talk about it privately so that I was able to check the story through other sources. It was confirmed.

Incident 2: This case was investigated and published by Commander Auriphebo Simoes, a well-known UAO researcher. He personally interviewed Captain de Beyssac, the chief witness, and printed his report in the Jan.-Feb. 1958 issue of "The Flying Saucer."

I will quote from it: "Jean Vincent de Beyssac is an airline captain actually working for Varig Airlines. Formerly he was a copilot for Cruzeiro do Sul Airlines.

"Do you know what happened to him on the morning of November 4th, 1957?

UFO HOSTILITIES AND THE EVIL ALIEN AGENDA

His flight that day started about midnight on November 3rd. He took off from Porto Alegre in Southern Brazil on a trip to Sao Paulo and Rio aboard a C-46 cargo ship. During the day a cold front had passed and the sky was starlit over Porte Alegre. About 1:20 a.m. while flying over Ararangua, Santa Catarina, at 7,000 feet, he flew over a layer of stratus. Suddenly, blinking his eyes in disbelief, he sighted an impossible red light to the left of his aircraft. He watched it curiously and joked with his copilot, suggesting that they were at last seeing an authentic flying saucer. When the THING grew larger, Jean decided to turn left and investigate. Just before he pressed his rudder, the thing jumped a 45 degrees arc in the horizon and became larger.

"Jean started the pursuit. He was about midway on his left 80 degrees turn when the thing became even brighter and at once he smelled something burning inside his ship.

"Yes, the ADF burned; the right generator burned, the transmitter burned, all at the same time! Then the 'thing' disappeared almost suddenly, while his crew looked for fire. Scared, Jean turned on his emergency transmitter and told the Porto Alegre control what had happened. He went back to Porto Alegre, where he landed about an hour later. After writing a full report he went home and got soused, just to scare the scare.

"On that same day, Varig Airlines issued an internal circular forbidding pilots to tell the press about their sightings of UAOs. This, according to the airline bigshots, was to prevent the public getting too wise about certain things happening to some pilots."

The evidence in this incident suggests that our visitors from outer space are dangerous when approached and definitely hostile when pursued. Positive proof is given - physical proof - that they have the means to interfere with any electronic instrument and make it useless. In other words, that they have a weapon - probably a high frequency radio-electric beam - with power enough to short-circuit anything within its range; any apparatus, or electric instrument, or motors with electrical systems. The technical aspects involved will be discussed later.

TOP SECRET MILITARY SIGHTING

It is printed here for the first time; it was never published before. It is a horror tale. Anyone, after reading the report, will understand the reasons why it has been withheld from the public under a curtain of absolute censorship. I am breaking this official secrecy because I believe it dangerous. I still believe that civilian scientists should be told. One of them may find a defense not yet discovered. Civilian scientists and technicians, working in every country, might help to find new weapons and defenses before it is too late.

On November 4th, 1957, at 2:00 a.m. (just forty minutes after Incident 2) something sinister took place at the Brazilian Fortress Itaipu. This fortress belongs to the Brazilian Army and was built along the coast of Sao Paulo state, at Sao Vicente, near Santos. It was a moonless tropical night. Everything was quiet. The whole garrison was sleeping in peace. Two sentinels were on duty on top of the military fortifications. They were common soldiers, they did not know that saucers existed. They were performing a routine task, relaxed because there was no enemy to be feared.

Then a new star suddenly burst into searing life among the others in the cloudless sky, over the Atlantic Ocean, near the horizon. The sentries watched the phenomenon. Their interest increased when they realized it was not a star, but a luminous flying object. It was coming toward the fortress. They thought at first that it was an airplane but the speed was strange - too high.

There was no need to alert the garrison, however. In fact, so tremendous was the object's speed that the two soldiers forgot their patrol just to observe it. It was approaching rapidly. In just a few seconds the UAO was flying over the fortress. Then it stopped abruptly in midair and drifted slowly down, its strong orange glow etching each man's shadow against the illuminated ground between the heavy cannon turrets.

It hovered about 120 to 180 feet above the highest cannon turret and then it became motionless. The sentries were frozen on the ground, their eyes wide with surprise, and their Tommy guns hung limply from their hands like dead things. The unknown object was a large craft about the size of a big Douglas, but round and shaped like a disk of some sort. It was encircled by an eerie orange glow.

It had been silent when approaching, but now, at close range, the two sentries heard a distinct humming sound coming from it. The strange object hovered overhead and nothing happened for about one minute. Then came the nightmare. The sentinels were startled, unable to think what to do about the UAO. But they felt no terror, no premonition, no hint of the danger.

Then something hot touched their faces (one of them thinks he heard a faint whining sound he could not identify at that same moment). In darkness, this would have been horrifying. But the UAO was bright and they could see that nothing had changed. Then came the heat. Suddenly an intolerable wave of heat struck the two soldiers. One of the sentries said later that, when the heat wave engulfed him, it was like a fire burning all over his clothes.

The air seemed to be filled with the UAO's humming sound. Blind panic yammered at him. He staggered, dazed, heat waves filling the air around him. It was too hot. He went stumbling and lurching, his whole conscious purpose that of escaping from that invisible fire burning him alive. He fought and gasped and beat the air before him. He was suffocating. Then he blacked out and collapsed to the ground unconscious.

The other sentry got the horrible feeling that his clothes were on fire. A wave of heat suddenly enveloped him. Horror filled him and he lost his mind. He began to scream desperately, running and stumbling and crying from one side to another, as a trapped animal. He did not know what he was doing, but somehow he skidded into shelter, beneath the heavy cannons of the fortress. His cries were so loud that he awoke the whole garrison, starting an alarm all over the place. Inside, the soldiers' living quarters everything was confusion. There was the sound of running footsteps everywhere, soldiers and officers trying to reach their battle stations, their eyes wide with shock.

No one knew what could explain those horrible screams outside. Then, just a few seconds later, the lights all over the fortress collapsed suddenly as well as the whole electric system that moved the turrets, heavy cannons and elevators. Even the ones supplied by the fortress' own generators. The intercommunications system was dead too. Someone tried to switch on the emergency circuits but these were dead too. The strangest thing, however, was the behavior of the alarms in the electric clocks, which had been set to ring at 5:00 a.m. - they all started to ring everywhere, at 2:03 a.m.

The fortress was dead, helpless. Inside it, confusion had changed to widespread panic, soldiers and officers running blindly from one corner to another along the dark corridors. There was fear on every face; fear of the unknown - hands nervously grasping the useless weapons. Then the lights came on again and every man ran outside to fight the unexpected enemy, who surely was attacking the fortress. Some officers and soldiers came in time to see an orange light climbing up vertically and then moving away through the sky at high speed.

One of the sentinels was on the ground, still unconscious. The other was hiding in a dark corner, mumbling and crying, entirely out of his mind. One of the officers who came first was a military doctor and, after a brief examination, he saw that both sentries were badly burned and ordered the men to take them to the infirmary immediately. They were put under medical care at once. It became clear that one of them was a severe case of heat syncope; he was still unconscious and showing evident signs of peripheral vascular failure.

Besides this, both soldiers presented first and deep second-degree burns of more than 10 per cent of body surface - mostly on areas that had been protected by clothes. The one that could talk was in deep nervous shock and many hours passed before he was able to tell the story. The nightmare had lasted for three minutes.

Next day the commander of the fortress (an army colonel) issued orders forbidding the whole garrison to tell anything about the incident to anyone; not even to their relatives. Intelligence officers came and took charge, working frantically to question and silence everyone with information pertaining to the matter. Soldiers and officers were instructed not to discuss the case. The fortress was placed in a state of martial law and a top-secret report was sent to the Q. G. (at Rio or Sao Paulo).

Days later, American officers from the U. S. Army Military Mission arrived at the fortress together with officers from the Brazilian Air Force, to question the sentries and other witnesses involved. Afterwards a special plane was chartered to bring the two burned sentinels to Rio. It was an Air Force military aircraft. At Rio, they were put in the Army's Central Hospital (HCE), completely isolated from the world behind a tight security curtain. Two months later they were still there.

I don't know where they are now. Three weeks after the incident, I was contacted by an officer from the Brazilian Army, a friend who knew about my interest in UAO research. He was at the Fortress of Itaipu the night of the incident. He was one of those who questioned the two sentries. He told me the whole story exactly as it was described above. His name was suppressed from this report in order to protect him. The reasons are obvious; he told me something he should not tell. As a matter of fact, this officer has asked me to forget his name and he wasn't laughing. He was too frightened.

I was aware, however, that the information was not enough, despite the fact that it had come directly from one of the witnesses. The case was too important. On the other hand, to get more information through the security ring built by Army Intelligence would be an almost hopeless task. The only way was to attempt to break the secrecy around the two soldiers under treatment in the Army's Central Hospital. As a physician, I might perhaps contact some doctors from the hospital and even examine the two patients if possible.

However, all my attempts failed. The only thing I was able to determine was the fact that two soldiers from the Fortress of Itaipu were really there under treatment for bad burns. Only that. The case remained in my files until two months ago, when the final proof that it was real was finally obtained.

Three other officers from the Brazilian Army who had been at the fortress on the night of the UAO were fortunately localized and contacted. They told the same story. They confirmed the report transcribed above in every detail.

UAO WEAPONS - COMMENTS ON TECHNICAL ASPECTS INVOLVED

The evidence at hand indicates that UAOs possess the means of creating, in the ignition system of internal combustion engines of cars and aircraft, secondary currents powerful enough to destroy synchronization of sparkplug action and so to stall the engines; that they can interfere at will with radio transmitters and receivers, with generators of electric current, with batteries, with telephone lines, and generally speaking, with all electric circuits; and that these "electric effects" are not merely side-effects of the powerful electromagnetic fields that exist around UAOs, but the result of purposeful interference, of a weapon used as means of defense and attack.

UFO HOSTILITIES AND THE EVIL ALIEN AGENDA

These effects are quite independent of the proximity or any movement of the UAO and sometimes (as in Incident 3) they appear to be provoked entirely by the behavior of the witnesses. Such a weapon is very efficacious, because the great majority of man-made machines are either electrical or depend on an electrical ignition system. In Incidents 1 and 2 it was used against two airplanes, but produced no biological effects on the crews inside them.

In other cases, however - chiefly in France - the witness reported that they were "electrified," "paralyzed by an electric current," or felt a "sensation of heat." But such a heat was not enough - in any case - to produce the biological effects described in Incident 3.

The evidence at hand suggests that such a weapon is not an alternating magnetic field in itself, but a high-frequency, long-range electromagnetic beam of some sort, i.e., a radio-electric wave concentrated into a narrow, powerful beam. After a careful analysis of the data, I came to the conclusion that this weapon might be a microwave ionizer - a generator of odd-shaped microwaves that ionize the air where they strike.

They would make air a high-resistance conductor, among other things. Nothing more than that. And if ionization can make air a high-resistance conductor, then an ionizing beam would make a high-resistance short between the power terminals of a battery. With the electric charge a battery carries, that short would get hot. So would the battery. It would get hot enough - given enough time - to boil the solution inside it. Which has happened in some cases (in some "stalled cars").

Besides, a microwave generator with power enough would short-circuit anything within its range; any apparatus or electric instrument (as in Incident 2, or motors with electric systems as in Incident 1); or it could momentarily paralyze every bit of electric equipment in a plane, ship, grounded vehicle, or military fortress (as in Incident 3).

Such a microwave device might be used as a scanner too. In this case, it might explain the so-called "spy beam" sometimes described in connection with UAOs. This appears to become visible near the focal point to radar, photography or the human eye. Jets have sometimes flown through such radar "ghosts," while others have appeared on film as discs, ovals or cones.

The "heat wave" which burned the two sentries in Incident 3 represents another problem. It was not a side-effect of the weapon which produced the "electric effects" - these came at least one minute later, when an all-over alarm had been caused by the soldiers' shouts. Besides, the heat produced by an electrical device would be diffuse, less intense, and similar to that obtained through diathermy. The witnesses would report also a tingling on the skin and a raising of hairs.

As it happened, it is clear that a weapon of a different kind was used against the soldiers. What was it? It is known that the temperature rise of any volume element of matter may be brought about by two different mechanisms: (1) a readily accessible surface is kept elevated in temperature and as a result of conduction there is heating of deeper parts. (2) Heat may be developed in the volume itself (in our case, a human body) by physical energy being conducted through it and converted into heat.

The first is exemplified by application of a heating lamp or hot packs, the second by diathermy. But in none of these cases would the heat produce the feeling of burning clothes; neither would the burns be worse on skin areas protected by clothes - as it happened with the two sentries. This effect is unique and can be termed "structural" heating. It can be produced only by ultrasonics.

This arises from the fact that the longitudinal ultrasonic oscillations are transformed into transverse waves (shear waves) at interfaces between mediums of different acoustic impedance as, for example, between clothes and skin. These resulting transverse waves are more rapidly absorbed than the longitudinal ones, with subsequent increased heat development at interface areas. This ability of ultrasound to produce a unique thermal effect, unduplicated by any other modality available, through differential heating at interfaces between different substances and differences in absorption capacity has been demonstrated and accepted.

An ultrasonic beam is the only thing that could produce the peculiar characteristics of the "heat wave" that struck the sentinels in Incident 3. There is no other choice.

(It is the contention of this office that the "charred roots" of the grass in the vicinity of the Desvergers incident in August 1952, should be considered within the scope of Dr. Fontes' proffered theory of the ultrasonic device. We refer the reader to page 242 of Capt. Edward Ruppelt's book, "The Report of the Unidentified Flying Objects," in which he describes the strange charring of grass roots. The blades of the grass were not harmed, except for the tips which had been bent over and touched the ground, Ruppelt said, and also claimed that the laboratory which did the analysis of the ground and grass samples could duplicate the condition described by heating the clumps of dirt and grass to about 300 degrees Fahrenheit. How it was actually done outside a laboratory, the technicians couldn't even guess. However, if we apply Fontes' postulation, we have at least an educated guess as to how the grass roots became charred. It is certain that they were not heated in a pan in a laboratory. - The Ed.)

An ultrasonic weapon can explain the sudden "heat wave" encountered by military pilots when pursuing UAOs. For example, early in 1954, one of the test pilots of the French Fouga Aircraft Company of Pau, in the Lower Pyrenees, tried to approach a UAO hovering near the town but was forced to turn away because of the intense heat that built up in his cockpit.

About two months later, a USAF "Starfire" was scrambled to intercept a UAO, but the crew bailed out because the cockpit had become unbearably hot. Anyway, this ultrasonic device seems to be a short-range weapon, used only at close range. It might also destroy aircraft if a powerful ultrasonics generator is used - through the phenomenon of resonance. If the driving frequency of the beam coincides with the natural one of the vibrating body (the metallic structure of an aircraft, for instance), then a maximum motion or vibration occurs.

Cases have been recorded in which such vibrations reached proportions where large structures were destroyed. In the case of an airplane, the molecular cohesion of its metallic structure would be suddenly disrupted; instantly all metallic parts of the plane would disintegrate into thousands of small fragments. The plane would explode as if hit by an invisible external force - an explosion without fire.

The nonmetallic pieces or objects wouldn't be affected by the sudden disintegration. The shredded condition of the plane would be the chief clue that such an ultrasonic weapon was used. Confirmatory evidence might be also found in the bodies of the crew members killed in the crash. An ultrasonic scanner, i.e., an instrument to meter the nature of the terrain below, might also be used by UAOs. The constant stream of reflections of the ultrasonic signals (microwaves or shortwaves might be used too) being sent out - channeled into proper computing devices at precisely the right time for comparative analysis - might give data enough to obtain a complete picture of the planet's outer crust.

Such a device could explain the strange behavior of animals and birds - chiefly dogs - when a UAO is sighted in their proximity. The ultrasonic vibration emitted from the UAO, which bypasses the ear and directly stimulates the brain, could play on that the way a musician plays on his instrument, creating emotional moods that would strike too deep for any untrained animal to resist.

Dogs would be especially sensitive for obvious reasons. At this point, I would like to emphasize that I can merely give you my technical interpretation of the available evidence. I cannot prove to everybody that such UAO weapons are, in fact, a microwave ionizer and an ultrasonic beam. But some of you will agree that my reasoning is sound. Others will feel that the conclusions are controversial. However, none of you can deny that the evidence included in the three incidents I have reported makes two things absolutely certain: (1) that UAOs possess weapons of a peculiar type which have been tested against planes, soldiers and a military fortress; (2) that UAOs are hostile - at least some of them.

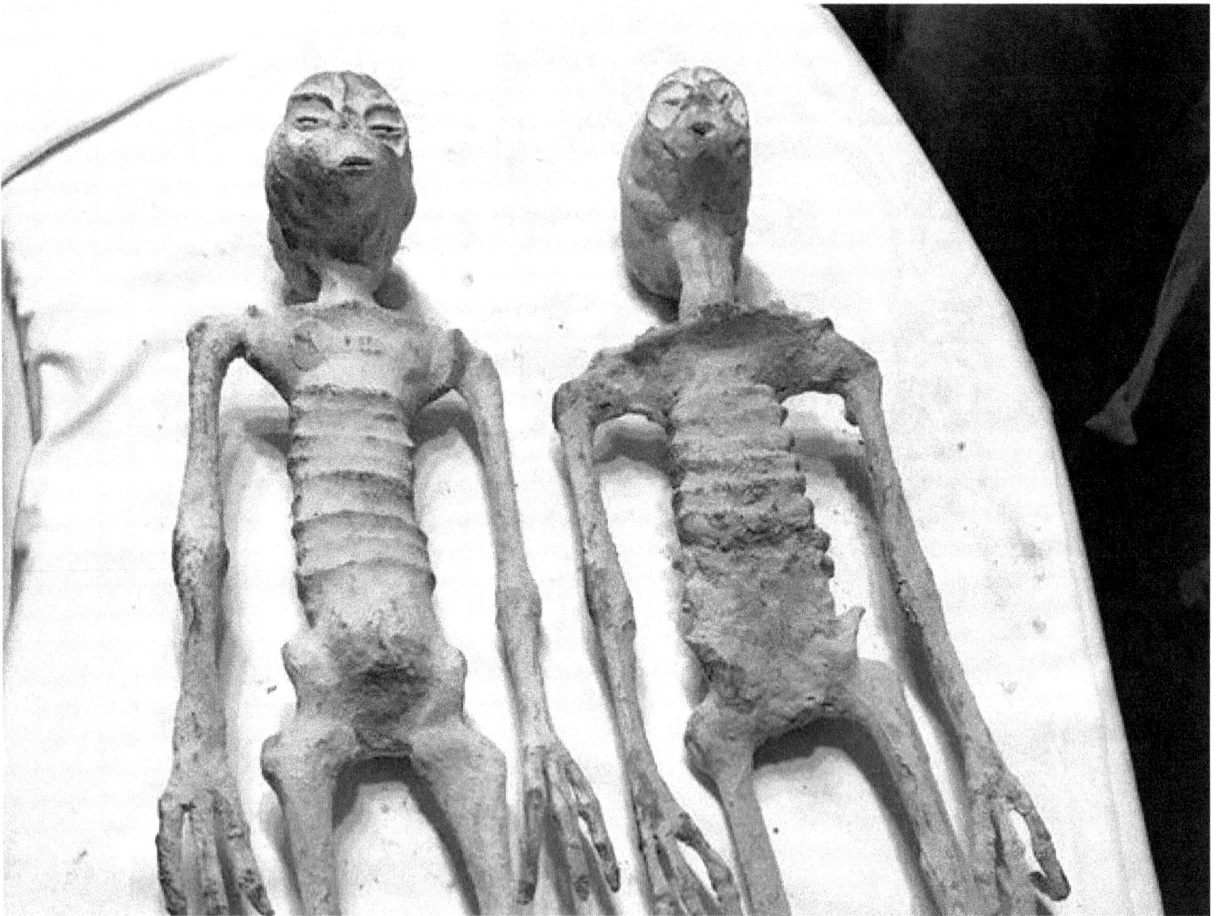

Chapter 15

BURNED BY A FLYING SAUCER
By George Sands

"You are the only man in the world who has come within reaching distance of a flying saucer." That was the thought-provoking statement which Florida newspapers claim was made recently by a serious-faced military officer to Scoutmaster J. D. Desvergers of West Palm Beach. The high ranking Washington official had hurried south to interview Desvergers, following the Scoutmaster's amazing experience in a dark woods nearby.

The incident occurred on Tuesday, August 19, 1952. The still night lay upon the land much the same as any other typically peaceful Florida midsummer evening. A million jewel-like stars glittered from their settings in the great velvet showcase overhead. The customary light southeast tradewind whispered through the pines and cabbage palms, launching an occasional misty cloud ship across the gleaming night skies. There had been a few scattered seasonal showers during the day and some restless summer lightning had played across the heavens around 9 P. M. There was nothing about the peaceful night to indicate that history was about to be made.

Following a meeting with his boys, Scoutmaster Desvergers was driving three of them home. The time was approximately 10:30 P. M. The scene was a lonely spot along Military Trail, about five miles west and a dozen miles south of West Palm Beach.

Suddenly, through the trees on his side of the road, the Scoutmaster saw what appeared to be a bright glare, as though something was burning. The glow seemed to be about 300 yards off the highway, at a place where there were no houses.

"Looks like a plane has crashed," he told the boys, quickly stopping the car and turning it around for a better view of the glare. "You fellows stay here. If I'm not back in ten minutes, go for help."

With that, he plunged into the dark pine woods. Used to moving through the Florida woods, where thick palmetto and other growths are often encountered, the Scoutmastser had remembered to take a machete with him.

The going here was fairly easy, however, and Desvergers was mainly concerned with avoiding the low hanging branches that threatened his face and eyes. Hurrying through the darkness, he was not acutely aware of the light ahead. It was there; he was going in the right direction, and it became brighter as he approached the scene of the supposed crash. It was not until he had nearly reached the site, however, that the perspiring Scoutmaster realized something was amiss.

There were no cries of injured people; no crackling sounds from burning wood or metal; no flickering shadows cast by leaping flames. Instead, the eerie, rosy glow that illuminated the pine woods was of a steady intensity. And it was coming from somewhere above.

The Scoutmaster looked up. There, humming and spinning a few feet over his head was a 30 foot diameter flying saucer.

"I saw the saucer in every detail as it hissed and hovered horizontally about ten feet from the ground," said Desvergers. "It was about three feet thick and rimmed with a phosphorus effect."

Startled, the Scoutmaster ducked and swung instinctively at the the strange machine with the machete he carried. No sooner had he done that than a misty flare blossomed from the bottom of the whirling craft and came floating down directly towards his face.

Too late, Desvergers put up his hands to protect his eyes. The ball of light struck him and he blacked out.

"It had a sickening, nauseating stench; worse than rotten eggs—more like burning flesh."

Meanwhile, back in the parked car, the three scouts had gotten worried. The ten minutes were already stretched into eighteen with no sign of their leader. Furthermore, they had each seen lights flare up on three different occasions; strange looking lights that glowed weirdly and caused the dark trees to stand out in stark silhouette each time before they died away again.

The boys, David Rowan, Bobby Ruffing and Charles Stevens, got out of the car and ran northward along the highway until they came to the home of J. D. Bryntson. Here, breathless and scared, they blurted out the story.

Bryntson telephoned Deputy Sheriff Mott N. Partin, whose home is at Boynton Beach, a half-dozen miles away. Racing to the scene, Partin spotted the abandoned car, its light still on, pulled off the highway onto the shoulder where Desvergers had left it. As he got out to inspect the car the sheriff heard sloshing footsteps stumbling towards him through a water filled ditch beside the road.

Partin swung his flashbeam towards the sound. It was Desvergers. "Here— over here," the Scoutmaster gasped. He lurched drunkenly, the machete thrust out before him. "His face was pale and he had the look of someone who was suffering from shock," the Sheriff said later.

Jerkily, the Scoutmaster blurted out the details of his strange experience. "When the thing finally knocked me down, I tried to scream and run at the same time," he concluded. "But I couldn't do either."

"Did it burn you?"

"Look." He held up his hands for the officer's inspection. The hair was singed on both forearms, as would be the case if he had held them protectively over his face.

The hair on Desverger's arm was severely singed.

"What happened then?"

"I don't know. The next thing I knew the lights from your car were shining in my eyes. I don't remember walking out here."

The Scoutmaster had been in the dark woods one and one-half hours.

After they had rounded up the three boys and had listened to their accounts confirming the strange flaring lights, Sheriff Partin found himself again staring thoughtfully at the darkened woods from which the Scoutmaster had stumbled. There seemed only one thing to do. Probing with his flashbeam before him, the officer went into the dark pine woods to investigate.

The woods were quiet, hushed and waiting for him. Desvergers' route was easy to follow and at every step the sheriff strained for some foreign warning sign. His sharp glance followed expectantly each new jabbing finger of light from the torch in his hand. There was nothing; only silence.

When he reached the small clearing where the Scoutmaster said he had encountered the hissing saucer, the sheriff studied the ground. "The grass seemed to be scorched or blistered," he admitted later.

He was about to turn away when a bright object caught his eye. At first he thought it must be a reflection of his flash beam off a stone or puddle of water. It turned out to be the Scoutmaster's flash light. It had been embedded into the soil so that only a small part of it was visible.

Back at the car the sheriff inspected Desvergers' cap. Three small holes, about the size of BB shot, had been burned into it. Two were on the top and one was on the brim. The hole in the brim was not burned all the way through.

The Scoutmaster did not appear to be in need of medical care so he was taken to the office of the West Palm Beach sheriff. Here he repeated the story. This version differed from the original in only a single detail. That single detail, however, fell like a bombshell upon the ears of his listeners.

The hissing saucers had contained beings, he said grimly, "...who were as afraid of me as I was of them. That's why they projected that ball of fire at my head."

The next morning, Deputy Sheriff Partin returned to the site of the incident with three Air Force officers from nearby Morrison Field. In the clearing they found a place where a body had clearly fallen.

The newspapers, meanwhile, were in a quandary about how to handle this seemingly fantastic story. They made a careful check of the Scoutmaster's background. A Marine and naval intelligence veteran of World War Two, the 30-year old hardware clerk had a spotless record. He was happily married and the father of three children.

No sooner had the story hit the news wires when telegrams, phone calls and letters began arriving from all parts of the country. People who might never have paid any attention to the harmless Scoutmaster now clamored at his door day and night. The public seemed to want to believe that the strange ship had

The scene of the encounter.

come from another world. And there was no evidence to say it had not.

The Air Technical Intelligence Center at Dayton, Ohio, quickly flew an officer to the scene. ATIC is charged with the task of analyzing reports of unusual aerial objects. Officials from Washington conducted inquiries and combed the wooded area with a Geiger counter.

"We cannot tell yet just what it was," an Air Force spokesman said. "We are going to continue to check and get all the facts together. The incident is under investigation."

Whatever those facts may have been, they were obviously not going to be made public knowledge. It was reported that the government officials had treated Desvergers like a Dutch uncle.

Desvergers, meanwhile, sleepless and harried after several consecutive days of contact with a curious and wearying public, decided to disappear for awhile. The parents of the three boys who been in the car that fateful night had also grown irritated with those who seemed to want only to misquote and ridicule their children. They grimly slammed the doors against further interviews.

Deputy Sheriff Partin, sadly, could not escape so easily. He continued to receive telephone calls and letters from far countries. One determined voice on the phone demanded that Deputy Partin make a statement for the British newservices.

Ten days passed and public interest gradually waned. Then, about 7:30 on the evening of August 29th, another saucer was sighted.

The incident occurred beside the same highway, only eight miles further north. Wendell Wells and his family of Belle Glade, a small farming community located 40 miles due west of West Palm Beach, were driving to a movie in the city when they noticed a big yellow-white light. It appeared to be drifting, slanting down from the sky.

"Then it got over into the woods on the side of the road and dropped straight down. We turned down Military Trail to get closer. Soon we could see where the bushes were all lit up. There are no houses around there, so it was clear that's where it landed."

The landing spot couldn't be reached by car and the family did not investigate because they did not want to leave their small children in the car.

By September 14th, still another eerie night visit occurred, this time a Belle Glade itself. Floyd Brown, a dependable employee of the Everglades Experiment Station, was enroute to the station's milking barn before daylight, when he heard a high-pitched whistling sound and was startled to observe a large, red light hovering about one hundred feet above the barn.

As Brown stared at the strange contraption, it settled to within about 40 feet of the ground as if to get a better look. Frightened by the whistling sound, the

The scoutmaster case received national press attention.

cows bolted. The saucer then quickly shot away. It was about 35 feet in diameter with alternate amber and red lights in rows along the outside rim on its underside. The glow from it was bright enough to light up the ground as it passed the Experiment Station, and an acrid, ammonia-like odor caused Brown's eyes and nostrils to smart and burn.

No sooner had Brown rounded up the cattle than the saucer whistled in from a different direction, causing the cows to bolt again. But this time the strange craft kept going.

Two weeks later, on September 28th, DeWitt Upthegrove and three adult members of his family spotted yet another of the strange sky craft. Upthegrove, a quiet mannered frog hunter, lives at the edge of the great Everglades Swamp, 20 miles north of the spot where Scoutmaster Desvergers encountered the original saucer. Like Brown, Upthegrove is regarded as a stable citizen, not the type to tell imaginative tales. He had been sitting on his darkened porch after dinner when the saucer appeared.

The sharp-eyed outdoorsman and his son-in-law, both experienced at piloting their airboats through the great swamp at night during long hours of frogging, were well aware of meteors and similar unusual lights in the sky. They tried to reach the saucer as soon as they saw it descend to earth about a quarter of a mile away. It must have been resting against the earth,, for they could see the tops of low mangrove trees clearly silhouetted by the strange rosy light cast by the craft. Yet, before they could cover the intervening distance, it zipped upward and disappeared.

What were these phenomena? Floridians, just as residents of other states, began to ask uneasy questions. The frequency and similarity of the appearances ruled out such theories as meteors, fie balls and swamp gas, or even hallucinations.

Desvergers, following his final interview with the "brass," had clammed up. He knew the government had no legal hold on him, yet he admitted they told him not to ever tell the whole truth of encounter. He never did.

"It is not foolish to say that it will determine the future of all of us someday," was the Scoutmaster's final comment "I know what it is and it is of vital importance. But it is better for me not to go any further because it may cause another "Orson Wells" panic."

Chapter 16

MAN KILLED BY DEATH RAY FROM FLYING SAUCER
By Tom Lingham
Reprinted from the October 1968 issue of "Flying Saucers - Mysteries of the Space Age"

****

Another victim has been mysteriously killed by a flying saucer. A New Zealand sheep rancher died a horrible death as a strange ray shot through his skull, melting skin and hair.

In early February, 1968, Amos Miller, 39, was killed under highly mysterious circumstances. His son, Bill, 17, witnessed the death, and an autopsy was performed by top medical men in Auckland, New Zealand. Based on the evidence at hand, police officials clammed up, made no charges and refused to release any further information to reporters or the public. What they had learned about the sheep farmer's death alarmed them enough to hush-up the case and keep it in a closed file.

It was found beyond a doubt that Amos Miller had been murdered by some unknown being or beings inhabiting a strange flying vehicle - in other words, by creatures in a flying saucer. It looked like a straightforward murder case in the beginning, and this Midnight reporter was able to get the full information about it before police officials put on the wraps.

On February 2, Amos Miller was found dead on his New Zealand farm following a phone call to police from his son, Bill. The boy was almost hysterical and claimed his father had been killed by a flying saucer.

When police went out to his farm to investigate, Miller was found lying in a gully with half the skin missing from his head. The son was taken into custody for questioning and held as a material witness. He stuck to his story that a UFO was the cause of his father's death.

"We were out in the fields repairing some fences in the morning," he told police. "Dad and I were working away when all of a sudden we hear like on a short-wave radio, sort of high-pitched like you hear on a short wave radio sometimes.

"We looked around and didn't see anything at first. But then, just to the right of us, where there's a wooded area, this thing suddenly hovered into sight.

"We dropped our tools and stared with our mouths open. It was about two hundred yards away and maybe about forty feet off the ground.

"It seemed to rest there, like a helicopter standing still, only it wasn't shaped like a helicopter.

"It was round and had a turret-like top on it. There were little portholes around it. We could see them as clear as anything.

"And all around it there was a glow lighting the thing up. It hovered there a while and then three prongs lowered out of the bottom of it. This was the landing gear, because right after that it settled down on the ground and stood on these three legs.

"I didn't know what to do, but my Dad said, 'Come on, Bill, we got to get a look at this thing.' Then he began running toward the ship.

"I tried to call to him to come back, but he wasn't afraid of anything. The spaceship was partly hidden by trees, and when Dad got about halfway to it, just where there's a little brook, he stopped.

"I noticed there was a brilliant light shining on him. It was the same light that surrounded the ship. Even though the sun was shining, you could see it, it was so bright.

"And then Dad slumped to the ground. I was so scared I couldn't move from the spot.

"Hardly had Dad fallen when the ship rose up in the air with the same humming noise, and then shot off into the sky so fast I could hardly follow it. It was out of sight almost immediately."

Bill ran to his father, he told police, and found him dead. Half the skin on his head was gone. The first reaction of the cops was to congratulate the boy on his imagination and then throw him into a cell.

An autopsy was held on the body while police made a routine check of the area where the death occurred. What they found gave them something to think about. At the spot where Bill claimed the saucer landed, the ground was singed in a circular area that had a radius of some 60 feet. There were three indentions in the ground around the edge of the burned area, suggesting an enormous weight had pressed into the ground at these spots. This area was immediately roped off by police and forbidden to reporters.

In the meantime the autopsy results were in. Cause of death unknown. Said

UFO HOSTILITIES AND THE EVIL ALIEN AGENDA

Dr. John Whitty, who headed the autopsy team: "It's the most unusual thing I've ever seen. Aside from the missing skin of the skull, there are no marks on the body.

"There's no evidence that the dead man was struck on the head, and I can in no way account for the disappearance of the head tissue.

"What is at least equally mystifying is the condition of the man's bones. There's a complete absence of phosphorus in them.

"What could have caused that condition, I couldn't possibly say."

The absence of phosphorus in the bones had some relevance in the case. Last year in Australia, following sightings of UFOs by dozens of people, a herd of cattle in the area was found dead. No marks were found on the carcasses. An autopsy revealed that the normal phosphorus in their bones had somehow disappeared.

Five days after he was taken into custody, Bill Miller was released. Not another word has been said on the case. The Miller family refused to talk to reporters about the incident, saying they've been told by police to keep mum.

****

PILOT AND PLANE VANISH — WITHOUT A TRACE
UFO IS DEEMED CULPRIT

PUBLISHER'S NOTE: While no direct link is overtly etched in stone, the death of the individual in the above "death ray" case brings to mind the disappearance of an Australian pilot who vanished years later, again under the well-established assumption that a UFO had "done him in."

On October 21, 1978, Frederick Valentich radioed Melbourne air traffic control at 7:06 PM to report an unidentified aircraft was following him at 4,500 feet (1,400 m) and was told there was no known traffic at that level. Valentich said he could see a large unknown aircraft which appeared to be illuminated by four bright landing lights. He was unable to confirm its type, but said it had passed about 1,000 feet (300 m) overhead and was moving at high speed. Valentich then reported that the aircraft was approaching him from the east and said the other pilot might be purposely toying with him. Valentich said the aircraft was "orbiting" above him and that it had a shiny metal surface and a green light on it. Valentich reported that he was experiencing engine problems. Asked to identify the aircraft, Valentich radioed, "It isn't an aircraft." His transmission was then interrupted by unidentified noise described as being "metallic, scraping sounds" before all contact was lost.

UFO HOSTILITIES AND THE EVIL ALIEN AGENDA

This is one of the better known hostility cases and it has received its due of attention in the media, especially on History Channel-like documentaries. The pilot's parents have been repeatedly interviewed and all they can do is shrug and suggest something strange happened to their son. No wreckage was ever found despite extensive searches. Where did he go? Heaven knows, but both plane and pilot are gone and a UFO was seen nearby. And, sadly, we know what that can mean.

THE AUSTRALIAN

'It's a long shape . . . coming for me right now . . . hovering on top of me'

UFO MYSTERY

Plane vanishes after 'loud metallic noise'

Chapter 17

CAUGHT IN THE MIDDLE OF BRAZIL'S "UFO DANGER ZONE"
By Bob Pratt
(As introduced by Timothy Green Beckley)

I cannot say for certain if I ever met Bob Pratt or not.

He was one of the top reporters for the National Enquirer headquartered in Lantana, Florida, at the time, and I was their main freelance stringer operating out of New York. I did visit their impressive spread upon several occasions and chit-chatted with numerous editors who had worked on my pieces to get them "into shape" for the paper (which usually meant cutting the original article way down and replacing my name with theirs - but, what the hell. I was getting paid some fairly decent coin for those days.) Not certain if Pratt was one of those I kibitzed with or not.

This was in the day when the tabloid was seriously fact-checking their articles and they were heavily into reporting on UFOs. I remember I ran into movie producer Tracy Torme ("Fire in the Sky - The Travis Walton Story") at a MUFON convention and causally mentioned that I had once interviewed his father "the Velvet Fog" Mel Torme about a UFO sighting the singer had had while walking his dog in Central Park. Tracy nodded and said, yes, his dad had said this was the most accurately quoted article anyone had ever written about him.

What I am attempting to say is that the Enquirer didn't make up stories back then. It wasn't the Weekly World News (which they did own). They made us writers present their fact-checkers with tapes of the interviews to make sure we had not over-hyped the story or blown it out of proportion. Certainly they got caught with their yellow journalism pants down upon occasion, but so has that great "grey lady," the NY Times.

ROBERT V. PRATT

Robert Vance Pratt died Monday, November 21, 2005 - age 79 - following a brief illness. He was born August 12, 1926. A journalist and UFO researcher, Bob Pratt, as his byline read for 50 years was a hard working newspaper and magazine reporter and editor. He studied at the University of California-Berkeley and American University.

During the early days of his journalistic career, he worked in various reporting and editing positions on the Alexandria (Va.) Gazette, the Charlottesville (Va.) Daily Progress, the Evansville (Ind.) Courier, the Buffalo (N.Y.) Evening News, the Miami (Fla.) News, the Philadelphia (Pa.) Inquirer, the Louisville (Ky.) Times and Courier-Journal. He worked for the Charlottesville newspaper three different times for a total of 10 years, twice as a reporter and the third time as managing editor, for seven years. After leaving the Louisville Times and Courier-Journal, where he had been executive assistant to the editor and publisher, he worked for eight and a half years as a writer and reporter for the National Enquirer. The last 17 years of his career he worked in computer page design and production for three other tabloid magazines, the National Examiner, Globe and the Sun.

After retiring, Bob returned to his favorite South American haunts several times to continue his quest. He also became an editor of the MUFON UFO Journal and co-author with Dr J. Allen Hynek of "Night Siege: The Hudson Valley UFO Sightings." Some of the material in this chapter has been excerpted and slightly abridged from his massive casebook, "UFO Danger Zone: Terror and Death In Brazil - Where Next?"

It should be noted that during his newspaper days he was a skeptic when it came to UFOs, but in May 1975, as a reporter for the National Enquirer, he came to believe UFOs were real. He revised his attitude after interviewing more than 60 people in one week who had seen something they could not explain, or had somehow gotten wrapped up in this grand celestial mystery.

From that moment on, UFOs became his major interest. In the years that followed, he interviewed more than 2,000 people who had what he believed were unexplained UFO experiences.

It all - kind of - started when his bosses at the Enquirer said they had heard stories of a wave of sightings along the Amazon, which might have involved the crash of an unidentified object. They sent Bob packing and off he went to the jungles of Brazil in search of adventure - and what he thought might be a wild goose chase.

Turns out it wasn't!

Bob ended up going to Brazil numerous times, learning more about this enigmatic topic on each visit. He recalls: "When I first began going to Brazil to look into UFO incidents, I had resources beyond the reach of most investigators. My first four trips were made as a reporter for a magazine that paid all my expenses. This allowed me to do many things that most researchers can't afford to do, such as spend weeks at a time on one case, travel wherever I needed to go, and hire cars, planes, boats, guides, interpreters and whatever else was needed."

It was during these early visits that Pratt began to hear about people getting hurt and killed by UFOs.

As Pratt noted: "UFOs (became) particularly grim news for people living in the small towns, farms and forests of central and northeastern Brazil. Their encounters, unlike those of most other experiencers in other parts of the world, often led to injury and even death."

UNDER ATTACK

It was dark and a light rain had begun to fall when Moises Campelo, a man in his thirties, began to make his way home.

He was alone and his rural surroundings might have made some big city slickers a bit nervous, but he was used to the terrain. Suddenly there was a bright light overhead, so bright that it began to hurt his eyes. But that wasn't the only thing menacing about the brilliant orb. The witness became paralyzed in its presence.

Pratt quotes the witness as he sat in his living room recalling the experience in horror, one snippet at a time:

"I thought the UFO was going to take me away!

"It lit up everything around me, and it was very hot. And then I felt like I was being sucked upwards.

"I got really scared...I was raised about one and a half meters. I couldn't cry for help, and I couldn't move. The light was very hot. I was terrified."

Pratt quotes the witness as having said that at this point they seemed to lose interest and let him down gently.

He thought his bizarre ordeal had ended and he would be able to make his way home without further incident.

Moises Campelo was not to be so lucky!

"I was crawling on the ground like a lizard because I couldn't walk. I got about a hundred meters and crawled under a little tree, where I rested for a moment. Then I went on again and just as I started to come out from the other side of the tree, they came back and got me again.

"My head hit the branches when I went up. I was paralyzed again, and, this time, I felt very cold. I couldn't move. I couldn't shout for help, and the light was hurting my eyes again. The thing was above me, going around and around and around."

They - whoever "they" are? - held him in the air and then dropped him really hard. "My left eye began to swell and bulge out of my head, and I couldn't seemingly see out of it by the time I got home."

Pratt noted that the witness was not able to sleep for several nights after the incident and he was blind for a few days. He had trouble with seeing for a great while afterward.

UFO HOSTILITIES AND THE EVIL ALIEN AGENDA

As Pratt probed the case further he discovered that, "Other people besides Moises had been pulled off the ground. Many more have resisted in one way or another. Some have even been snatched up by hooks, and a few have actually been taken away. Many have been chased by UFOs and still others have been burned, zapped and otherwise hurt. Some have died."

On a nightly basis, UFOs were zapping cresidents of Colares.

The astute, keenly aware researcher followed a trail of cases that consistently reminded him that he was not dealing with some benign type of entity but was instead facing an unknown terror.

Take the case of Januncio, who suddenly was confronted with a huge, dark apparatus which appeared just above his head the moment he had lit a match.

"It was like a big silo," he told Pratt. "It was at least twenty five feet tall and had a round bottom twelve to fifteen feet in diameter.

"A door opened on the bottom," he continued, spreading his hands apart to show what it looked like. "And I could see a man and a woman sitting in seats like a car. They were sitting still, alive, but very still and stiff. They never moved. The woman looked like she was wearing a dress.

"When the door opened, a lot of light came out, and I felt like I was being pulled up into the object. It was like a magnet. I grabbed a small palm tree and wrapped my arms and legs around it. The light was very hot and I was terrified. . . This happened five times," he said, raising and lowering his arms. "Up and down, again and again. My chest was scraped raw."

The worst part of the ordeal was when the occupants of the ship saw he was not about to let go of the tree he was holding onto as a life support; that was the only thing keeping him out of their grasp, and his only connection to the solid ground below.

"I began to cry and thought I was going to die. That's when the man and woman saw I was not going to let go, and so they dropped something like hot oil on me to make me let go of the tree. I felt like I was between two big fires. I couldn't move. It burned my arms and hurt very much, but I was too afraid to let go. I almost died I was so scared. "

He finally made it back to his house, where he remained sick for two days. "I couldn't eat. I had a bad headache, my chest was scratched and red, and my arms had burns on them like cigarette burns."

Januncio's case held great importance for Bob Pratt for the remainder of his career in UFOlogy. "I have talked to other people who have had worse experiences, but for many years this was the one case against which I measured all others. Few match it for sheer horror, brute force and crudeness."

But let us turn over the "mighty pen" to Bob Pratt himself as he takes us on a creepy journey through the back roads of Brazil, where UFOs have often pitted themselves in a hostile manner against the occupants of several rural, coastal communities.

TGB

Pratt examines the boat used to sail to Crab Island.

UFO HOSTILITIES AND THE EVIL ALIEN AGENDA

PERMANENTLY ZAPPED

For seventeen years, an old man lay in a small room at the back of a large house in Ceara, helpless and needing a full-time nurse. In a small, humble, mud-walled house one-hundred-thirty miles to the north, a younger man sits all day long in a wheelchair, unable to walk or speak clearly. Both were victims of UFOs.

The older man, Luis Fernandes Barroso, had been a businessman and rancher in Quixada, a city in the central part of the state. From late 1976 until he died in April 1993, Luis had been in a vegetative state, totally dependent, able to say only three words and recognizing no one but his wife, Teresina.

His sad story began several hours before dawn on April 23, 1976, when he hitched a donkey to a two-wheel carriage and set out from the family home in the city to go to his farm ten miles east of town. He never made it, at least not on his own.

Around seven o'clock that morning, a cowboy named Joao Francisco found him sitting in the carriage on the side of a highway three miles from the farm. Both Luis, then fifty-three, and the donkey were in a daze. Joao took them to Luis's farm. Later in the day, when Luis became coherent, he told his wife, Teresina, what had happened.

Two hours before daybreak, he said, a big, lighted object came down from the sky and hovered over him and the donkey. A door opened on the bottom, and a beam of hot light hit him and the donkey. He passed out and doesn't know what happened after that.

Before the day ended, Luis was very sick, vomiting and suffering from nausea, diarrhea and headaches. Teresina took him to Dr. Antonio Moreira Megalhes in Quixada. The doctor, then forty, had known Luis most of his life and listened sympathetically to his story about the UFO. He then gave him some medicine.

HAIR TURNS WHITE

Luis didn't get any better, though, and Dr. Megalhes suggested he see a psychiatrist. Teresina took Luis to Fortaleza, one hundred miles to the north, and over the next two months he was examined by a total of twelve psychiatrists and psychologists. None could say what was wrong with him although some thought he might have a lesion on his brain.

Not one of the doctors put any credence in Luis's story about the UFO, and because of this he stopped talking about it long before the end of his stay in Fortaleza. The only people who believed him were Teresina and Dr. Megalhes, and Dr. Megalhes was ridiculed by fellow physicians because he did.

By the time Luis returned to Quixada, his speech was beginning to deteriorate. Three months after the UFO incident occurred, his hair turned white almost overnight. By the end of six months, he had lost all his mental faculties and had regressed to the age of a one-year-old child.

UFO HOSTILITIES AND THE EVIL ALIEN AGENDA

From that moment on, the only words he ever uttered were mamae, medo and da. In English, they mean "mama," "danger" and "give." He didn't react to any stimuli, with one exception. When anyone took a photograph of him with a strobe light, as I did once without realizing what would happen, he would scream when he saw the flash.

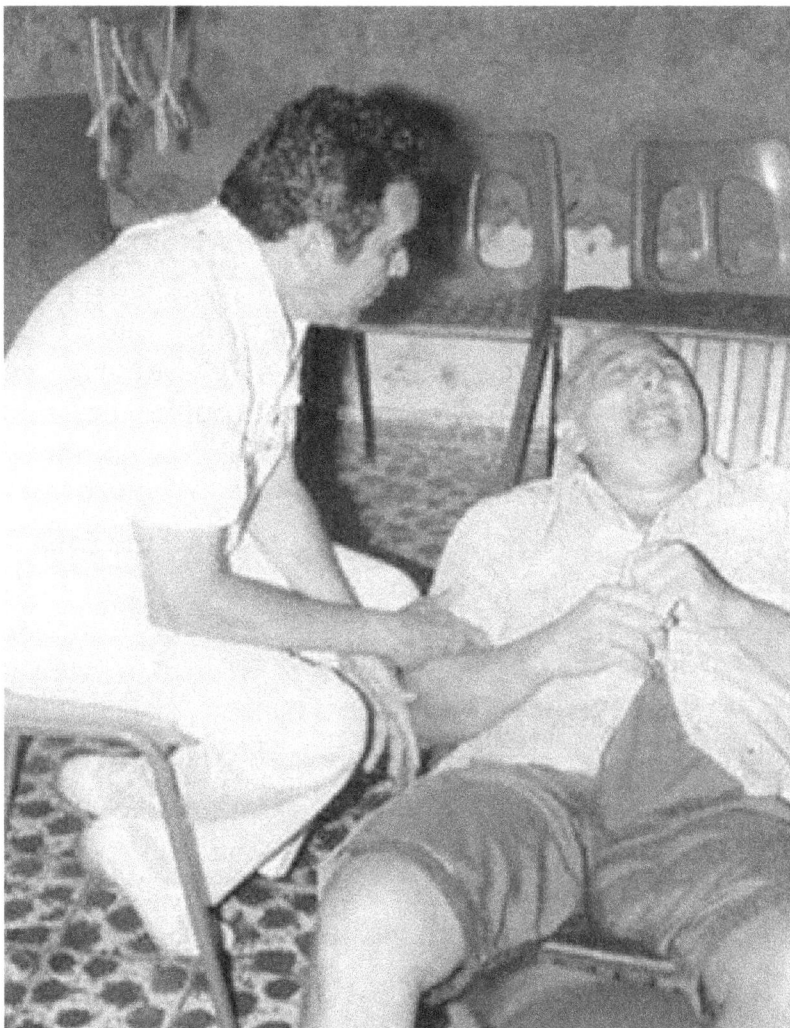

Luis Barroso Fernandes, years after his experience. Beside him, the doctor Antônio Moreira Magalhães, who accompanied him until his death.

Reginaldo Athayde was the first investigator to study this case. In 1986, we went to Quixada together and talked to Dr. Megalhes and Teresina Barroso, both of whom told me the story of what had happened to Luis.

Dr. Megalhes even went with us to the Barroso home, where Luis was being watched over by a nurse. Luis sat in an easy chair, staring straight ahead and moving his eyes from time to time but apparently seeing nothing. All his days for seventeen years were like that, with Luis either sitting in the chair or lying in a bed. One by one Dr. Megalhes raised Luis's arms and legs, and Luis would slowly lower them by himself. This, Dr. Megalhes said, showed that Luis still had full control of his arms and legs and hadn't suffered a stroke.

Luis may have suffered some undetected mental condition at the same time the UFO hit him with a beam of light, and the whole affair could have been just a strange coincidence. However, the donkey that was pulling his carriage that morning was also affected by the light. Teresina said the animal seemed to be in a stupor for about a week after the incident and wouldn't eat for several weeks, but then recovered without any further effects. Forever after, though, it was skittish and easily frightened.

UFO HOSTILITIES AND THE EVIL ALIEN AGENDA

Luis died on April 1, 1993, of pneumonia. Fortunately for him, Teresina was able to carry on the family businesses and could afford full-time nurses to take care of him for the seventeen years he lived after the UFO left him a helpless invalid.

ANOTHER ZAPPED

The other victim crippled by a UFO, Jose Vonilson Dos Santos, has not been as lucky. Vonilson has been an invalid ever since the day he, too, was hit by a beam of light from a UFO.

It happened shortly after dark one evening in May 1979 as he climbed over a wooden fence while walking across a farm on his way to a religious meeting. He was twenty-two at the time, worked as a farm laborer and lived with his widowed mother on a farm near Carnaubinha, west of Fortaleza.

"When I was almost over the fence, I was hit by a beam of yellow and green light," Vonilson said. "The UFO was about ten meters above me. The beam hit me on the right side of my neck. I was halfway over the fence and was putting one foot on the ground at the time, and I tried to hide in the bushes. But I fell down. I got up and fell down again. I couldn't move. I was paralyzed."

It was when he first fell to the ground that he looked up and saw the round, silent object above him.

"It was about two meters across, and it had a small hole on the bottom," he said. "I saw some lights on it, but I don't know how many there were. After about five seconds, it went toward the ocean and disappeared."

HELPLESS FOR THREE DAYS

The Atlantic coast is five miles north of the spot where Vonilson fell to the ground. As he watched the UFO disappear, he discovered he couldn't move either leg and had only slight control over his arms. He lay where he fell and, incredibly, stayed in the same spot for nearly three days and nights, helpless. He was in a field, and the nearest house was a mile away.

In the beginning, the ground was still wet from rain that had fallen the day before. He was bothered at times by mosquitoes but escaped getting sunburned because he was in the shade of a tree. Unable to control himself, he also wet his pants.

"I was scared but I didn't feel any pain," he said. "I was very thirsty and hungry. I was awake most of the time. There were so many stars, so much wind."

No one realized he was missing. This occurred on a Sunday evening, and his mother had gone to Fortaleza and didn't return home until the following Wednesday. The people at the farm where he worked assumed he, too, had gone away for a few days.

Vonilson was discovered on Wednesday by twelve-year-old Luis Perreira, a farm boy who happened to be walking by. Luis ran and got his father, Joaquim, who took Vonilson home. Vonilson's mother, Regina, wasn't there so they took him to another farm where he was given some herbal tea and brandy. When Vonilson's mother returned, she took him home and also gave him tea and brandy.

He didn't get any better and, some weeks later, she took him to a doctor who examined him and said he had suffered a stroke. Regina, who has ten other children living elsewhere, told us Vonilson had been healthy, spoke normally and had no disabilities before the UFO incident.

Vonilson, who never went to school but learned to read, now has no way to support himself. He and his mother are very poor and live on a small pension she receives in a simple house on the outskirts of Caucaia.

In one way, Vonilson is like Benedito Bogea, the farmer who wants to be zapped again. "I think the UFO did something to me," Vonilson said. "Now I want someone from the UFO to come back and fix me."

What happened to Luis and Vonilson is tragic. Luis lived seventeen more years and never recovered, while Vonilson is still alive and probably will be confined to a wheelchair for the rest of his life.

Other victims of close encounters, though, haven't been even that fortunate.

DEATH ON CRAB ISLAND

Jose Sousa was twenty-two the day he died. He was healthy and had no known ailments. What killed him is a mystery.

The day started out sunny and hot as he and three other men sailed an old, weather-beaten boat from Sao Luis to Crab Island fifteen miles to the south in Sao Marcos Bay. They arrived early in the afternoon, anchored in a stream well inside the island and spent the rest of the afternoon cutting down trees and shaping them into poles. They planned to sell the poles for use in constructing simple buildings.

The island is twenty-five miles long and seven wide. It's a desolate, swampy, uninhabited place infested with mosquitoes and covered with scrub brush and trees. People go there only to get wood or catch crabs. With Jose were two of his brothers, Apolinario, thirty-one, and Firmino, thirty-eight, and a cousin, Auleriano Bispo Alves, thirty-six. They worked all afternoon cutting and stacking poles on the bank. They quit at six o'clock as the sun was setting and ate a supper of beef and rice. The tide was out, and the boat was sitting in the mud of the empty stream. They chatted until eight o'clock, then went to sleep inside the boat, covering the hatch with a piece of canvas to keep mosquitoes out. A small louvered window at the back of the cabin allowed a little air to circulate. A lantern with the wick turned low hung on one side of the cabin.

The men planned to wake up around midnight when the tide came in, load the poles onto the boat and sail back to Sao Luis on the outgoing tide. Jose,

UFO HOSTILITIES AND THE EVIL ALIEN AGENDA

Apolinario and Auleriano had made this trip at least a hundred times before and had never failed to wake up when the tide came in. The rocking motion of the boat as the stream filled plus the sound of the rushing water hitting the hull was as good as any alarm clock.

Firmino was the only novice. The regular fourth man in the crew was sick and Firmino, a farmer, asked to take his place because he needed poles for an addition to his house in the tropical forest. It was his first trip, and he was to regret it.

Something went terribly wrong after they went to sleep. By midnight, Jose was dead and Firmino and Auleriano were badly injured, but no one would know what happened or why. No one knew then. No one knows now.

SHOCKING DISCOVERY

Instead of waking up at midnight, no one awoke until five the next morning as the sun was rising. Apolinario, who had slept on a mat on the cabin floor, heard Auleriano calling for help in the front of the boat. Apolinario was puzzled, because Auleriano had gone to sleep in a hammock at the back of the boat, four feet behind Apolinario's mat.

Apolinario scrambled forward, ducked under another hammock where Jose had gone to sleep, and threw back the canvas covering the hatchway. With the cargo area suddenly visible in the early light of dawn, Apolinario saw Auleriano lying in several inches of water in the bilge. He asked what was wrong, but Auleriano didn't know. He was in pain, couldn't stand up and didn't know how he got there.

Apolinario helped Auleriano climb through the hatch onto the deck and discovered he was burned on both shoulder blades. Auleriano then pulled his shorts down and discovered he also had a burn on his left buttock. Strangely, his shorts were not burned.

Apolinario began fixing tea for Auleriano, and then he heard someone moaning back in the boat. He went down into the cabin, again ducked under Jose's hammock, and found Firmino lying on the floor under Auleriano's hammock. This was another surprise, because Firmino had gone to sleep in the front of the boat, where he'd found Auleriano. But Apolinario's surprise turned to shock as he examined Firmino.

"Firmino was all burned and swollen and the skin had come off," Apolinario said later. "I tried to talk to him, but he didn't answer. His eyes were closed, and I tried to open them but couldn't. I got really scared."

Desperate, Apolinario turned to Jose's hammock to get him to help, but as soon as he touched him, he realized Jose was dead. Apolinario was horrified by the latest discovery and checked Jose's pulse. But there was no beat. Jose's body was cold and stiffening with rigor mortis. One leg hung over the side of the hammock. Grief-stricken, Apolinario felt he had to put the leg back in the hammock, but it was a struggle.

He was overwhelmed and wanted to cry, but he was the only healthy man on board, and he'd have to get everyone back to Sao Luis by himself. There was no medicine or first-aid kit aboard, and he couldn't do anything for the burned men. Worse, the tide was now out and the boat was sitting in the mud again.

GOD HELPED ME

He had to wait more than eight hours for the tide to come in again. About two o'clock in the afternoon, he began sailing back to Sao Luis. It was a difficult journey because normally it takes at least three men to handle the sail and rudder of the forty-foot boat, and Apolinario had no help. Jose was dead, Firmino was unconscious and Auleriano was in too much pain. All the way back, Firmino rolled from one side of the cabin floor to the other as the boat rode the heaving waves in the bay.

"God helped me," Apolinario, a small, thin man just over five feet tall, said simply. "We would all have died without God's help."

The sun was setting when they arrived at the Port of Itaqui near Sao Luis, but Apolinario's nightmare was far from over. The only people at the small, deep-water port were two security guards, and they weren't able to help him. He had to walk six miles into Sao Luis, tell the police what had happened and then walk home to get his oldest brother, Pedrinho. The two returned to the port in a car at nine o'clock and took Firmino to a hospital. Although Auleriano was suffering from his own burns, he stayed with Jose's body.

The police didn't get to the boat until one A.M. They took Jose's body to the Medical Legal Institute, and only then was Auleriano able to go to a hospital for treatment. His burns were to leave scars, but he was able to go home that night.

'EMOTIONAL SHOCK'

Firmino was in a coma for a week, and he stayed in the hospital for more than a month. Second-degree burns covered much of his body. The most serious were on the left side of his rib cage, the inside of his left arm and on his forehead. The arm muscles were so badly damaged that the fingers on his left hand were left permanently curled and virtually useless.

No autopsy was performed on Jose. Sao Luis is near the equator and, after more than twenty-four hours in the heat, his body was badly decomposed. The doctor who examined him for the Medical Legal Institute said in his report that there were no cuts or bruises on the body. The death certificate stated that Jose had suffered a "cerebral vascular accident caused by arterial hypertension as a consequence of an emotional shock." The cause of death was listed as "emotional shock."

There was no explanation of what that "emotional shock" was. I spent one month in the Sao Luis area investigating this and other cases and, for much of that time, tried to find the doctor. With Monica Carneiro and other interpreters, I tracked him from place to place, leaving messages everywhere, but when we finally found him, he refused to talk to me and wouldn't say why. However, I discovered that when he submitted his report on Jose's death, his boss strongly criticized him for his conclusions.

The police couldn't determine what happened on Crab Island. Investigators went to the island, examined the area where the boat had been anchored, inspected the boat itself, and talked to the survivors and people who knew them. There was no evidence that the men had been drinking or taking drugs, suffered from food poisoning or toxic fumes or had been fighting. The police found no sign of a fire on the boat or on the island. The only conclusion they did reach was that the three survivors truthfully did not know what happened.

None of the three men can recall the smallest detail of that night, not even under deep hypnosis. A burn has to be one of the most excruciatingly painful injuries anyone can suffer, yet two men were severely burned before midnight and neither knew anything about it, one not until the next morning and the other not until he came out of a coma a week later.

How could these things happen and the men not have any memory of how they got burned? What or who could inflict such injuries and then completely blot the painful experience out of the minds of the victims? How could a healthy young man like Jose simply die in his sleep without any apparent cause?

MEDIA ATTENTION

These are a few of the questions that puzzled Maranhao police, and they have never found the answers. There is no direct evidence that a UFO was involved in the incident. The men saw nothing unusual. It occurred on the night of April 25, 1977, during a period of numerous UFO sightings throughout the area. The newspapers and radio and TV stations in Sao Luis immediately jumped on the story, and most blamed a UFO for what happened because of the mystery surrounding the case and because so many UFOs had been seen. Despite the media attention, the Crab Island incident was not publicized outside of Sao Luis. I learned of it only because Roberto Granchi, son of veteran Rio de Janeiro UFO investigator Irene Granchi, went to Sao Luis in early 1978 to repair some electronic equipment on a boat at the Port of Itaqui. While there, Roberto heard about the case and managed to talk to Auleriano. He told Irene what he'd learned, and I heard it from her. In late November 1978, I went to Sao Luis. There is a remarkable similarity between Firmino's serious burn and coma and what happened in a case investigated by Hulvio Aleixo in the Valley of the Old Women. In Florestal one afternoon, an elderly woman was found unconscious in her backyard with a bad burn on one arm. She was taken to a hospital, where she later recovered. The burn was so severe that she needed skin grafts, and it took three months to heal. No one knew what caused the burn, and she didn't have any idea how it happened. For some days before this occurred, people living in the area had seen strange balls of fire flying through the sky. Some people thought there was a connection between her burn and the fireballs.

It is an old, colonial city on an island at the mouth of a huge bay, with narrow, hilly streets and buildings in pastel shades of green, pink, blue, yellow and other colors, many covered with ornamental tiles. It has miles of beautiful beaches. At that time it had a quarter of a million people, but the city grew rapidly in the 1980s and, by the mid-1990s, was approaching a population of one million.

One of the first persons I talked to about the Crab Island case was Clesio Muniz, chief of criminal investigation for the Maranhao police.

"I saw these people with those strange burns, and I do not believe they were burned by ordinary fire," Muniz said. "I don't believe in UFOs, but this is a strange phenomenon that I have no explanation for. I had heard reports of the 'fire ball' having been seen in cities around Crab Island and west of here. A lot of people had seen the 'fire ball' when this happened, both before and after.

"From reports I received, the 'fire balls' do not seem like falling stars. They go up or down or to the left or right, horizontally or vertically, slowly, fast, or very slow and then very, very fast. It is an unusual phenomenon, and I do not know what it is."

Another investigator told me he believed lightning caused the death and burns. His theory was that lightning struck the sand or mud near the boat, bounced back up and then flew horizontally into the cabin, striking three of the four sleeping men.

HE SAW A 'FIRE'

Two doctors from the Medical Legal Institute who examined Firmino in the hospital also thought lightning was the cause. One was Dr. Carneiro Belfort, then director of the Institute and later a professor of medicine at one of the universities in Sao Luis.

"I wanted to see Firmino because the newspapers were saying UFOs caused it, and I wanted to see for myself," Dr. Belfort said. "I've never seen a UFO, and I don't believe they exist. The burns were characteristic of lightning, but I cannot definitely say it was lightning. If it wasn't lightning, I don't know what it could have been. The man told me he had seen a 'fire' before passing out."

That last remark - that Firmino in his delirium had mumbled something about "lithe fire" - was the only discernible link to a UFO. O fogo or "lithe fire" is a common term for a UFO throughout Brazil.

The other doctor favoring the lightning theory was Jose Oliveira, then a member of the staff of the Legal Medical Institute.

"Firmino had many second-degree burns and could have died," he said. "In my opinion, it was lightning. But if lightning was the cause, then the boat should have had some damage or burns and the one who died should have been burned."

Neither doctor saw the boat or Jose's body, but the death certificate stated there were no marks or lesions on the body.

As we talked, Dr. Oliveira examined the institute's records on the injured men. Regarding the burn on Auleriano's buttock, he said: "It is likely that if he had been struck by lightning his clothing would have been burned as well." Both Auleriano and Apolinario said Auleriano's shorts were not burned.

Clesio Muniz, the chief criminal investigator, strongly disagreed with the lightning theory, as did Sergeant Atenor Costa, an Air Force meteorologist at the Sao Luis airport. The airport, fifteen miles northeast of the island, is served by four national airlines, two regional airlines and several air taxi companies. The meteorology station's records show no lightning or violent weather between five P.M. April 25 and six A.M. April 26. There was a light rain at eleven P.M. and again at midnight, but otherwise the night was clear and quiet. "There is no way lightning could come down, hit the sand and bounce back up and then go sideways into the boat," Sergeant Costa said. "It just doesn't do that. If it had, it should have burned the canvas curtain. Lightning wouldn't hit two or three men at the same time because their positions in the boat were so different. The lightning would have to be like a winding road to do that.

"Furthermore, it is highly unlikely it could kill the one man without burning him. It just is not possible for lightning to burn two men and kill the other without leaving a mark on him."

Natalino Filho, director of the weather station, said lightning could have hit the water and passed through it to the boat, since water is a good conductor of electricity. "If that had happened, however, Apolinario should have been killed because he was lying on the floor nearest the water," Filho said.

THE SWAMP FROM HELL

There definitely weren't any burns on the boat. I personally inspected it myself, and it was a hellish experience. Firmino was then living in the forest some distance south of Itauna, the western ferryboat terminal across Sao Marcos Bay from Sao Luis. With Ana Teresa Britto and her sister, Leila, as interpreters, I went to find Firmino and take him back to Sao Luis. When we reached his home, we learned that the Maria Rosa, the boat used by the four men for their trip to Crab Island, was anchored in a nearby stream. I had been searching for it for days but far away in the Sao Luis area.

We had to wait for Firmino to get ready to go to Sao Luis with us, so Ana Teresa, Leila and I set out to inspect the boat, with Firmino's wife, Maria, showing us the way. We drove to a small village, parked and started walking down a path into the forest. Five minutes later we came to a swamp where the path disappeared under the water for about seventy-five yards. Maria said there was no other way to get to the boat.

Thoughts of piranhas and other vicious creatures gave me a headache as I studied the dark water. I couldn't see a thing under the murky surface, and we had to go through it barefoot or lose our shoes in the muck. I wanted to cry.

Maria assured me it was only knee deep, but I didn't want to go through it barefoot or otherwise no matter how shallow it was. However, I had no choice if I wanted to examine the boat. All three women laughed at me as I stalled. Then, hating every second of it, I plunged in and sloshed across, with Maria in front and Ana Teresa and Leila behind me. But nothing happened, and we got to the other side with all our toes intact.

A few minutes later we reached the boat. The tide was out, and it was sitting in the mud. It was made entirely of wood and had a single huge sail. It was old, and its paint was so faded that I could just barely make out the name, Maria Rosa.

There was no one around. As the three women sat on a log and waited, I walked up a plank onto the deck. The only entrance to the cabin and cargo hold below the deck is through a square hatch just behind the mast. I spent about thirty minutes looking the boat over thoroughly inside and out. There was no sign of any fire or violence anywhere. I took a number of photos, and then the four of us went back through that same swamp again.

We took Firmino to Sao Luis because I had arranged for Dr. Silvio Lago to come to Sao Luis from Niteroi, near Rio de Janeiro, to hypnotize the three men. Dr. Lago was a physician and professor of medicine who, at that time, had used hypnosis in his practice for forty-five years. The three men agreed to the sessions because they had been depressed ever since the incident and hoped he could help them.

Several years later, Ana Teresa remarked: "You know, that was dangerous."

Firmino showing the scar the UFO left years after his experience.

UFO HOSTILITIES AND THE EVIL ALIEN AGENDA

MENTAL BLOCK

Dr. Lago spent a total of sixteen hours with the men, six hours talking to them individually and together about their lives and what happened at Crab Island and the other ten hours in individual hypnosis sessions. When he was through, he was convinced the men were telling the truth, but he hadn't gained any clues about what happened that night. "They were unable to remember anything that happened to them after they went to sleep that night," Dr. Lago said later. "I am not accustomed to seeing this kind of mental block. This is a very strange and complicated case."

Emotion alone would not be enough to cause the mental block, he said. "It was something physical and psychic but not common. A very strong emotion could cause amnesia, but it doesn't seem likely that it was their emotional reaction that caused the mental block. It is possible that before or during the experience they had some kind of hypnosis, a very deep one, preparing them not to remember whatever it was after they experienced it."

Another thing that puzzled him was that Apolinario, who had no apparent injuries, had the same kind of mental block as the other two.

"One hypothesis is that Apolinario would have to have had a very strong emotion that would cause the block," Dr. Lago said. "I can't imagine what that would be unless he had seen whatever happened. Whatever imposed this mental block was much stronger than his grief at seeing his brother dead, because he remembers everything before and after but nothing in between, and I cannot believe there is any greater emotion than seeing a brother dead and two men injured. It is very strange."

Still another part of the mystery is the fact that Auleriano went to sleep at the very back of the boat and awoke in the front, while Firmino, who had gone to sleep in the front, was found at the back near Auleriano's hammock. Neither man had any recollection of changing position during the night.

Some people familiar with the case believe a UFO plucked the men out of the boat, did whatever it did to them, and then put them back but mistakenly placed Firmino in Auleriano's location and Auleriano where Firmino had been.

Whatever happened that night aboard the Maria Rosa occurred between the time they went to sleep at eight P.M. and midnight, when they intended to awaken. Three of the men were accustomed to waking up when the tide came in, but no one awoke until the next morning. This suggests that all four were unconscious before midnight.

Whatever or whoever burned Firmino and Auleriano most likely was also responsible for causing Jose's death. Just exactly when these events occurred can't be determined, but probably before midnight. Jose's body was getting stiff, and Apolinario had difficulty putting his leg back in the hammock between five and five-thirty A.M. Normally, rigor mortis begins setting in three to four hours after death and takes about twelve hours for complete stiffening of the body.

NOW IN POOR HEALTH

The terrified experiencer, Apolinario, undergoes hypnotic regression and unfortunately suffers flashbacks of the incident in question.

When I interviewed the three survivors, I was hoping that the mental blocks that suppressed their memories of that night would eventually weaken and they'd remember.

But that may never happen. I went back in 1981 and talked to Auleriano and Apolinario, and again in 1992, when I talked to all three. None of them has ever remembered anything.

Interestingly, the two who were burned, Firmino and Auleriano, are now in robust health but Apolinario, who suffered no apparent injuries that night, is in poor health. A year and a half after the incident, he began to feel a weakness in his left arm. By 1981, the year he turned thirty-six, he could no longer hold anything in his left hand without dropping it. By 1992, at the age of forty-six, he had little strength in his left arm and hand, suffered from severe headaches, and walked with an odd, stiff-legged gait. He doesn't know why. He's never had any crippling accidents or illnesses. When he can work, he makes charcoal.

Firmino, who lost weight and did little for several years after the incident - and even acted a little silly at times, according to his wife, Maria - is husky and mentally sharp once again. He now does light labor work despite his twisted left hand. He and Maria also now own and operate a small grocery in one of Sao Luis' poorer neighborhoods.

Auleriano's scars have virtually vanished. Two years after the incident, he began going to Crab Island to get wood once again and continued going there until 1991, all without anything unusual happening. But he gave that up and now works as a security guard for a construction company. Neither Apolinario nor Firmino has ever gone back to Crab Island.

UFO HOSTILITIES AND THE EVIL ALIEN AGENDA

ANOTHER CRAB ISLAND DEATH

That is not the end of the Crab Island story. Virtually the same thing happened nine years later to another crew, with one man dying, another being burned and two others mysteriously stricken.

On April 28, 1986, the four other men sailed in a similar boat to Crab Island to get wood. They worked for two days cutting more than three hundred poles and stacking them on the bank next to the boat.

On April 30, they finished working at six o'clock, and one of the men, Juvencio, twenty-two, began cooking supper. Verissimo, twenty-one, said he wasn't feeling well and asked Juvencio for garlic to rub on his arms to make him feel better, but Juvencio suddenly became dizzy and fell to the deck unconscious. In quick succession, the other two men, Anselmo and Lazaro, both in their forties, also passed out.

No one knows what happened to Verissimo. Lazaro regained consciousness at noon the next day, eighteen hours later, and found Verissimo lying on the deck dead. There were no marks on him, but a little blood had trickled from his mouth.

Anselmo awoke two hours later, and Juvencio revived about five o'clock, almost twenty-four hours after he passed out. The right side of his head was burned and swollen. Anselmo and Lazaro tried to load the wood onto the boat but gave up after getting less than thirty poles on board. They began sailing back to Sao Luis, but it was difficult because all three were sick and nauseated.

LOUD CRASHING NOISE

The second Crab Island death wasn't news outside of Sao Luis, either. I happened to go to Sao Luis five months after it happened and learned about it from Monica Carneiro and Ana Teresa Britto, the principal interpreters during my investigation of the first case. They helped me find Juvencio, who told me what had happened.

As in the first case, none of the three survivors knows what happened that night, except that all three got dizzy and passed out. Port authorities questioned them and told me they believe the men were telling the truth. The men were certain food poisoning was not to blame. They hadn't yet eaten and were feeling well until they became dizzy. Authorities also discounted the possibility that any kind of poisonous gas seeping from the swampy land could have been the cause. Juvencio said no one smelled any unusual odors before getting dizzy.

No autopsy was performed on Verissimo. As in the first case, by the time the boat reached port, his body was badly decomposed. Verissimo's death certificate simply lists the cause as "undetermined."

The UFO connection in this case is also tenuous. One unusual thing happened shortly before the men passed out. They heard a loud crashing noise in the brush somewhere near the boat. In the darkness, they couldn't see anything, and they don't know what caused the noise.

The only way to get to the island is by boat or helicopter, and the men weren't aware of anyone else being on the island with them. UFO buffs may see the crashing noise as a clear indication that a UFO landed, crushing trees or bushes in its way, while debunkers might claim it was just a tree falling down. There is no way to prove either is right, but the men would have recognized the sound of a falling tree.

When Monica, Ana Teresa and I interviewed Juvencio in his home, a number of neighbors gathered around to listen. One man in the crowd said he'd had a UFO encounter in a similar boat not far from Crab Island one night in 1983. His boat was anchored in a stream on the western side of the bay when a big bright object came down and hovered overhead, shining a light down on the boat. The man and his companions dived overboard and hid in the bushes until the UFO went away. He said people in several other boats in the area also had UFO encounters that year.

Both Lazaro and Anselmo were in the interior the two times I've been in Sao Luis since the second incident, and I've never talked to them. However, I saw Juvencio again in 1992. He said he was in good health, but Anselmo and Lazaro both now feel numbness in their legs, while Lazaro sometimes has dizzy spells and headaches.

The two cases are strikingly similar, except that none of the men in the first incident felt dizzy at any time. It's very possible that UFOs were not involved in either case, since the men remember nothing unusual, and there were no other witnesses. But if UFOs weren't to blame, then some other phenomenon just as bizarre was responsible. Either way, it is all part of a strange mystery that injures people and leaves some dead.

VAMPIRES IN THE SKY

UFOs have been deadlier in Colares than perhaps any other place in the world. Colares is one of thirty villages at the mouth of the Amazon River where for more than a year people weren't safe from UFOs even inside their own homes. UFOs would beam down rays of light that would penetrate roofs as if they didn't exist. Sometimes the beams would strike someone and, at other times, they would snake around as if searching for something. Terrified inhabitants would jump out of the way, but some got burned anyway.

This happened in a broad area only fifteen to thirty miles north of the busy metropolis of Belem. Just how many persons were injured will never be known, but in Colares alone approximately forty were burned and two of them died. A third person died three years later, possibly as a result of the UFO burns she suffered.

In addition, a pregnant woman lost her baby after being hit by a beam of light, and a dog died after a UFO shined a light on it when it was barking at the UFO.

The discs were seen so often that the Air Force conducted an official investigation in the area for months. A captain and a number of sergeants from the Belem air base questioned hundreds of people who'd had sightings and harmful encounters. They also reportedly administered drugs to help calm people down. The information that the team gathered, plus numerous photographs of UFOs that the investigators themselves took, were sent to Air Force headquarters in the nation's capital, Brasilia, but none of the findings or photos have ever been officially released.

Sightings and encounters began in July 1977, were particularly heavy in the last four months of the year, and continued sporadically through November 1978.

UFO HOSTILITIES AND THE EVIL ALIEN AGENDA

UFOs were reportedly seen in thirty communities, with Colares, Mosqueiro Island and Baia do Sol seemingly getting the brunt of the visits.

DOCTOR TREATS BURNS

With the results of the military's investigation still secret, the principal source for the number of injuries and deaths in Colares is Wellaide Cecim Carvalho, who was the only doctor on the island when the sightings began in 1977. Then twenty-four, she was in charge of the small state-run hospital in Colares, serving from December 1976 to December 1977. She's now a doctor of public health for the state of Para, dealing with children, cholera and AIDS.

Wellaide lived through the worst months of the UFO attacks, and she discussed them during an interview in her Public Health office in Belem on July 15, 1993, with Daniel Rebisso interpreting for us.

Q. How many people did you treat for injuries connected to the sightings in Colares in 1977?

A. Approximately forty people, mostly adults.

Q. What kind of injuries were they?

A. Mostly burns on the chest, like sunburns, near the face, the throat and chest.

Q. Most of the people were like that?

A. Yes. The burns usually covered an area of ten to twenty centimeters.

Q. Almost as big as a soccer ball?

A. Yes... The skin peeled off. These burns healed quickly. Usually it takes about seventy-two hours for burned skin to peel. UFO burns begin to peel almost immediately.

Q. Peeled off and healed quickly?

A. Yes. Very interesting cases. I could see two small puncture wounds in the center of these burns.

Q. In all of these cases?

A. Yes, in all. All had irritation, swelling, redness. Very red.

Q. Almost always in the chest and throat and face?

A. Yes.

Q. A ray of light hit them?

A. Yes.

Q. This is what people told you?

A. Yes. I know of two deaths.

Q. Two deaths?

A. Yes, a man and a woman. The woman was taken to Belem ... Eight hours later, after I treated her for burns, she died here in Belem of a heart attack, and that morning she had a large burn on her chest.

Q. Do you remember her name?

A. No. I remember another woman who had her hands burned by a light from a UFO and a great burn on her chest. She was working.

Q. But not the same woman?

A. No. This other one was sewing at home at night when a UFO arrived and shined a light through the garden. She was on her veranda. Her hands were badly burned, and she said the cloth was burned, too.

Q. Do you remember the month the woman died?

A. September 1977.

Q. You also said a man died?

A. And then the man died. He was a fisherman.

Q. One month later?

A. More or less, yes.

Q. Do you remember the approximate age of the woman or man?

A. She was forty-four to forty-five years old and the man was younger, thirty-two.

Q. What kind of injuries did he have? Was he burned?

A. The same burns on the chest. The document from the government didn't specify the cause of death. There was no autopsy. The Air Force didn't allow an autopsy.

Q. Was the Air Force there at the time these two people died?

A. Yes. They arrived in Colares in August or September 1977.

Q. Hollanda?

A. Yes. Hollanda's team. They had drugs to calm people down because many people were very frightened. All of my patients were burned. I did blood tests on them, and all had low levels of hemoglobin.

Q. This is not normal?

A. No.

Q. The woman who died, she was burned and then she went to the hospital and died of a heart attack. The man who died, did he go to the hospital or did he die at his home?

A. They both died the same day they were burned. He died at home in Colares, the same day, about two hours later after I talked with him.

Q. What was his occupation?

A. He was a fisherman, and the woman was a domestic worker.

Q. Did you believe these people (who had been burned) when they came to you and told you what happened?

UFO HOSTILITIES AND THE EVIL ALIEN AGENDA

A. Not at first, I didn't believe them.

Q. When did you start believing?

A. At first I thought they were crazy, but after about the fifth case, I began to take it seriously ... For a long time the sheriff, the priest and I were the only professionals in Colares. When the UFOs came to Colares, many people left. Only three professionals remained there. No stores were open. We had little to eat except eggs and farinha. The fishermen wouldn't fish because they were afraid.

Q. You treated about forty people over how long a period?

A. About three months. Many people left the area.

Q. How long did you go without food, a week, two weeks?

A. September, October and November, no food as no fishermen would go out. Only eggs and farinha.

(This indicates anemia, but low levels are common in people who are malnourished, and most people living in Colares at that time didn't have balanced diets, according to Daniel Rebisso, a biologist.)

Q. You saw a UFO yourself?

A. Yes, I saw a UFO in November 1977. It was a cylindrical thing. About six o'clock in the evening. My secretary was with me. She fainted.

Q. What did you think about the UFO?

A. I think the UFOs come from another place in the universe. It is stupid to think we're the only ones in the universe.

Q. What colors did you see on the UFO?

A. Metallic, silver. Part of it was a lighter color.

Q. How close was it?

A. About forty meters.

Q. For how many minutes?

A. A few minutes. I don't know how long because I was fascinated by it. More than ten minutes. It was very beautiful.

Q. Several other people saw this?

A. Yes. Everybody else was afraid, and they ran home. I was the only one who stayed on the street. Everyone yelled for me to run, but I stayed.

Q. This was near the beach?

A. About two hundred meters. People were shouting at me to run, but I didn't. I was too fascinated ... One time a UFO flew very low. It was going to land. People shot at it and threw stones to drive it away, but Hollanda's team arrived at that moment and shouted: "No! No! No! Don't do that!" But people were very frightened ... Many people said they could see people in UFOs like me, with blond hair, long blond hair. Many people said this. They said I was like the extraterrestrials.

UFO HOSTILITIES AND THE EVIL ALIEN AGENDA

I was the only woman there with blond hair ... The UFO I saw circled and made rings in the sky. It was beautiful, beautiful, beautiful! The Air Force knows about this case, but the Air Force told people not to talk about these cases.

Q. Hollanda?

A. Hollanda's team. But I don't forget. She said she began keeping a record of the people she treated but later destroyed her notes. Brazil was then under a military government, and she got worried about official reaction to the record keeping.

No autopsy was performed on the woman who died, she said, and the death certificate stated simply that the woman had suffered a heart attack. It made no mention of the burns on her chest. Wellaide said she thought that was strange because she had seen the burns that same morning.

UFOs GO INTO RIVER

I have learned about the Colares-area cases in December 1978 and have gone to Colares three times, in February 1979, July 1981 and July 1993. On the first visit, people said they'd seen UFOs going into and out of the Amazon River near Colares. Some had also seen glowing blue objects moving about under the surface of the water. They were frightened and felt threatened by them.

Rosil Aranha De Oliveira, thirty-six, who owned a store on the beach at Colares, told me: "I often go fishing at night, and I get out to this spot, and we can see these lighted things coming at great speed, and when they get close, they just stop. Sometimes they go into the water and sometimes they don't."

He had seen UFOs going into the river three times, twice in 1978 and the third time on February 14, just two days before the interview. At three o'clock in the morning, he said, he and his brother, Sebastiao, eighteen, were in a boat when a blue light went into the river about one-hundred-fifty yards from shore.

"Some men were fishing in other boats nearby, and they seemed to be frightened by it," Rosil said. "I could hear them shouting: 'The Thing! Here comes the Thing!'

"I have seen a blue spot moving around in the water, and I've seen them come out of the water. It just goes up and away, a blue light, going north toward the ocean. Once I saw one sitting on another beach south of here for about fifteen minutes. I tried to figure out the shape, but all I could see was just the lights. Then it took off and went north, going up and down in a wavy motion as it went away."

During that same visit to Colares, Marcelo Do Nascimento, who was in charge of the village's electricity plant, said: "Many times we would see UFOs very high, and they'd be shining red beams of light down on the houses. It was like an airplane light pointed to the ground, like a streak of reddish light. Many people saw the UFOs, and the beams of red light would go inside the house and circle around as if they were searching for something." He knew of at least three people who

Health care worker Wellaide Cecim Carvalho attended to many of the stricken.

had been burned, but we couldn't locate any of them then, so I went back to Colares again in July 1981.

THE AIRSTRIP THAT WASN'T

You can get to Colares in one to two hours by car or boat but the quickest way is by plane. It's only twenty minutes by air from Belem, but it's risky because Colares has no airport or landing strip. Instead, your plane touches down in a churchyard and hurtles a hundred yards down a narrow lane with bushes smacking the wings before coming to a halt.

Taking off is scarier because by now you know - or should know - that the lane wasn't meant to be a runway. I've flown in there twice, but it wasn't until the second trip that I realized that.

The first time, in 1979, it was raining and a veteran jungle pilot was at the controls of our Cessna. He was concerned only with how muddy the ground was. He simply swooped down for a quick look, came around again and landed without hesitation. The bushes weren't so lush then, and they posed no threat as we squished along in the mud of what I assumed was a landing strip.

On the second trip I learned the truth. With me this time were Hollanda, the officer who had led the Air Force investigation, and Charles Tucker, a UFO investigator from Indiana. Our pilot was twenty-three-year-old Carlos Montenegro, who'd been flying for only three years. In Hollanda's earlier visits to Colares, he had gone by car or helicopter. I was the only one of the four who'd flown there in a plane.

On this day the weather was sunny and dry. Carlos took one look down at that lane and quickly vetoed any landing there. I assured him it could be done, but the wide dirt road leading into the village looked much safer, and Carlos started to land on it. However, he changed his mind when he spotted a bus coming toward Colares. Rather than have the driver and a bunch of passengers mad at him for blocking the road, he headed back to the churchyard.

As soon as we hit the hard, packed dirt in front of the church, things got very scary. Unlike a car, a plane has no undercoating to deaden the noise and virtually no springs. With the engine screaming at high pitch just a few feet in front of us, the wheels smacking every bump and rut, setting up a horrible racket, with each jolt instantly transmitted to our backbones, and with the bushes whacking our wing tips as if they were trying to rip them off - with all this, the din was deafening. This wasn't at all the way I remembered it. In seconds, however, Carlos had braked to a halt just before we reached the end of the lane, spun around and taxied back to the churchyard.

Even before the propeller stopped turning over, we were surrounded by at least a hundred kids and adults. The circus had come to town.

On this second visit, we found one of the burn victims, Claudiomira Rodrigues, who was working at a nearby farm when we walked up to her house. One of the boys who had greeted our plane ran to the field to bring her back.

While we waited, her husband, Manoel, told us a twenty-three-year-old woman named Domingas had just died, three years after getting burned by a UFO. He said Domingas' health had been bad ever since she was burned, and her family believed the burn caused her death.

It would be impossible to land on that lane today. Over the years it has been washed out by rains and recreated by cars and wagons countless times, and neither the lane nor the ruts go in a straight line any longer. Electricity wires also cross the lane. Colares itself has grown enormously and has paved streets. The highway leading to it is also now paved, and the Colares beach is one of many in the area that attract huge throngs from Belem on weekends.

SUGGESTED READING

UFO DANGER ZONE - Bob Pratt

SURVIVING CATASTROPHIC EARTH CHANGES - Prof G. Cope Schellhorn

THE WHITE SANDS INCIDENT - Dr. Daniel W. Fry

UFOS AND ABDUCTIONS IN BRAZIL BY IRENE GRANCHI

SERPENTS OF THE SKY, DRAGONS OF THE EARTH - F.W. Holiday

Available on Amazon or Horus House Press, Box 55185, Madison WI 53705-8985

Examining The Grievous Alien
Agenda And The Evil Underbelly
Of Hispanic Ufology—A Trilogy

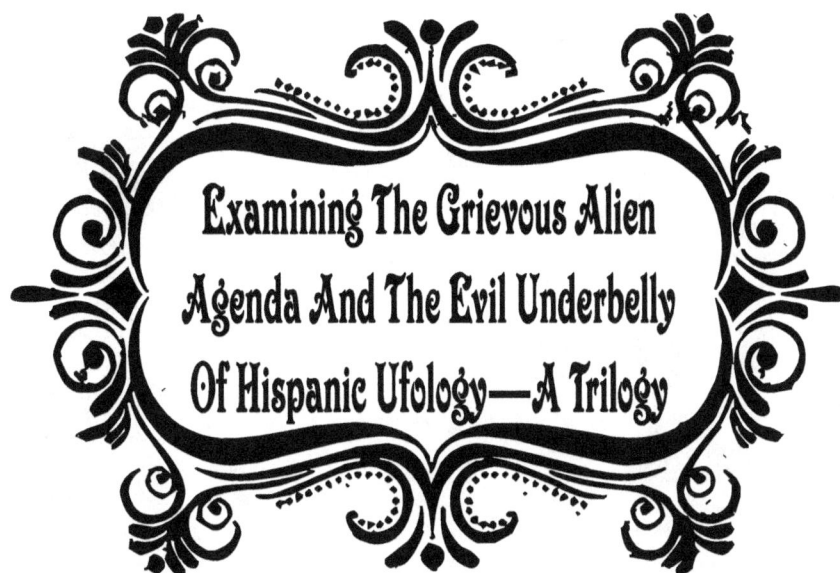

By Scott Corrales

Chapter 18

ALIEN SHOCK
By Scott Corrales

If we could divide the ufologically-turbulent Nineties into sections, we could perhaps argue that the early part of the decade was devoted to allegations of subterranean bases, MJ-12, and secret government projects with names like SNOW-BIRD and AQUARIUS; the middle part of the decade belonged squarely to the abduction phenomenon, with the release of ground-breaking books such as John Mack's Abductions and the discussion "alien abductions" in public forums; the remainder of the decade has belonged to the Roswell enthusiasts and their antagonists and those who are bent on wrestling whatever information the government clings to.

Unfortunately, the importance of abduction research overshadowed conventional encounters with non-human entities, the so-called "traditional cases" which usually involved a nocturnal encounter by a roadside, the accidental encounter with a landed saucer and its occupants, and other forms of human/non-human contact that did not involve the clearly-defined parameters of the abduction phenomenon. It is perhaps of interest to investigators that this "traditional" type of case continues to occur today, often far beyond our borders.

VICTIMS OF ALIEN PERFIDY?

Books and magazine articles dealing with the very real perils, both mental and physical, suffered by experiencers of the UFO phenomenon are commonplace today. Distinguished ufologists like David Jacobs openly state that the involvement of non-human intelligences in human events may not be as sanguine as many had firmly believed in earlier decades - that UFO occupants were here to help us take the next evolutionary step or eventually render assistance in solving humanity's most pressing problems. The downright eerie experience of a hapless Mexican ceramics technician should have served as an early warning to investigators when it occurred over forty years ago.

In 1972, researchers Jorge Reichert and Salvador Freixedo looked into the experiences of Heriberto Garza, who had allegedly had repeated encounters with otherworldly entities. Garza, a tall slender man who lived in the city of Puebla with his only son, had been unwilling to go public with his paranormal experiences out of fear of being ostracized by the conservative residents of his community.

His experience began when he was getting ready to go to bed on a given night. After turning off the light and getting between the sheets, he heard an unusual noise in the living room. Fearing that a break-in was in progress, he promptly went to investigate and was surprised to find a tall man with distinguished, almost feminine facial features. Taken aback, Garza demanded to know how the figure had entered his apartment. The entity told him in perfect Spanish that it could obviate physical obstacles and go where it pleased - but the reason for its visit was to grant Heriberto Garza "an experience that many would wish to have." His involvement with creatures from an improbable world known as Auko was about to begin.

Garza claimed to have subsequently been taken aboard a spacecraft where he met other beings similar in appearance to his original contact. One alien took his left hand and drew blood from his ring finger before returning him to his apartment, a return trip which he did not remember. He suddenly found himself sitting on an easy chair back home, with the door to the outside hallway open.

Strange phenomena began to occur soon after this experience. One morning, while shaving in front of the bathroom mirror, Garza saw his reflection vanish, only to reappear as he heard alien voices ringing in his ears, bearing a message that he was unable to understand. He would soon be subjected to intense telepathic communication with his non-human "friends," the consequences of which led him to seek psychiatric advice.

During a follow-up visit with researcher Ian Norrie, Reichert was perplexed by the change in Heriberto Garza's demeanor. The once-articulate man spoke sluggishly and did not appear to be himself. At one point, Garza said: "I want to show you what is happening to me," and proceeded to unbutton his shirt. The researchers were astounded to see a number of nipples growing randomly across Garza's abdomen, some of them small, others larger and with abundant hair. Reichert and Freixedo concluded that something had been injected into Garza which tampered with his DNA. Detailed study of the case became impossible when the experiencer "disappeared." Visitors to the humble apartment building in Puebla were angrily turned away by Garza's son, whose father appears to have

Salvador Freixedo has been involved in a variety of perplexing cases, including the case of Heriberto Garza, who grew a third nipple following some sort of blood work onboard an alien ship.

become an early casualty of tampering by uncaring non-human forces.

THE INSANITY RAP

Luis Ramírez Reyes may not be one of Mexico's most visible UFO researchers, but he is certainly one of the more thoughtful ones to have emerged from that country's rich ufological tradition. A journalist and radio announcer, Ramírez's non-doctrinaire position has made him accessible to individuals who would have otherwise chosen to remain silent.

This was precisely the case with a young man known only as "Pedro," who made an appointment to meet with the distinguished author one day to tell him his story.

During a weekend in December 1988, Pedro and a friend had gone to play an early morning game of tennis at the clay courts facing a large auto assembly plant on the outskirts of Mexico City. While waiting for other colleagues to join them, the two men suddenly felt that "the sun was rising behind them." Turning around, they were absolutely floored by the sight of a descending circular ve-

UFO HOSTILITIES AND THE EVIL ALIEN AGENDA

Luis Ramírez Reyes is a qualified Mexican researcher.

hicle that irradiated formidable amounts of white light, illuminating the entire area. The saucer-shaped craft touched down on a nearby field.

Suppressing a strong urge to flee, Pedro and his companion forced themselves to remain and see what further incredible developments would occur. Their courage and patience were rewarded with a glimpse of two creatures, described as clad in tight-fitting grey outfits and standing some four feet tall. Pedro added that "the creatures didn't look like you ufologists describe them," indicating that their heads had normal proportions, had small mouths and noses and slanted eyes.

Pedro estimated that the riveting experience lasted some twenty minutes, after which the diminutive aliens returned to their craft, which rose into the air and disappeared "like they do in the cartoons."

The friends decided not to speak further about the matter. The following day, Pedro returned to his job at the car assembly factory feeling confused and dejected. He told investigator Ramírez that he feared that his co-workers would take him for "a lunatic or a drug user" if he related his story.

While carrying out his duties, the UFO witness was suddenly gripped by unexplained seizures, convulsing on the assembly line. He was whisked off to a medical facility, where the doctor on duty decided to send him to a psychiatrist, given that Pedro "ranted about aliens during his seizures."

The psychiatrist determined that while he could find nothing wrong with Pedro, his disclosures of the sighting and the aliens might indicate schizophrenia. The hapless experiencer was sent to a mental health facility where he claims he was injected with a substance that made him "look like a nut," thereby making it easier for everyone around him to dismiss him as hopelessly insane. Despite the drug's influence, Pedro tried telling his parents that he wasn't crazy, but he was not believed.

The UFO witness was cast into an insane asylum where he witnessed the most atrocious abuse of the inmates by their keepers. One of the asylum's order-lies suspected that Pedro was clearly not insane, and told him to "behave like a paranoid" to avoid further problems during his stay at the institution.

Fortunately for Pedro, his companion at the tennis court had chosen to disclose the UFO experience in its entirety, despite having promised to conceal it. This ultimately proved to be the key that secured Pedro's release from the mental health facility.

"But upon my release," he told Ramírez, who included the harrowing experience in his book Contacto: México (1997). "I was still not free from criticism by my fellows. People clearly did not believe me or my friend, to the extent that I was refused employment in [the car assembly plant] or in other area factories."

LESS THAN HUMAN

Luis Ramírez was also made privy to an even more sensational case - one which, due to its very nature, has been kept under wraps, since it may well involve a suspension of disbelief greater than any audience might be willing to concede.

In early 1993, an anonymous young woman was driving between Mexico City and Poza Rica, Veracruz, as part of her regular route as a cosmetics sales-woman. Upon reaching the Teotihuacán archaeological site, she became aware of an object she thought to be a UFO in the clear blue skies.

The next thing she knew, she had arrived at her destination. Perplexed and afraid, she glanced repeatedly at her wristwatch and noticed that it indicated the very same time at which she'd left Mexico City and was driving along the expressway that runs past the Teotihuacán pyramids. Her pulse racing, she pulled into an alleyway in Poza Rica to steady herself. A passerby informed her that the time was now 2:00 pm - three hours later than the time on her wristwatch. The mechanics of how she had been able to traverse the 300 kilometer distance without being aware of it eluded her completely.

She conducted her business transactions nervously, haunted by the experi-

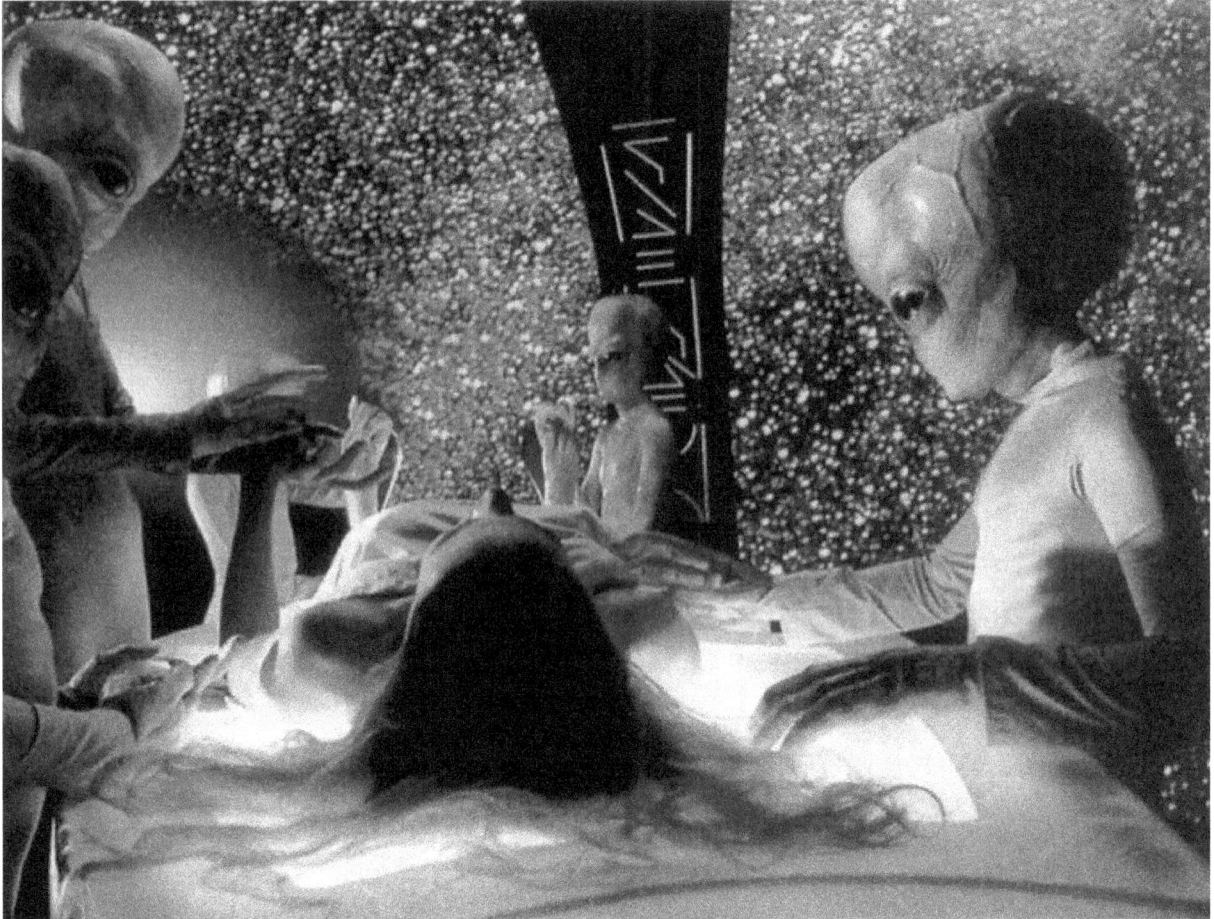

ence she had undergone. In the weeks which followed her "missing time" experience, the cosmetics saleswoman began to experience lassitude and nausea to the extent that she went to see a doctor. The physician dutifully informed her that she was pregnant — a statement that astonished her, since she was still a virgin and did not even have a boyfriend.

Seven months later, at a private clinic whose name and location Ramirez has kept confidential, she gave birth to a strange creature having double-membraned eyes, thick frog-like lips, joined fingers and hard, shell-like features on its skin which were similar to a tortoise's shell. Panic spread among the delivery room doctors and nurses, and only stern admonitions from the clinic's director kept the story from circulating any further.

The bizarre newborn remained inside an incubator for three weeks after its birth in September 1993. A physician's report indicated that the "baby" would not drink any formula or dairy products, but appeared to crave herbs. Other peculiarities included its inability to withstand light, preferring "infrared light sources," and the development of scales along its spine.

Photos of the creature were shown to an analyst who has also requested anonymity. His expert opinion was that the newborn belonged to a "saurian or reptilian species'' of some sort. The researcher's sources claim that the mother is raising her "child" alone, and that the latter is growing and developing into a full grown amphibian reptile "horrible to behold within our notions of beauty."

Is this reptilian infant merely a throwback to the very beginnings of the evolutionary trail? A human child deformed by unknown radiation or toxicity? Or can we actually believe that it is the offspring of a human mother and a clearly non-human father during a "missing time" experience? If so, the case would clearly be Exhibit A in the case presented by believers in reptilian aliens from nameless planets in space. This successful hybridization case - if true - represents the furthest possible limit of "high strangeness." Beyond it lies only madness, of the kind described in the works of H.P. Lovecraft.

ACROSS THE SEA TO SPAIN

Spain's first recorded UFO abduction was that of Próspera Muñoz in 1947 on the outskirts of Jumilla, a town in the southern province of Murcia, well known as a wine-producing region. While on a farm belonging to one of her uncles, Muñoz and her sister witnessed the presence of a "circular automobile" from which descended two diminutive, large-headed beings who cautioned the girls that very same night "they would return for one of them."

The little aliens made good on their threat and took Próspera to an enormous disk-shaped craft where she was examined by the occupants and allegedly had a "micro device inserted into her neck." The Muñoz experience, which was not made known until thirty years later, would simply be the introduction to a number of cases involving contact between humans and supposedly non-human entities in the Iberian Peninsula.

Few are the occasions in which a UFO investigator gets to see an unexplained celestial phenomenon that he or she can classify as a "UFO" with any degree of certainty. Far fewer are the occasions when an investigator manages to get a terrifying glimpse of alien intruders.

In 1991, researcher Josep Guijarro travelled from his home in Barcelona to the island of Gran Canaria (largest of the Canary archipelago) as part of his continuing investigation into the experiences of Judith, a nurse at one of Gran Canaria's hospitals, who had undergone a number of abduction experiences. Her first experience had occurred the previous summer, when she drove into a dense fog bank aboard her Renault and was found unconscious at the wheel the following morning by another motorist; subsequent experiences had included a number of disturbing "bedroom visitations" by supposedly alien entities.

Guijarro and Judith worked out a plan by which they would try to catch one of these unknown quantities at work: the ufologist would sleep in a bedroom next to that of the experiencer and would try to document "the source of her phobias."

But let's yield the floor to the researcher himself. "That night," he writes in

his book Infiltrados (Sangrila, 1992), "Judith and I spoke until well into the night, when suddenly her pet dog stood to attention and the TV set's volume control began increasing and decreasing of its own accord. We exchanged a knowing look. When everything appeared to have calmed down, we began hearing the sound of chanting. I cannot deny that I began to feel scared. With a look of fear still etched on my face, I suggested that we go to bed straightaway. If the Visitors existed, if they were not a figment of our imaginations, this night had all the makings for catching one."

Ufologist and experiencer vanished into their separate chambers. The former readied his camera and tape recorder, lying down in bed with his eyes firmly glued to the open doorway, expecting something to happen. In the darkness, Guijarro claims having heard all manner of creaking and squealing sounds, which he attributed to the structure of the house. At around 3:00 A.M., the dog began to howl as steps could be heard on the staircase.

"It was then that I saw it with stunning tranquility," Guijarro writes. "The outline of a short creature with a large head had just gone past my bedroom's doorway. My reaction to it was equally surprising - I made no movements whatsoever beyond taking a deep breath and falling asleep."

The following day, the ufologist told Judith about his experiences, realizing that while he may have worked himself into a highly suggestible state, that night he had lived the anguishing experience that affected not only his present subject, but tens of thousands of others worldwide.

Aside from the obvious fact of having "witnessed" what could have been one of the large-headed Greys, Josep Guijarro's account is significant due to the occurrence of high strangeness phenomena bordering on the paranormal: the fluctuations in the television set's volume control, the defensive attitude of the household pet and its subsequent howling, and the unnerving sound of "chanting," which prompted both individuals to retire to their rooms...incidents that should give boosters of the ETH (extraterrestrial hypothesis) food for thought.

A UFO FRIGHT NIGHT?

Ever since Flying Saucer Review's Gordon Creighton and Charles Bowen began focusing upon Argentina's seemingly inexhaustible supply of UFO and high strangeness incidents, the material has been the subject of fear and wonderment around the world. Argentina, the world's sixth largest country, boasts a population of only 30 million, with a fifth of its inhabitants tightly clustered in the communities surrounding Buenos Aires. To the west lie the majestic Andes; north and east are dominated by the plains and grasslands collectively known as the Pampa, and the south is occupied by the barren plateau known as Patagonia. While not exclusive to these remote open areas, the bulk of Argentina's UFO case histories have occurred in such lonely reaches.

During the 1965 UFO flap (one of the largest ever experienced in the southern hemisphere), Rialto Flores, an investigator for Argentina's defunct CODOVNI organizations, visited the locale of Corrientes to interview Carlos Soriou, at the time a high-school senior and hapless experiencer of one of the most terrifying high-strangeness events ever recorded in that country.

One night in February 1965, Soriou and his older brother went on an armadillo hunt accompanied by the farmworkers of their father's estate. Upon returning home from the hunt, they noticed short, unusual forms lurking in the field under cover of darkness. According to Soriou, the forms were no larger than three feet in height. Their shortness prompted one of the farmworkers to say to Soriou's brother: "They're midgets, patroncito. Let's cut them down with our machetes!"

Drawing his cutlass, the farmworker proceeded to act out his aggression upon the silent bundles. But the unexpected happened: the farmworker's arm was momentarily paralyzed as he was about to deliver the first blow, and the "midgets" increased in size to a height well in excess of seven feet.

Soriou's brother quickly fired his .22 caliber automatic rifle and was dumbfounded to see that no bullet had exited the muzzle. Replacing the bullet with others only had the same effect - no projectile would issue from the barrel to strike the now-towering forms. Helpless against the unknown entities, the hunters broke into a mad dash to a nearby barn, where they bolted themselves in.

But the wooden structure would afford little protection against whatever forces had been stirred up by their reckless behavior. Beams of light poured in throughout the wood, lighting up the barn's inside with an actinic glare; Soriou himself was hysterical with fear, and the others had to cover him with boxes and saddle blankets to keep him from seeing the unearthly light that poured in.

The glare stopped after a while, prompting the farmhands to believe that the worst was over and the "critters" had gone. The older brother courageously decided to venture out into the night once more to start a pickup truck that was kept nearby, hoping to leave the area and get help, but halfway through his sortie he was surprised by the entities who seemingly "appeared" out of nowhere. Propelled by sheer adrenaline, the man ran back toward the barn, where the farmhands refused to open the door lest the "critters" gain entry. Soriou's brother screams prompted them to unbolt the door just as one of the "critters" seized him, encircling his waist with its unearthly arms. The human broke free and made it into the safety of the barn.

Many hours later, the terrified band of hunters made it to the safety of the pickup truck and drove off to another field owned by the Soriou family without being harassed by the entities. Subsequently, many of the farmhands refused to return to the field in which the incident had occurred and one of them had to be dismissed from his position due to his fear. During his conversation with Rialto Flores, the younger Soriou believed that the gigantic presences were perhaps sitting when his group came across them, which would account for the mistaken impression that they were dealing with "midgets." The witness was adamant about the sheer horror of the event, and about the fact that at no point was a vehicle or UFO seen anywhere in the vast open area. The appendage that encircled his brother was not of a humanoid type. Rather it appeared to be "made out of hair or something similar," which he could not explain.

CONCLUSION

It became very trendy in the late 1990's to seek the comfort of an earlier age in ufological history - the sunny period of the contactees and the benign, all-caring, all-wonderful and even all-powerful space brothers who peddled redemption from their saucers. This author has been taken to task for "accentuating the negative" rather than concentrating on sweetness, light and other new ageisms. Wishing for something does not make it so: the negative effects of human involvement - albeit involuntary - with the UFO phenomenon have been documented to satiety both in the U.S. and abroad, to the extent that it is well possible that said negative experiences outweigh the positive ones even if not by a wide margin.

UFO HOSTILITIES AND THE EVIL ALIEN AGENDA

It is neither sensationalistic nor exploitative to dwell on these aspects when the aim is to provide the reader with all the facts rather than capriciously-worded summaries of events. Not even the most hardened contactee or channeler can dispute the unwholesomeness of Heriberto Garza's metamorphosis, the Mexican human/reptoid hybrid, or the creatures that ambushed the Soriou brothers in the Pampas. Eminent authors of the field, such as Keel, Steiger, Vallée, Freixedo, Creighton and many others have cautioned us about alien perfidy for decades. When will we listen?

Chapter 19

SUSPENSION OF DISBELIEF: BAFFLING PARANORMAL CASES
By Scott Corrales

The location: The Canary Islands in the Atlantic Ocean. The year: 1979. The protagonists: two youths involved in the paranormal. The responders: Elements of Spain's Guardia Civil (state police). The outcome: Mind-bending.

Two fifteen-year-old boys, enthusiasts of flying saucers, contacteeism and the occult in general, had decided to contact "alien intelligences" by means of a Ouija board - the vehicle of choice in such matters, it would seem - and had sustained a prolonged relationship with entities claiming to be captain this or commander that, spouting the same shopworn rhetoric about the environment and spirituality. But things were about to take an unexpected turn in the summer of 1979 as the non-human intelligences purportedly from distant planets upped the ante: were the two young seekers of the truth ready for a close encounter?

An increasing number of individuals are trying to make contact with aliens via the Ouija Board - often with traumatic results.

One evening, with their fingers on the planchette, the boys were told where and when to report for their meeting: a day in August of '79, and in one of the most remote and desolate locations on Grand Canary. Out where the buses don't run, as they say. Armed with courage and decent footwear, the seekers walked the sun-blasted expanses encouraged by the long-awaited meeting with the alien masters.

UFO HOSTILITIES AND THE EVIL ALIEN AGENDA

But there was no one there, as might have been expected. Late in the afternoon, possibly showing signs of heat prostration, one of the young men - unable to move — asked his companion to go for help, walking the hard distance back to the village of San Nicolás, fifteen kilometers away. Loath to leave his companion, the healthier of the two set off on the three-hour walk, arriving late in the evening. In the early morning hours, the would-be contactee returned to the remote location with a doctor and some of the concerned residents of the tiny island village, hoping it wasn't too late to offer assistance.

"They found nothing of the fellow but ashes," writes Atienza, "which Guardia Civil officers had to collect with shovels, as they disintegrated at the slightest touch. The coroner's verdict was death by intense heatstroke. The survivor was committed to a mental institution a few months later."

One is unsure as to which is the more monstrous of the two events - the bizarre manner of passing of the stricken contact-seeker, or the coroner's dismissal of his ashes as "death by acute heatstroke." Although no UFOs were reported or seen, and one of the pair never came into contact with anything, much like the Barra de Tijuca contactees in the 1950s in Brazil (readers will remember the two men found with strange lead masks), they were summoned to a place by unknown forces. Had both young men fallen sick, perhaps rescuers would've found two piles of ashes.

Such a case should appear in all the UFO case histories, but it doesn't. It isn't even mentioned in UFO chronicles of the Canary Islands. A call for help was placed to Alfonso Ferrer, author of "Las Cronicas del Fin del Mundo," a resident of the Canary Islands who kindly made inquiries into the case on our behalf. He approached José Gregorio González, whose books on Canarian ufology and mutology enjoy a wide readership. Ferrer provided the researcher's verbatim reply:

"This case is a sort of urban legend, but with a factual background. As far as I know, there were two individuals, and the succinct description given [in Atienza's book] is correct. The problem consists in locating the survivor, given that the other party died. The survivor was apparently accused of murdering his friend and was not exactly treated with kindness. He did some jail time, but had mental issues, as one can imagine. I was able to contact him a few years ago - we exchanged letters and a few phone calls, but I had the impression that he wasn't fully healed. I agreed to meet him in Las Palmas on two separate occasions and he never showed up. He has never replied to my calls or correspondence since then - I think that his family realized I was contacting him [about this matter] and decided to intervene. To protect him, I suppose."

The trail goes cold at this point, and the reader is left to wonder whether it is just another "UFO tall tale" of many that abound in the 70 year history of writings on the subject. We may never find out what happened on the Canary Islands in August 1979, but it is as good a preface as any to other cases in which innocent humans involved with the paranormal have come to an unfortunate end.

Researcher Javier García Blanco gives us a case with not one, but two, unidentified objects.

There is a paranormal aspect associated with many UFO incidents in the Hispanic world as in this case with a suspected ET lifted off into the air.

In the summer of 1980, Luis Gonzalez and his wife Bienvenida, residents of the Spanish village of Mediana (Zaragoza), hopped into their old station wagon at three o'clock in the morning for the long trip to Mercazaragoza, a large wholesaler from which they made regular purchases to stock their small family business. As they left the little town, something was waiting for them.

According to their daughter Ana, her parents left home along the road, and upon reaching a hilltop, found themselves suddenly flanked by two strange objects, which Bienvenida would later describe as "a pair of silver bells, with very strange colors. Inexplicable, but very beautiful." The couple was terrified, but there was no point in turning back - they continued driving toward Zaragoza with their unwelcome, unworldly escorts. Within a few kilometers, the silvery bells blinked out of existence as if they had never been. The journey continued without further comment and the supernatural occurrence was soon forgotten. But seven years later, the couple would die from a blood disorder that would baffle physicians. Coincidence, or a result of their exposure to unknown radiation emanating from their strange escort that morning on the way to market?

The trustworthiness of sources plays a crucial role when writing about matters of high strangeness, especially when harm to humans is involved. Noted researchers like T. Peter Park and Chris Aubeck have done their level best, in this case, to find the whereabouts of John Macklin, the author whose little compilations of ghostly tales fired the imagination of many young readers in the late 1960s and early 1970s with accounts that had a distinct ring of truth, but with a signal lack of footnotes or sources. I am nonetheless including the following account -

taken from Mr. Macklin's "A Look Through Secret Doors" (Ace Books, 1969) as an example of the mysterious situations that often involve unlucky experiencers and law enforcement.

In May 1951, a young woman named Clarita Villanueva found herself at the center of a paranormal mystery that was supposedly witnessed by hundreds of people one muggy afternoon in Manila, the capital city of The Phillipines. Aside from the dumbfounded onlookers, two trained observers would also have their names attached to this case: Dr. Mariana Lara, a medical officer attached to the Manila

BITTEN BY DEVILS

BY LESTER SUMRALL

THE SUPERNATURAL ACCOUNT
OF A YOUNG GIRL
BITTEN BY UNSEEN DEMONS,
DOCUMENTED BY MEDICAL DOCTORS
& HER MIRACULOUS DELIVERANCE
THAT WOULD BRING
RIVIVAL TO A NATION

Above left: The Canary Islands would seem an unlikely place to encounter UFOs of any variety, but actually it has a rich history of encounters with pilots and craft, both good and bad. Some witnesses say they have actually been able to see inside the craft.

Above center: The impressive Canary Island sightings circa 1976 had a negative effect on a number of witnesses who saw it coming out of the sea.

Left: From approximately May 9 to May 18, 1963, a young woman named Clarita Villanueva in the Phillipines was apparently attacked and bitten by two strange beings that only she could see.

UFO HOSTILITIES AND THE EVIL ALIEN AGENDA

Police Department, two officers from the selfsame department, Arsenio Lascon, Mayor of the City of Manila, and prominent journalists of the time.

According to Macklin's account, a police cruiser responded to a disturbance near the port of Manila. Upon reaching the waterfront, the officers found many dozens of onlookers clustered around the form of a woman on the ground, screaming at the top of her lungs as she wrestled with an unknown assailant: "Get it off me! Please! I can't stand the pain!"

No one dared to lend the poor soul a hand, unsure if she was insane or possessed by demons. Any forces attacking her were completely invisible to the human eye, yet the police officers were able to make out bite marks and bruises appearing on the woman's arms and neck. Resolutely making their way through the crowd, the law enforcement agents grabbed the writhing, screaming woman and managed to get her into their vehicle for the trip back to the station. But there was nothing anyone at the precinct could do for her, as they suspected Clarita Villanueva was either inebriated or drugged. Dr. Mariana Lara opined that the woman was going through some form of epileptic fit; placing her in a holding cell was the best remedy anyone could think of.

With Clarita safely behind bars, the policemen lent a deaf ear to her pleas for assistance and not being left alone with "whatever" seemed to be accosting her. She eventually passed out and the station went about its normal business - but only for a few minutes: The woman in the holding cell woke up a few minutes later, shouting that "the thing had found her again, and was coming at her through the cell's iron bars." The terrified woman described her invisible assailant as having the general shape of a male human, but with large, bulging eyes and the sartorial detail of wearing a cape pinned to its shoulders.

The guards entering the cell could only see fresh bite-marks on her arms. At this point they decided to summon not only Dr. Lara, but also to place a call to Mayor Lascón for advice. According to Macklin, the assembled authorities concluded that there was an external cause at work, as Ms. Villanueva "could not possibly inflict bite-marks on her own back."

But even this vote of confidence didn't help Clarita's situation. She was remanded to court the following morning to face charges of disorderly behavior in public. But even in the hallowed halls of justice, the woman wailed that the creature had returned, and even as she did so, her police escort noticed how deep bites manifested on her skin during a brutal attack that lasted five minutes before she dropped to the floor in a dead faint. Another medical specialist pronounced her wounds as "genuine," adding: "she was not the cause of the bites".

The local press soon turned the Villanueva mystery into the talk of the town. Mayor Lascón and Dr. Lara swore under oath that they had seen the bite marks forming, especially when the mayor gallantly offered to be at Ms. Villanueva's side aboard the ambulance taking her to the general hospital. The fifteen-minute ride to the medical facility, remarked the mayor later on, "felt like twenty-four hours in hell."

Once hospitalized, the attacks by the invisible creature drew to a close and Clarita Villanueva was on the road to recovery. Dr. Lara, skeptical of the paranormal, was quoted as saying: "What happened to Clarita Villanueva is a total mystery. She was attacked by something with sharp and invisible fangs. We'll never know what it is, but I'm not at all hesitant to admit that I was never so frightened in all my life."

Accentuating the positive - and downplaying the often negative aspects of the unknown - has been a trademark of communities involved with the paranormal (abductees, contactees, Ouija board users, etc.) but cases such as the ones presented here suggest that such involvement is less than wholesome. The baffling cases are sent to the bottom of the drawer, waiting for their moment in the sun.

Chapter 20

LOW INTENSITY WARFARE: UFOs VS. EARTH
By Scott Corrales

Captain Charles Wendorf's orders were straightforward enough: fly his B-52 Stratofortress to the Saddle Rock Mid-Air Refueling Area to meet a KC-135 tanker. The clear skies over the Mediterranean coast of Spain made Saddle Rock a particularly suitable refueling site. The giant aircraft, an element of the 68th Bomber Squadron out of North Carolina, was in the middle of a long patrol of the Atlantic Ocean, coming as close to the USSR as they dared. But Cold War tension would be the very last factor to affect the B-52's fate.

At 10:22 a.m. on January 17, 1966, at an altitude of thirty thousand feet, Captain Wendorf's nuclear-warhead laden Stratofortress sighted the KC-35 some 15 miles ahead in the refueling zone. The B-52 carefully jockeyed into position behind the tanker to connect with its refueling mast - a complex but efficient operation that did not involve any loss in speed on the bomber's part and in which remarkably small amounts of fuel were lost.

But something went wrong. An unseen force bumped against the bomber's underside, pushing it upward and causing the KC-135's starboard wing to graze the B-52's cockpit. The bomber's crew felt another terrible jolt as their plane rammed into the tanker's fuselage

To observers on the ground witnessing the refueling maneuver, the tanker exploded into a ball of orange flame while both military aircraft disintegrated high above the earth. The long-range bomber's crew managed to jump clear of the explosion and were later rescued by Spanish fishermen

after having miraculously survived their high-altitude jump. But four hydrogen bombs now lay at the bottom of the shallow coastal waters, and the efforts to retrieve them before lethal gamma radiation spread throughout the sea made headlines worldwide.

Witnesses to the explosion claimed having seen three objects in the sky at the time of the explosion, although only the downed bomber and the disintegrated tanker should have been in the area. Suspicions arose among the Spanish military elements assisting with the rescue efforts that the USAF's frantic search for the missing warheads was, in fact, a thinly-veiled excuse for finding the elusive third "airplane" - the UFO which had caused the destruction of the aircraft and then disappeared without a trace.

The USAF had good reasons, perhaps, to worry about a force inimical to its interests somewhere over the Mediterranean: Eight days before the Palomares debacle, a colossal fireball of unknown origin had flown over the Italian cities of Capri and Naples, causing a general blackout. Nearly four years later, in October 1969, two jet fighters would disappear without a trace during NATO exercises held off Crete. The previous year, the French air force had lost two Mystére IV fighter-bombers on routine patrol over Corsica. The result of the military inquest was that both planes were lost due to "undetermined causes."

Incidents such as this one are legion: UFOs - whether interplanetary or interdimensional - have engaged in a sort of "air supremacy" struggle with military aircraft of all nations since the first foo-fighter encounters during the Second World War. Fighters, bombers and transport planes have been intercepted and fired upon by unidentified objects, while others have been destroyed during efforts at getting closer to the mysterious intruders that fly over sovereign airspace with impunity.

The concept of an interplanetary war in our times, a thought which admittedly smacks of the Cold War's "flying saucer" scare, has been seriously invoked by Gen. Douglas MacArthur in the 1950's and by President Ronald Reagan in the 1980's. NATO allegedly prepared a document evaluating the threat posed to ex-

isting conventional and military forces in the 1960s, conceding that "we virtually have no defense against their advanced technology." Serious researchers like the late Dr. Olavo Fontes analyzed the possibility that northern Brazil would be occupied by a UFO invasion force in the wake of the attack on the Itaipú garrison in 1965. But the hostility between unidentified objects and human air forces is not limited to our own air force nor to our own times, as can be seen in the following paragraph.

During World War II, both Allied and German aviators had close encounters with unknown aircraft. The following incident appeared in La Nature, a French scientific magazine:

"It took place in Le Mans at the start of the summer of 1941...The Luftwaffe had occupied the old Le Mans airfield at the La Sarthe circuit and was stationing many Messerschmitt 109 fighters - not very fast vehicles, but highly maneuverable ones. The German pilots were kept in constant state of readiness, and I had the opportunity to witness a number of Messerschmitt flight squadron maneuvers in broad daylight.

"The weather was splendid, and as far as I can recall, it was a Sunday. At around 1300 hours, the skies were clear and only a few large cumuli could be seen at quite a distance from each other. Toward that time, the entire city heard the roar of the airplanes flying at full speed, and I was able to ascertain at the time that the cause for alarm appeared to be within one of the large cumulus clouds, which at the time was slowly passing over the airfield. One could see the Messerschmitts flying around the cloud, diving into it, shooting up out of it, or emerging from it sideways before engaging in the same maneuver again. It was a spectacular performance to watch, but it must have been terribly dangerous for the pilots. Was this an exercise ordered by the base commandant or an alert? I never found out, but it was interesting to see the looks of terror on the faces of the German soldiers and officers that comprised the city garrison as they followed the spectacle from the windows of their homes.

"I was then able to observe how the cloud, whose base upon its arrival was 900 feet high to a maximum height of 3000 feet, began to grow, and when it cleared the airport to move away from the city, had acquired the shape of a very tall pyramid with perfectly clear outlines. Its sharp apex must have been at an altitude of some 10000 feet, judging by its visible height over the horizon and its probable location over the terrain. My vantage point was some 2 miles away from the airfield, and the cloud was some 3 miles away while I watched its apex."

The Korean War produced a number of incidents which became classics of ufology and stressed the inimical nature of whatever appeared to be sharing the skies with our own aircraft.

A B-29 flying over the Korean Peninsula during the 1950-53 conflict had its

UFO HOSTILITIES AND THE EVIL ALIEN AGENDA

In the skies above Korea, B-29 crews encountered Chinese Mig-15s and swift saucer-shaped craft. No one ever identified the origin of the UFOs.

otherwise uneventful mission interrupted by the sudden appearance of a 20 inch wide glowing disk within the cockpit, which remained immobile and suspended in mid-air to the shock of the flight crew. When one of the airmen approached it with a fire extinguisher, the strange object vanished, only to reappear later on. The bizarre "probe" inspected the bomber's interior before returning to its materialization point and disappearing for good.

A squadron of fighters flew into a cloudbank similar to the one assaulted by the German Messerschmitts during the previous war, but this time, whatever was inside the cloud won: the fighters never emerged from the seemingly innocuous clouds.

The air and ground forces of the former Soviet Union were by no means immune to UFO attacks. In 1961, a giant UFO escorted by a host of lesser craft challenged the grim defenses of the Moscow Anti-Ballistic Missile defense system, one of a kind in the whole world. The commander of the Rybinsk battery opened fire on the intruder, but to the surprise of the assembled misslemen, the projectiles detonated well short of their targets. Now it was the UFO's turn: all the electrical systems of the Rybinsk battery were stalled until the small flotilla of intruders was safely out of its range.

Apparently, the elements of the elite Soviet Rocketry Forces were unable to avoid the temptation of taking potshots at passing UFOs. In 1969, one such detachment was in charge of the nine brigades of SAMs (Surface-to-Air Missiles) which defended the North Vietnamese capital of Hanoi against American bomber raids. The Soviet artillerymen, stationed at some distance from Hanoi proper, observed a battleship-sized discoidal UFO (its diameter was in excess of 1000 feet, according to calculations) approach silently and suddenly, oblivious to the urgent IFF ("Identification Friend-or-Foe") requests transmitted by the ground installations. The battery commander phoned his superiors for instructions, and after a brief delay, was given the order to open fire. Three out of five missile brigades launched a total of ten SAMs against the aeroform, but as had occurred in Rybinsk years before, the salvos detonated well short of the target, to the dismay of the ground personnel.

The UFO fired a needle-thin blue beam against one of the SAM launchers, turning the entire unit, which included three launchers, as well as radar and guidance stations, into a smoking heap of metal. According to the report (which has been classified as a hoax by many researchers), two hundred lives were lost in the process before the UFO resumed its trajectory and vanished from sight.

However, the Petrozavodsk UFO of September 1977 seized the initiative when, at four o'clock on the morning of September 20th, it terrorized the city's inhabitants. The medusa-like craft emitted golden shafts of light that pierced holes in windows and pavements. The intruder returned five or six times within a month.

Another much-publicized incident dealt with the destruction of a Cuban MiG-21 "Fishbed" by a UFO. According to a specialist stationed at Florida's Boca Chica Naval Station, some 90 miles off the Cuban coast, he noticed that a strange object was approaching the island nation at some 600 miles an hour at an altitude of 32,000 feet. Two Cuban MiG's intercepted the object and attempted radio contact. The vehicle was described as a large, featureless sphere. When orders to destroy the object were issued from the ground, one of the MiG's suddenly exploded and disintegrated (in other retellings of the story, it was "zapped" by the UFO) while the UFO climbed suddenly to 100,000 feet and continued its southward trajectory unchallenged.

Other incidents, like the following, have bordered on the unreal - and in fact, there exists no way of proving its authenticity. Author Luis Anglada Font, a World War II fighter pilot who later devoted himself to UFO research in the early days of the phenomenon, included the following highly dramatic account (as told to him by physicist Raymond Harvey) in one of his books, La Realidad de los Ovnis A Través de los Siglos. The case was allegedly "hushed" by ATIC at Wright Paterson AFB and never included in the Blue Book Files. Anglada simply refers to it as

LUIS ANGLADA FONT

LA REALIDAD DE LOS OVNI A TRAVES DE LOS SIGLOS

HORUS

KIER

"The Steinbeck Affair."

According to the author, this chilling incident took place sometime in 1953. Steinbeck - no rank given - was a thirty-year-old aviator who was considered among "the best and brightest" due to his bravery and considerable sang froid. He had been summoned to an unnamed air base near Las Vegas to test a one-of-a-kind fighter prototype: a sky-blue interceptor armed with atomic cannon.

A lengthy briefing session ensued, after which the test pilot, technicians and three high-ranking officers went on to inspect the interceptor itself, which was kept in a hangar guarded by an armed infantry platoon. Steinbeck and the officers had to present their credentials to be allowed in.

Once inside, a German engineer met the visitors and proceeded to show off the interceptor. The test pilot climbed into the cockpit and gave the instrumentation a thorough examination. Steinbeck was to fly the plane at four o'clock the following morning and perform a certain number of routine maneuvers before opening fire with his atomic arsenal against an airborne target.

The story relates that Steinbeck and the others met with the press later that evening at the officers' lounge. In conversation, the test pilot revealed that he had studied stenography (!) and that a mechanized steno pad had been installed on the right-hand side of the cockpit so that he could note every single step of the test flight. The steno pad's housing was equipped with a parachute that would land it safely in the event of a malfunction.

All went well the following morning. Steinbeck took to the dark desert sky in the prototype without any incident while the unidentified German technician followed his progress from the control tower. At one point, witnesses in the tower saw the foreign expert blanch as he stared at the radar screen, speaking into the microphone with a sense of urgency: "Please, come back to base! There are unidentified machines to your right! Land quickly!"

UFO HOSTILITIES AND THE EVIL ALIEN AGENDA

The radar had in fact detected an object that crossed the field at breathtaking speed, followed by others: a total of eight unidentified vehicles were closing in on Steinbeck's prototype, sending the ground control personnel into a frenzy. The entire base was placed on full alert.

Steinbeck allegedly radioed back that the vehicles were surrounding him and that he was going to open fire on them with the interceptor's advanced weaponry. Two bursts from the atomic cannon [sic] were fired against one of the intruders, and moments later, the prototype crashed into the desert sand at high speed some two and a half miles downrange from the control tower.

Rescue crews found a twisted heap of metal along with the test pilot's carbonized remains. The engines and fuel tanks, however, were intact and unexploded; more importantly, the steno pad in its aluminum housing was recovered intact a few hundred feet away from the downed fighter, and it told the astounding tale of how a long, flaming "tube" had been aimed at Steinbeck's plane. The pilot's final desperate shorthand read: "They're attacking, I think they're attacking...God have mercy on me...I'm shooting back."

Anglada's "atomic cannons," the most questionable detail in the account, could refer to small warheads expelled by means of a launcher-

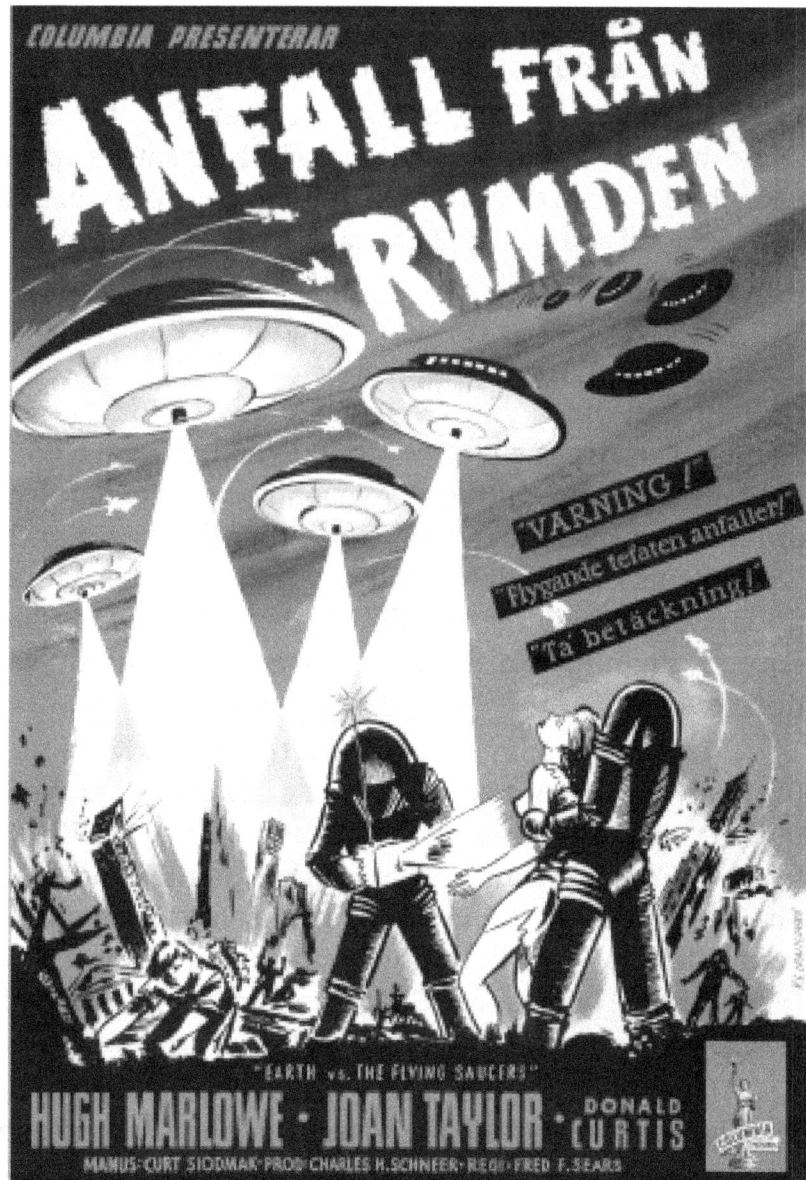

Movies like "Earth vs. The Flying Saucers" were very popular even in Spanish-speaking countries, as this cinema poster shows.

like arrangement rather than science-fictionish "blasters." It is a fact that the military experimented with different permutations of nuclear ordnance, ranging from atomic hand grenades and bazookas (like the "Davy Crockett") to huge atomic railguns. If true, the case would represent one of the most significant unprovoked attacks ever recorded.

Heat weapons appear to be the weapon of choice in aerial engagements between UFOs and terrestrial aircraft. One case involving fighters "scrambled" to intercept UFOs ended with the object pursued firing a heat weapon at the fighters which made their wings incredibly hot, forcing them to break off their mission and return to base.

The Walesville, NY incident, which occurred in July of 1954, is perhaps the best known of the heat-weapon attacks: An F-94 Starfire scrambled to investigate a UFO tracked by the radars at Griffins AFB was greeted with a terrific blast of heat issuing from the unknown vehicle. Pilot and navigator bailed out while the fighter rammed into a house, causing five deaths.

Carlos Alejo Rodríguez, a trainer pilot for the Uruguayan Air Force, also found himself at the receiving end of a heat weapon of some sort when he approached a large discoidal UFO flying over the Curubelo Air Force Base. The sudden surge of suffocating heat caused the startled pilot to break off his pursuit and return to the base.

Helicopters have, on occasion, had cause to defend themselves from intruders: In January 1974, the crew of a Marine UH-1E helicopter from the 3rd Marine Aircraft wing flying from Quantico, Va. to Beaufort, S.C. found itself having to stand tall against a silver-white UFO heading on a collision course toward the helicopter. The pilot engaged in evasive action only to discover that the featureless object was able to match his dodging effortlessly and even pull alongside until it filled the cockpit's view. Disregarding the rules of engagement, the gunner fired tracers point-blank at the UFO with no apparent effect. But to the surprise of all aboard, the huge object peeled away with a slight wobble before receding into the distance.

Lest the reader suspect that these incidents belong to the "heroic past" of UFO research, it is well worth remembering that the most graphic cases of UFO-military aircraft interaction occurred in the late Eighties, when the US Navy lost two F-14 Tomcats while intercepting a giant triangular object over Cabo Rojo, P.R. The magnitude of the incident apparently forced all military flights over the island to take off with their full complement of missile pods, and prompted PRANG (Puerto Rico National Air Guard) to replace its ageing fleet of Grumman A-7's with F-15's.

The final question is always: does Earth stand a chance against a bona fide extraterrestrial attack? There is evidence that terrestrial air forces have tried to develop deterrents against this unknown threat since the 1950s. It has been suggested that the formation of the USAF's Space Command in 1985 (ostensibly in charge of all satellite operations) was the first step in providing a capability against an attack from space. Upon close inspection, the SDI program (alternatively known as "Star Wars" or the "Peace Shield") applications appear better suited to repel a space-based threat than an atmospheric one. In September 1985, an F-15 launched an 18 ft. long missile dubbed "Tomato Can" at P871, a solar observation satellite, striking it at 30,000 miles per hour and nearly pulverizing it. While "Tomato Can" accomplished its mission without the benefit of a warhead, one can imagine its capability against "hostile unknowns" if endowed with one.

Model kit hobbyists may recall Aurora's "Ragnarok," a prototype of a U.S. Air Force nuclear-powered and laser-equipped interceptor specifically "developed to protect the earth from any hostile alien invasion." The detailed 1/200 scale kit provided an informative rundown on the ultimate orbital interceptor, which was 255 feet long and with a weight in excess of 330 tons (heavier than a B-52) and capable of achieving its cruising altitude of 200,000 feet in a startling 8 minutes.

The information went on to state that "protection from alien craft" was provided in the shape of five nuclear rockets and an ingenious manned support craft (a descendant of the D-21 reconnaissance drone carried piggyback by the SR-71?) armed with laser weapons capable of piercing two feet of solid steel. The kit's info blurb ended with a price tag of 100 million dollars for each "Ragnarok." While the model kit may just have been a fanciful take on the aviation developments that followed one another in the 1960s - U-2s, SR-71s, XB-70s - the fact remains that modeling companies have often "scooped" the Air Force's official acknowledgement of the existence of certain craft, as was the case with the F119 "Stealth" Fighter and the B-2 Bomber. Another modelmaker, Testors, recently made the pages of TIME magazine with a disturbing model kit of a UFO, described as "a scale replica of one of the ships, based on descriptions from Bob Lazar." We can certainly feel confident that there are no Ragnaroks lurking in secret warrens under the Nevada desert, but at some point, a design similar to Aurora's must have been on the drawing boards as a viable response to the threat posed by UFOs, despite Project Blue Book's insistence that "no UFO reported, investigated and evaluated by the Air Force has ever given the indication of being a threat to our national security."

UFO HOSTILITIES AND THE EVIL ALIEN AGENDA

About Scott Corrales:

Scott Corrales became interested in the UFO phenomenon as a result of the heavy UFO activity while he lived in both Mexico and Puerto Rico. He was also influenced by Mexican ufologists Pedro Ferriz and Salvador Freixedo, a former Jesuit priest who advocated a paranormal, interdimensional interpretation of the phenomenon. In 1990, Scott began translating the works of Freixedo into English, making the literature and research of experts and journalists available to English-reading audiences everywhere. This led to the creation of the SAMIZDAT journal in 1993 and his collaboration with Mexico's CEFP group, Puerto Rico's PRRG and the foremost researchers of Spain's so-called third generation of UFO researchers.

In 1995, Corrales documented the manifestations of an entity popularly known as the Chupacabras in three works: "The Chupacabras Diaries," "Nemesis: The Chupacabras at Large," and "Chupacabras and Other Mysteries." In 1998, the SAMIZDAT bulletin was replaced by Inexplicata: The Journal of Hispanic Ufology as the official publication of the nascent Institute of Hispanic Ufology. In addition, Scott has been a guest on numerous radio shows and his articles have been featured in several national publications.

Hide Your Men, Women And Children! UFO Related Kidnappings Are On The Rise!

Chapter 21

THEY NEVER CAME BACK
ACTRESS CAROLE LOMBARD'S DEATH DUE TO UFOS

PUBLISHER'S NOTE: To this day a cloak of mystery surrounds the death of actress Carole Lombard and includes the specter of synchronicity and a possible UFO connection.

I was tipped off to these astounding historical events that fit right into our presentation by script writer and Edgar Allen Poe aficionado Cynthia Cirile. It's not that the whole plane she was traveling in vanished, but she certainly never came back!

According to researcher Kyle J. Wood, "officially" Ms. Lombard's doomed flight was running late and the pilot was cruising at too low an altitude, resulting in the death of all twenty-one passengers. Says the on-air contributing guest to E Entertainment's "Mysteries and Scandals": – "Unofficially, some still believe it to have been the work of fate (via 'supernatural' design). In short, the number three, a reputed 'bad luck' numeral at the time, seemed to be everywhere. The Lombard party was supposed to travel back to Los Angeles via train after the War Bond rally in Indianapolis. Their group of three (consisting of Carole, her mother Elizabeth and friend Otto Winkler) only managed to board the plane because of three last minute seat cancellations. Carole was 33 years old at the time. It had taken three days to reach Indiana (her home state) by train. The plane was a DC-3, designated 'Flight 3.'"

An FBI document filed at the time and released under the Freedom of Information Act tells of mysterious lights seen over a nearby mountain range, which some say may have contributed to the plane's crash in some form or another, perhaps drawing undue attention by the pilot. Journalist Kyle J. Wood summarizes: "Much has been forgotten about the 'peculiar lights in the sky' that were noted in regards to the crash of Lombard's flight. 'Hanging like a suspended lantern' above the mountain range, the mysterious unexplained light was first theorized to have possibly been part of a sabotage plot to lure Flight 3 to its doom (because of the 15 Army Air Corps ferry pilots onboard). This theory was quickly discounted after an airway beacon mechanic came forward and reported to the FBI about seeing an 'identical type light' hovering in the sky above Baker, California, just three days earlier. The mechanic was engaged in recovery efforts following the Lombard crash when he by chance learned of the similar light from an area ranch owner who had seen both the light and the plane's fiery explosion." The case has also been tied in with the Battle of Los Angeles UFO sightings which occurred around the same time. –https://welkerlots.wordpress.com/

Writing in true Hollywood gossip style, scandal monger Robert Matzen seems to concur with our UFO driven speculation about the late actress: "As I mentioned to a reporter in Fort Wayne, I have now finally accessed the FBI files on the plane crash and, I kid you not, UFOs were seen in the Flight 3 airway on nights leading up to January 16. No, seriously, UFOs. Many people logically dismiss UFOs as Cold War paranoia, but we're talking sightings of odd lights in the sky *before* the era of Roswell and the 'flying saucer' by eyewitnesses that include a Civil Aeronautics Authority man. A fed. A trained observer equivalent to today's FAA investigators, who saw spherical lights in the sky that were not aircraft. (For the record, I don't believe that UFOs had anything to do with the crash of TWA Flight 3. I am remarking on how bizarre it is to find UFOs in the official FBI investigation.) Then there's the other incident that I'm still working on that's no less odd. Neither solves the mystery of Flight 3. On the contrary, both make answers all the more elusive and demonstrate how sometimes evidence and logic go right out the window, and, suddenly, you're in that other dimension where people are dead and nobody can figure out how they got that way."

The FBI "Top Secret" file, dated Feb 2, 1942, has been summarized best by Jack El-Hai, author of "The Nazi and the Psychiatrist" (2013), "Non-Stop" (2013), and "The Lobotomist" (2005). We cover the part only about the unidentified specters in the sky that might have pulled the actress and other unfortunate souls to their demise.

"The FBI file tells how agents did follow up on another lead: UFOs, reported by several eye-witnesses, may have played a role in the crash. One of those witnesses [name redacted], a CAA employee, explained in a letter dated February 23, 1942, that he saw 'a strange light' on the night of January 12 or 13.

"At about 7 p.m. that evening, he and another employee were driving a CAA truck towards Baker, California, on the Death Valley Highway. 'I glanced to the west and we both noted a light above the crest line of the mountains, which was about 15 miles distant,' he wrote.

""This light was a white bright light similar to an 18-inch course light, stationary and suspended against the sky as a background, and never moved or varied as long as we could see…. It looked round, more like a ball. I am satisfied it was not a star because we drove back to the station approximately an hour later

Left: Glamorous Hollywood actress Carol Lombard.

Below: Carol Lombard and Hollywood Leading man, Clark Gable.

and the light was gone.' In addition, the light was much larger and brighter than any star. 'We dismissed it from [our] minds because we could give no reasonable explanation for it.'

"The letter goes on to explain that a few days later, while helping in the search for the plane wreck, the CAA employee recalled the sighting when he met a local rancher named Willard H. George, who described a similar light he saw just a few minutes before the crash.

"The file contains a copy of a long letter George wrote to a CAA panel that was investigating the cause of the accident. George, a native of Las Vegas, had owned a cattle ranch in the nearby hills for 15 years. "[I] have ridden this territory [on] horseback both when I was a boy and in later years...and know the territory in which this plane fell probably as well as anyone in that country," he wrote.

On the evening of January 16, the letter explains, George and his wife were driving home from the El Rancho Hotel in Las Vegas. About six miles west of the city, they saw above a mountain ridge "a reddish-yellow glowing light which seemed suspended in the air."

The Georges continued to see the light as they drove west. "We first thought it was a big bonfire in the mountains, but looking more closely at it saw that there was no flicker to it. . .."

Above: Recovery crews at the scene of the DC3 crash.

Left: LA Times headlines tell the story in lurid detail.

UFO HOSTILITIES AND THE EVIL ALIEN AGENDA

The mountain ridge where the DC3 went down, killing all aboard.

COLD CASES

It's an undisputed fact that every year hundreds – actually thousands! – of people vanish into thin air – some right smack dab in front of friends and family.

Some turn up again. Oftentimes with a "strange story" to tell, or with a lack of memory that prevents them from discussing where they were or what they were doing during the period they were "missing."

Others are out there somewhere in limbo never to be heard from again!

Historian Charles Fort kept tabs on the missing. But we're not just talking about cases where some poor soul was murdered and the body hidden in a shallow grave or cemented somewhere down in a dark corner of your neighbor's basement.

No, we're talking about the guy next door who went to buy a pack of smokes and his family never sees him again. There was the famous case of the very dapper Judge Crater who became the "missingest man in New York City" when on August 6, 1930, he vanished into thin air amongst the four corners of the world, Time Square. He was on his way to have dinner with a lawyer friend and a sassy showgirl and then attend the play "Dancing Partner," for which he had scored tickets. For some reason his evening adventures ended prematurely as he sauntered by other Manhattanites on their way home from work never to be seen or heard from again.

A massive manhunt was undertaken but to no avail. It was speculated that he had split from his wife, packed his bags and headed to Mexico with some young chickadee. There were reported sightings of the Judge over the years, but none of these sightings panned out, and, finally, he was declared dead ten years later after every means to find him had been exhausted. Some say he was abducted by the mob and "laid to rest" under the Coney Island boardwalk, but even though the boardwalk was dug up at the designated spot, no bones or remains were ever found.

The Judge had been vanquished to the Neverland of the missing – thank you very much.

Fort also wrote about Ambrose Small, another dapper gentlemen complete with well-coiffed mustache and known as the toast of the Toronto theater district.

Small owned a chain of successful theaters, but he was a notorious gambler and was always looking for ways to get out from behind his debts. One day in 1919 he concluded a business deal which netted him a million dollars and he was last seen in public in a joyful, frolicking mood at the local opera house.

It was a bitter cold, snowy night and eventually Ambrose knew he had to duck out and head for home. It was his usual custom to stop at a newsstand on the nearby corner of Adelaide and Yonge and pick up the late edition coming in by train. But this evening the proprietor informed the theatrical promoter that, due to the inclement weather, the paper had not arrived as the locomotive had gotten snowbound south of town.

Mr. Small was fit to be tied that his usual habit of picking up the paper had been disrupted and he started fuming and cursing, a state he was never known to have been in before.

Eventually, he gave up the ghost – and became one!

It took several days before anyone realized Ambrose had gone with the wind. But, when the authorities discovered that he was missing, they put the full force of the city into looking for him.

NOTHING!

They looked high and low and found no trace of Small. He had stepped into the snowy night and seemingly melted away, despite an offered reward of $50,000 for information as to his whereabouts. Catchpenny sleuths were called in and psychics were consulted. No stone went unturned. There were a number of sightings of the entrepreneur but they were challenged and proven unreliable.

Sherlock Holmes creator – and psychical researcher – Sir Arthur Conan Doyle expressed interest in the case and said he was willing to lend a hand in the search, but was never officially approached by the police or the family. So eventually Doyle returned to England, leaving everyone involved no wiser in the case

A death certificate was eventually issued and the episode was officially closed!

Poor Small was reduced to nothing.

Poor theatrical community which experienced a reduction in ticket sales as their chief impresario was no more!

BREAKING NEWS

Today you can hardly spend an evening in front of your smart TV without being offered up a variety of programs that deal with those who have gone missing.

There is "Dateline," "20/20" on the networks. "Vanished," "Gone," "Disappeared" on cable's ID Channel and a host of other "one shots" and solo season programs that dot Showtime, HBO, Netflix and Amazon prime.

My favorite show that delves into those that have "gone astray" for one reason or another would be "Crime Watch," which is a daily syndicated program which you kind of have to hunt for but which is worth the search. Nancy Grace used to devote a lot of time to those whose lives were shortened by being abducted or running off into the night either on their own or at the hands of some masked avenger. But than her production dropped this popular – to me – aspect of the program in lieu of repeat performances of the Mendez Brothers and Lacy Peterson crimes. Boring! Let's help the disadvantaged and those who are seeking their loved ones' whereabouts, not glorify those who the hangman should have found a long time ago.

Most of the cases investigated on these shows turn out to be homicides that are eventually solved. But once in a while you will hear some rural sheriff or constable make a statement like – "It's as if they were picked up by a spaceship," when trying to account for the sudden, and at least temporary, disappearance of an individual, individuals, or in some cases an entire family.

"PEOPLE ARE NOT SUPPOSED TO JUST DISAPPEAR"

It was June 7, 1992, and the above is a direct quote attributed to local authorities in Springfield, Missouri, who after all this time have no explanation as to what might have happened to three woman who disappeared out of a quiet suburban home, leaving behind cash, credit cards, car keys and leaving nothing out of place. Nor were they any indications of forced entry, a struggle or any form of violence in the case. Sherrill Levitt, her daughter Suzie Streeter, and friend Stacy McCall have never been seen since. The two young women were friends who had graduated from Kickapoo High School the day before, and had apparently gotten ready for bed when the unexpected took place – whatever the unexpected might have been.

The media in and around Springfield, such as the Kansas City Star, still brings up the disappearance from time to time as it still weighs heavy on the minds of local residents who kept their doors locked day and night for a long time afterward.

It was June 7, 1992, when the three women vanished. The three haven't been heard from since that day when friends showed up at Levitt's home on East Delmar Street and found a broken porch light. Little else seemed out of place.

"All three of the women's cars were parked out front. Their purses and keys were inside the small white home. A smoker, Sherrill, 47, had left her cigarettes behind. And Stacy, 18, who battled migraines, hadn't taken along her medication.

A lack of evidence or any real sign of foul play has frustrated a long line of detectives who have taken their turn at trying to solve the mystery.

"'How do you wrap your head around three people literally disappearing? With no idea where they went?" said Sgt. Todd King, who started at the police department in 1994 and remembers as a rookie taking reports from people who had information they thought would be helpful. "In a lot of cold cases, you can look back and say this is probably what occurred, you just can't prove it. With this case, it's anything goes. Anything could have happened.

"You don't have anything that says they were abducted, they were harmed. … It's this big mystery."

THREE WHO DID RETURN

I like this account in particular as it tells of an event involving three women who disappeared and were not seen any more. It's strikingly similar to an incident that took place in Liberty, Kentucky, where three women literally "went up into the night" but did return to their homes to tell the tale. I spoke with one of the ladies at the time and I had no reason to doubt her story, which was vivid and lucid.

January 6 was Mona Stafford's 36th birthday. She was joined by her best friends, Louise Smith and Elaine Thomas, to celebrate the event and just have a friendly dinner together. The three good friends took their dinner plans to the Redwood restaurant, which was located south of Lancaster, Kentucky, on U.S. 27, thirty-five miles from their hometown of Liberty. As the three finished their dinner, they started what they thought would be a nice, pleasant drive back to their houses. Little did they know what lay in store for them in what would be a night they would never – could never – forget.

Upon leaving the restaurant, the women came across what they described as a bright red object in the sky. Mona Stafford thought it was an airplane on fire, but as it approached they saw it was not that. According to their tale, the object stopped a few hundred yards in front of them, an object bigger that "two houses" which rocked back and forth a few times and moved off.Perhaps at this point the women felt their unusual sighting was over, but it seems the strangeness had only

UFO HOSTILITIES AND THE EVIL ALIEN AGENDA

just begun.

All three women would later recall the same thing. What they were looking at was an enormous, metallic, disc-shaped object with a dome on top, and a ring of red lights around its midsection. The women all saw it close enough to see a yellow, blinking light. However, after they had driven about a quarter of a mile, a blue light appeared through the rear window of the car. At first they thought it was a highway patrol car with its lights flashing, but soon they realized that the flying object had circled around and had come up behind them. Suddenly, something wrested control of the car away from Louise Smith. The car accelerated, even though Mrs. Smith took her foot off the accelerator, and the speedometer was soon on 85 mph. Mona Stafford, in the front passenger seat, tried to help Louise regain control of the car, but it was not possible. The women began to feel a burning sensation in their eyes. The ignition lights lit up on the instrument panel, an indication that the car's engine was stalled, but they were still speeding along. They saw a wide, brightly lit road ahead of them, and then, seconds later, the scene became Highway 78 and they recognized their surroundings. They were on the outskirts of Hustonville, a full eight miles from where they had just been. Checking the time, they found that, incredibly, an hour and twenty minutes had passed.

Upon arriving at Louise Smith's trailer the women found they all had

red, burn-like marks on the back of their necks. Their eyes were also irritated, and Louise Smith reported the hands on her watch we moving at a much higher rate than normal.

ENTER THE MUTUAL UFO NETWORK

Picking up the story, ufocasebook go into some detail regarding the psychological issues the women were having post-sighting.

Mrs. Smith was having difficulties in performing her everyday duties as an assistant for the Casey County Extension Office. Mrs. Stafford was not only suffering from an eye inflammation, she was desperate to know what happened during the missing time. The three women were assured that they would be able to undergo regressive hypnosis, and uncover their missing time, therefore alleviating some, if not all of their emotional stress. Several things were evident to the investigators at this first meeting. The three women witnesses were sincere about what they had experienced, they were suffering from the so-called "beam," and there were obvious physical scars from the encounter. Smith lifted her hair up, and showed a mark on the nape of her neck. It was a roundish, pinkish-gray blotch, the size of a half-dollar.

Dr. J. Allen Hynek, and Walter Andrus from MUFON heard of the case, found the story compelling, and went to interview the women as well as give them regressive hypnosis. I am personally dubious regarding the efficacy of this hypnosis, but the tale which Mrs. Stafford (the only woman to initially agree to be hypnotized) tells is both strange and interesting.

In a slow, cautious manner, Stafford began to recall the events of her night of terror. She was able to relive her perceived experience of seeing what she thought was an airplane crashing. She was not able to go any further detail at this first session, as tears rolled down her cheeks and exhaustion set in. After the session, hypnotist Dr. Leo Sprinkle stated that Stafford was still in a posthypnotic state, and that she should be questioned very carefully and cautiously. After the first hypnosis, the investigators continued to ask questions of Louise and Elaine. Mona moved away from the others to rest. Another interviewer began to show Stafford some drawings of aliens. The word "alien" had not been mentioned in the case before this time, out of respect to the three, and also to not coach or "lead them" in any way. Mona sat and silently looked at the pictures, and then, in a dramatic fashion, she proclaimed, "This looks like the light I saw...It was shaped like that head!" pointing to a specific alien. Again, Mona sat for a time, thinking about that night. Then she added to her previous statement, "Yes...I can see the face now, but it doesn't seem solid. It comes and goes... I mean, fades and reappears like in a fog. Its eyes are far apart and at the bottom... the chin... is like that drawing."

The fact that Dr. Leo Sprinkle from the University of Wyoming became involved as the hypnotist working with Dr. Hynek is enough to convince me of the validity of this encounter.

In so many cases, those that vanished were never heard from again or seen in any shape, manner or form. It's an easy opinion to form that those that came back to us with little or no memory or a discombobulated understanding of what took place could just as easily have been whisked away by a UFO and its terrorizing crew. But what of those who were silenced forever, regardless of if they had ultimately been taken elsewhere still alive, or cannibalized in some non-fiction horror movie like those of rock maniac/filmmaker Rob Zombie?

www.UFOcasebook.com www.UFOEvidence.org

MISSING 411 PART TWO
By Timothy Green Beckley
THE RETIRED COP WHO PUT THE *OOMPH* INTO MISSING

Before retired police officer David Paulides came upon the scene with his series of "Missing 411" volumes, it could be said that there was hardly any life in the cases of people who had gone missing throughout the country. Outside of TV's Nancy Grace, such cases of those who had ventured off – or were openly abducted – and were never seen again, received little or no attention. Sure there were the "Milk Carton Kids," their photos plastered on the side of milk cartons. While those incidences were horrific, they still had a "logical" explanation. Some predator had ventured out of the woodwork to create havoc and went off with their prey of a young boy or girl to molest and kill before discarding the body.

And while David has sympathy for the families of those so taken, his main concern is those unfortunates who took a few steps into the beyond and went missing in some space netherworld. Most of the disappearances David deals with have taken place in National Parks in the U.S. and Canada.

David, who was with the Freemont and San Jose Police Departments for sixteen years and upon retiring became a respected private detective, Bigfoot hunter, and author, has become a frequent and exceedingly popular guest on a variety of podcasts and talk shows. For example, Coast to Coast AM, where he has been interviewed multiple times by hosts George Noory and George Knapp regarding his multiple book series. "Missing-411" is unarguably the first comprehensive research about people who have disappeared in the wilds of North America. It's understood that people routinely get lost and some want to disappear, but David's investigation is about the unusual. Nobody has ever studied the archives for similarities, traits and geographical clusters of missing people, until now.

A tip from a national park ranger led to this decades-long and 9000-hour investigative effort into understanding the stories behind people who have vanished. The books chronicle children, adults and the elderly who disappeared, sometimes in the presence of friends and relatives. As Search and Rescue personnel exhaust leads and places to search, relatives start to believe kidnappings and abductions have occurred. The belief by the relatives is not an isolated occurrence; it replicates itself time after time, case after case, across North America.

The research depicts 28 clusters of missing people across the continent, something that has never been exposed and was a shocking find to researchers. Topography does play a part into the age of the victims and certain clusters have

a specific age and sex consistency that is baffling. This is not a phenomenon that has been occurring in just the last few decades; clusters of missing people have been identified as far back as the 1800's.

The research has become so extensive that it has now been split between a half dozen books, which David sells exclusively on his web site:

www.canammissing.com/ bypassing even Amazon.

Some of the issues that are discussed in each edition:

· The National Park Service attitude toward missing people

· How specific factors in certain cases replicate themselves in different clusters

· Exposing cases involving missing children that aren't on any national database

· Unusual behavior by bloodhounds/canines involved in the search process

· How storms, berries, swamps, briar patches, boulder fields and victim disabilities play a role in the disappearance

· The strategies of Search and Rescue personnel need to change under specific circumstances.

David keeps up on all the recent cases of mysterious disappearances and his involvement, if any, can be found on his very active web site. One such recent case involves lost Toronto firefighter who had search-and-rescue teams combing Whiteface Mountain for a week and who eventually found himself clear across the country After a brief stop in the Adirondacks for police questioning, Constantinos *"Danny"* Filippidis returned to Canada, according to Frank Ramagnano, the president of the Toronto Professional Fire Fighters Association.

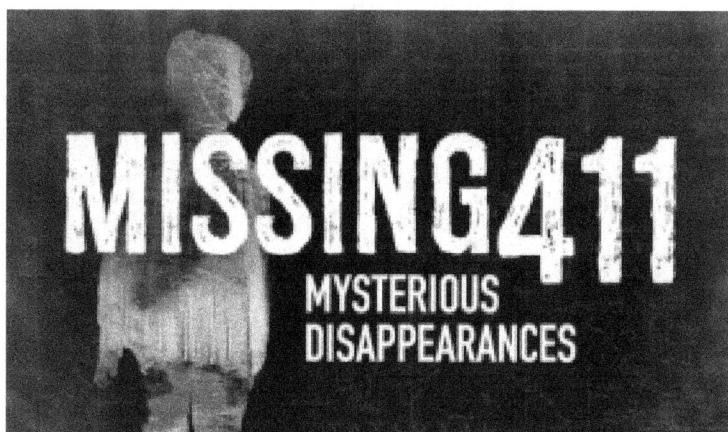

According to reporter Griffin Kelly, writing in the Adirondack Ledger,

Ramagnano said he spoke to Filippidis briefly but didn't ask too many questions because he didn't want to over-stress him.

Filippidis went missing at Whiteface Mountain Ski Center in Wilmington on Wednesday, Feb. 7. Organizations such as state police, the Department of Environmental Conservation and the Olympic Regional Development Authority that runs the mountain set up a search-and-rescue initiative. They utilized helicopters, dogs, drones and dozens of volunteers. The only problem was that Filippidis wasn't on Whiteface. He wasn't even in New York.

Filippidis was eventually found on the other side of the country in Sacramento, California, when he called his wife from there Tuesday morning

Sgt. Shaun Hampton of the Sacramento Sheriff's Office said Filippidis called police at 9:30 a.m. after his wife told him to. Police found him by the car rental terminal at Sacramento International Airport wearing a winter jacket, snow pants and a pair of Sorel boots. Hampton said Filippidis still had his skiing helmet, too.

Filippidis told police he was a passenger on a big rig truck, but couldn't remember much about the trip or the driver because he possibly suffered a head injury, according to Hampton. Filippidis also told police he got a haircut from an African-American male while in California.

Calls to numerous barbershops near the Sacramento airport did not turn up the person who cut Filippidis's hair.

Despite his apparent memory loss, police deduced that Filippidis arrived in Sacramento Monday and stayed outside overnight on the sidewalks before calling his wife the next morning, Hampton said. She told him to call 911, which he did.

Sacramento police are not entirely sure how Filippidis got to California but they don't believe drugs were involved.

"There is nothing that leads us to believe he was under the influence of drugs," Hampton said.

"I can't confirm how he got [to California]," Ramagnano said. *"I don't know if he even knows how he got there."*

State police Public Information Officer Jennifer Fleishman said she could not provide any information from Filippidis' interview with state police since the case is still under investigation.

/www.adirondackdailyenterprise.com/

SAMUEL BOEHLKE
MISSING: Oct. 14, 2006
Crater Lake, Oregon

THEY WILL NEVER COME HOME

David Paulides prides himself on a thorough investigation, leaving no stone unturned. In particular, he has concentrated on visitors to national parks who disappeared a few scant feet from the trail, often with their family members a few footsteps behind. Crater Place in Oregon seems to be the center of a great deal of David's focus. A feature article under "High Strangeness" in www.Sott.net lays out a small portion of what DP has found himself confronted with within the confines of Crater Lake, said to be one of the most beautiful places to be found anywhere in the world.

"Ten years ago, 8-year-old Samuel Boehlke darted away from his father near Cleetwood Cove at Crater Lake. He had a mild form of autism and feared loud noises and bright lights, so search crews couldn't use the customary air horns and whistles to find him.

"At the time, searchers told KOIN 6 News the family said Sammy liked to curl up in small spaces."The circumstances behind his disappearance and the subsequent inability of the park service to find him is unusual," Paulides said.

"They brought in canines, canines couldn't pick up the scent. They brought in air support, they couldn't find him. A multi, multi day search couldn't locate the boy.""Eight years before that, another 8-year-old boy, Derrick Engebretson, vanished from a densely wooded mountainside above Upper Klamath Lake, not far from Crater Lake. Search and rescue crews spent more than 10,000 man hours looking for Derrick, but didn't find a single clue."When Derrick was out there with his dad and his grandpa, somehow or another he just walked around, didn't go far," Paulides said. "There was snow on the ground. They should have been able to track him, and he vanishes.""Scott Lucas, a search and rescue coordinator with the Oregon Office of Emergency Management, said only one or two percent of the missions they launch don't get answers. They average about 900 searches a year."Oregon has its share of miraculous survival stories too. Cody Sheehy was just 6 years old when he wandered off deep into the Wallowa wilderness."Search crews using two helicopters with forward-looking infrared technology couldn't find a trace of him in the rugged woods, dampened by snow and rain."In almost all these cases they bring helicopters with FLIR to look for heat signatures on the ground," Paulides said. "They can't find a heat signature. That's unusual."

Despite that, the next morning, 15 hours and 20 miles away, Cody walked up to a house and asked for help."I was physically at the end of my rope that next morning," Sheehy said. "And if I hadn't been in a situation where people found me at that time, I don't know how I would have done for another night out there. I

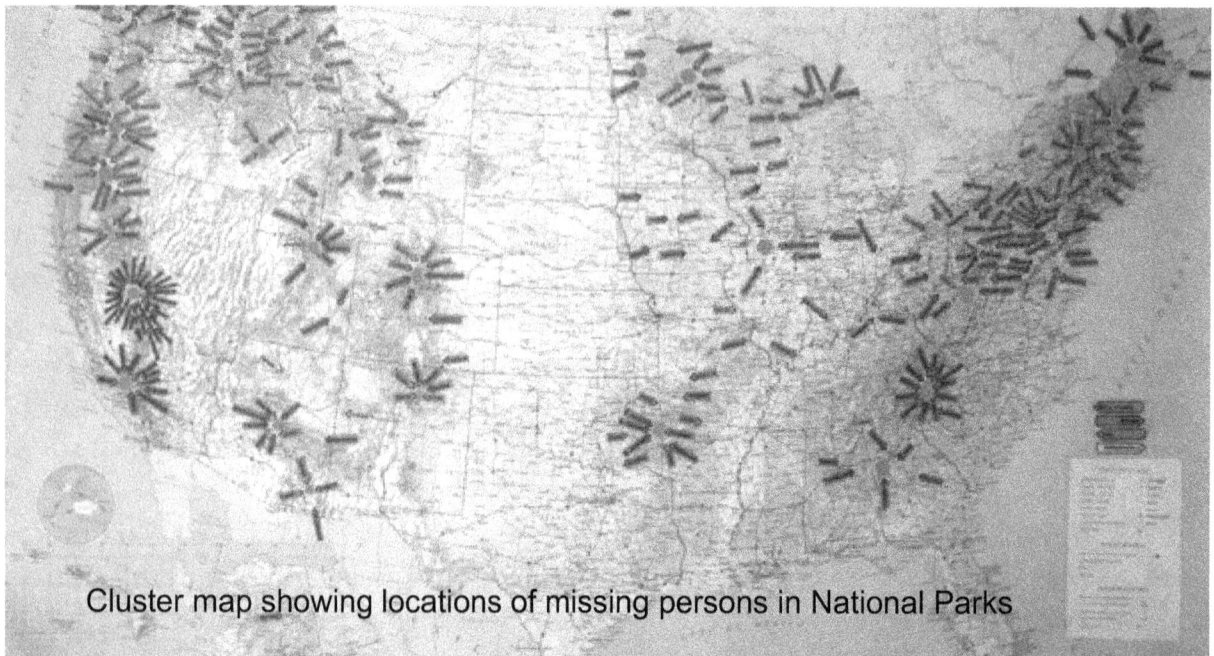

Cluster map showing locations of missing persons in National Parks

could easily have died."More than 30 years before Sheehy's harrowing experience, another astonishing story unfolded near Pendleton, where 2 1/2-year-old Keith Parkins ran and stumbled over a dozen miles of snow-covered timberland and mountains before he was found 19 hours later. He was stiff and cold, but alive."I mean, to me as a parent, I can't see my two-year-old climbing over two mountain ranges in the dark," Paulides said. "That's pretty hard to believe, and there are some cases where little kids are alleged to have walked up to 20 miles a night, or climbed phenomenal heights, over 3,000 or 4,000 feet, and those are facts and it's highly hard to believe."Sheehy says he's not an "overly mystical person.""I didn't encounter anything that was unusual," he said. "Other than the fact that that situation was extremely unusual."Sheehy is alive, but the fate of others, like Derrick Engerbretson and Sammy Boehlke, are why Paulides keeps digging for answers.

ANY CONCLUSIONS?

DP hesitates to speculate as to what is actually happening to these individuals that disappear while their loved ones are literally blinking a few feet away. Predators like Bigfoot or a Wildman have been offered, interdimensional portals a person can unknowingly step into without being alerted, and even abduction by aliens is not off the table.

Best summing it up, one hearty outdoorsman who has spent many days and nights on the trail boldly confided: "I find it very hard to believe there was an unknown force in the wilderness that is capable of taking people from right under the nose of others who are only a few feet away." – But that seems to be what we are talking about!

So curl up with a good Missing 411 book available from the author's site. Watch the movie version (Missing 411 available on Amazon) and listen closely to what David has to say when he updates his findings on C2C AM.

In the meantime, it's time to check in with our own reporters to see who has stepped into oblivion and won't be coming for dinner.

****

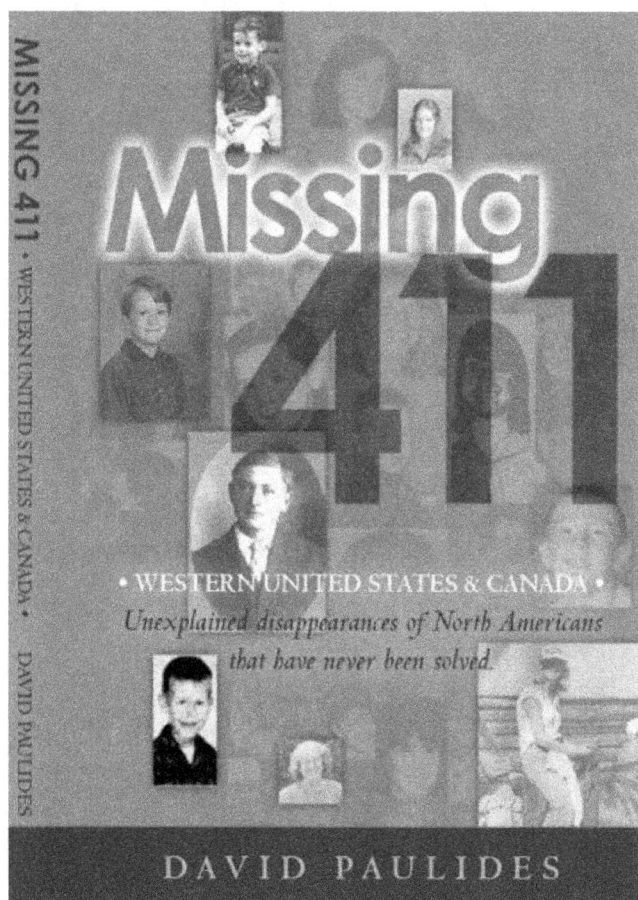

MISSING 411 • WESTERN UNITED STATES & CANADA • DAVID PAULIDES

Missing 411

• WESTERN UNITED STATES & CANADA •
Unexplained disappearances of North Americans that have never been solved.

DAVID PAULIDES

Chapter 22

VANISHED FOR GOOD
B. Ann Slate

The true story of how saucer sightings have been linked to mysterious disappearances of hundreds – maybe thousands – of men, women, and children. The whole horrifying story was told in the headlines.

OREGON FAMILY OF 4 MISSING IN FOREST, Oregon Journal, Sept. 5, 1974;

NATIONAL GUARD HUNTS LOST FAMILY; ABDUCTION FEARS GROWING. Oregon Journal. Sept. 7. 1974;

$1,000 REWARD OFFERED FOR DATA ON FAMILY, The Oregonian, Sept. 7, 1974;

THEIR CAMP READY, FAMILY WENT AWAY. Yakima Herald-Republic, Sept. 8, 1974,

"Copper, Oregon. (AP) Their food was on the picnic table, a pan of dishwater had been set out, and dishes and utensils were neatly laid out near a camp stove. Clothes were folded on a cot. But Richard Cowden and his family, who were camping there, have been missing since September 1st."

And finally: SEARCH CALLED OFF, Oregon Journal, Sept. 8, 1974.

Here are the details: Richard Cowden, 28, his wife Belinda, 22, and their two children, David, 5, and infant daughter Melissa, had completely disappeared from the face of the Earth, dressed only in bathing suits.

The family was camped for the Labor Day weekend at the Rogue River National Forest Campground, one mile from the town of Copper, Oregon, which is near the California border. The last time any of them were seen was on Sunday morning at nine A.M., when Richard Cowden went to the store in Copper with his son for a quart of milk. The family was expected at the home of Belinda Cowden's mother for dinner that night in Copper. They never showed up.

On Monday, the Cowden's basset hound, Droopy, scratched on the door of the Copper store, indicating something was radically wrong. The owner of the store, Guy Watkins, and Ruth Grayson Smith, mother of the missing woman, called

the authorities.

"That camp sure was spooky," Trooper Lee Rickson commented upon investigation of the site. "Even the milk was on the table."

It was as if the family planned to come back – but couldn't. Everything in camp was in order. A stove and cooking utensils were placed on a stump being used as a table. Washed dishes were on the table nearby, along with bananas and other foodstuffs. Richard Cowden's wallet and his wife's purse were found, and both contained small amounts of money. Their fishing rods were leaning against a tree and the family's Ford pickup was parked just above the campsite. There was no sign of a struggle or disturbance.

Nothing seemed logical about the disappearance. Investigators were baffled. "We can't find any motive, any reason, for their being lost," said Lt. Tom Phillips of the Oregon State Police. Nor was there anything to suggest foul play. The missing woman's mother believed they had never left the camp on their own but might have been forcibly abducted. Had they been taken for ransom?

Belinda Cowden's father didn't think so. "We are not wealthy," said Carl Grayson. The missing man wasn't affluent either. Richard Cowden drove a logging truck for the Steve Wilson Logging Company. No one ever responded to the $1,000 reward for information leading to the whereabouts of the family.

Oregon Governor Tom McCall authorized the use of the National Guard to aid state police and forestry personnel in the search. More than 100 specially trained Search and Rescue Guardsmen of the First Battalion, 186th Infantry, combed the rugged Siskiyou Mountains. They were joined by members of the Alpine Res-

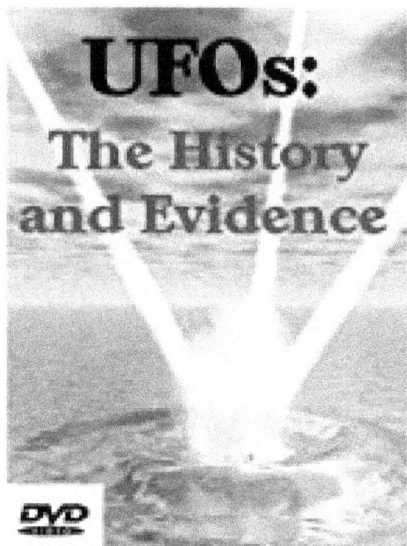

cue Club, friends of the family, and members of a motorcycle gang called the "Hessians" who were in the region for the holiday. Even a psychic went along.

However, not a single clue was found in the 25 mile radius covered, nor did high resolution aerial photographs reveal a thing. The area is a wilderness of deep gullies and canyons with heavy forests. It's also crisscrossed with logging roads and

CIRCLES indicate needle marks left by aliens. Maria Molero, left, and her sister Emma flank their little brothers Carlos, below, and Jorge.

honeycombed with old gold mine shafts. Most of the known shafts were searched, but the Cowdens knew the region well, having camped there many times before, so the chances of the whole family falling down an abandoned shaft was considered remote.

The intensive ground and air search was officially called off six days after the family vanished, although volunteers who carried on were coordinated by officers remaining at the site. The Cowden family had simply vanished into thin air. Only their basset hound Droopy knew what really had happened to them. State Police Sgt. Ernest Walden said there had been no indication that the family was kidnapped, but added, "although we've all had the thought."

Kidnapped for what reason? And by whom? Checks with the Oregon police three months after the mysterious disappearance indicate their fate is still unknown.

Ever since men took to the sea in ships and into the air with wings, there have been disappearances. Whether wreckage is found or not, it is presumed some natural disaster or mechanical failure was the cause and the oceans, mountains, or deserts have claimed the remains, But what about those frequent incidents when just the people themselves seem to vanish from planes and ships or right off the ground as if they were picked up off the face of the Earth?

The late Frank Edwards, news commentator and author, was intrigued with UFOs as well as those baffling events that all of man's knowledge cannot explain. In his books, "Stranger Than Science" and "Strange Happenings," he documented

UNSOLVED

In Loving Memory of
Gene
Felix Eugene Moncla, Jr.
1st Lt. United States Air Force
Born October 21, 1926
Disappeared November 23, 1953
Intercepting an UFO Over Canadian
Border as Pilot of a F 89 Jet Plane.

THE KINROSS INCIDENT

Pilot Lt. Felix "Gene" Moncla Jr. and
Second Lt. Robert Wilson were dispatched to check out an
unidentified object flying over Lake Superior. They seemed
to hone in on it. Then they just disappeared.

the following bizarre and unsolved mysteries: After the first atomic bomb tests in the Pacific, the island trading ship, Joyita, was found wallowing in light seas near Fiji, its crew and passengers having disappeared into thin air. A topside section was burned out in the form of a semicircle.

In 1858, 650 French colonial troops were marching toward Saigon, Indochina, when, as they walked across open country some 15 miles from the city, they vanished as completely as if they had simply walked off the Earth.

In 1953, the abandoned ship, Holchu, was seen drifting in the Indian Ocean. Everything was in order; there was plenty of food and water and ample amounts of fuel. A meal for five had been prepared in the galley but it was untouched by the crew, which had simply vanished.

The Northwest mounted police investigated a report from trapper Joe Labelle regarding the disappearance of an entire Eskimo village in November 1930. They had not taken their prized guns, their dogs or their clothing. Pots of food hung over fires which had long since grown cold. They had not gone across the lakes as the kayaks had also been left behind. Skilled trackers could find no footprints of the 30 Eskimos leading out of the village. Something had made them all rush out of their huts, stopping whatever they had been doing in the course of their normal daily lives. And none of them had ever got back from whatever had attracted their attention.

In 1939, with the Japanese conquering China, it was imperative to stall the aggressors as long as possible. Almost 3,000 Chinese troops were placed on the outskirts of Nanking to block the advancing enemy. When they didn't respond to radio calls, a colonel's aide investigated. He found their guns neatly stacked beside cooking fires, but the troops were gone. They couldn't possibly have deserted across the open country without being detected.

The midair mystery, which occurred in 1942 when two experienced men in a U.S. Navy blimp, the L-8, took off on a routine flight from their base on Treasure Island and were never seen again. The blimp was later found in proper order. The men could not have fallen into the sea as every move the L-8 made was observed by the crews of two patrol boats beneath it.

Ruling out some of the more fantastic concepts of "time-warp," "other dimensions," "materializing-dematerializing space cannibals," and "cosmic spacial whatzits," the finger of suspicion has long pointed to UFOs as being behind some of these blatant kidnappings. Admittedly, in some cases, it is a rather circumstantial finger. But in other instances, eyewitnesses and radar equipment attest to their presence.

There is the Kinross Affair of Nov. 23, 1953, as reported in Maj. Donald Keyhoe's book "The Flying Saucer Conspiracy."

At the Kinross Air Force Base in Michigan, radar operators announced the "paint," or blip, of an unknown object flying over the Soo Locks. No aircraft was scheduled to be near that strategic location so an identification had to be made. An F-89 was scrambled immediately and streaked toward the UFO. The jet was piloted by Lt. Felix Moncla, Jr.; and Lt. R. R. Wilson was the radar operator.

As the jet moved in closer, the UFO changed course, heading toward Lake Superior. The F-89 raced after it at over 500 miles per hour. At Ground Control Intercept, the radar operators watched their scopes intently, almost breathlessly, as the second "paint," which was the F-89, came closer to the UFO. Suddenly, the two blips merged into one! For just one moment, the huge blip remained on the scope and then it quickly flew out of range.

No trace of the missing men was ever found, nor any wreckage, although Search and Rescue planes and boats crisscrossed a 100 mile area. The UFO had simply swallowed up the interceptor, taking both Moncla and Wilson.

Initially, the Air Force admitted what had really happened in a statement to the Associated Press which read: "The plane was followed by radar until it merged with an object 70 miles off Keweenaw Point in upper Michigan."

This report appeared in one early edition of the Chicago Tribune and then was quickly yanked. Then the Air Force claimed the UFO had been a Royal Canadian Air Force C-47 and the F-89 had crashed into the lake and the radar technicians had read their scopes incorrectly. Yet the RCAF denied it was one of their planes, the radar operators had distinctly seen the two blips merge with the larger one then moving off the scope, no wreckage or bodies were ever found and those men manning the scopes that night were highly-trained experts!

And Eugene Metcalfe of Paris, Illinois, swears that he saw a bell-shaped UFO "swallowing" an Air Force bomber on March 9, 1955!

That wry master chronologist of the weird and unexplained, Charles Fort, never doubted that space explorers might be behind many of the strange disappearances he documented. His most quoted statement has become a classic: "I think that we're fished for. It may be that we are highly esteemed by super-epicures somewhere. It makes me more cheerful when I think that we may be of some use after all. I think that dragnets have often come down and have been mistaken for whirlwinds and waterspouts ... I think we're fished for..."

And from his book, "Lo!," Fort ventures, "It may be that if beings from somewhere else would seize inhabitants of this Earth, wantonly or out of curiosity, or as a matter of scientific research, the preference would be for an operation at sea, remote from observations by other humans of this Earth."

Or, perhaps, getting lured out of the sky to land because "they" don't want the plane, just the passengers! From Fort's "Wild Talents": "Upon July 24, 1924, at a time of Arab hostility, Flight Lieutenant W. T. Day and Pilot Officer D. R. Stewart were sent from British headquarters, upon an ordinary reconnaissance over a

UFO HOSTILITIES AND THE EVIL ALIEN AGENDA

Frank Edwards, author of "Flying Saucers: Serious Business," often dealt with close encounters between planes and UFOs. We wonder what he might have discussed with President Truman?

desert in Mesopotamia.

"According to schedule, they would not be absent more than several hours. The men did not return and they were searched for. The plane was soon found in the desert. Why it should have landed was a problem. There was some petrol left in the tank. There was nothing wrong with the craft. It was, in fact, flown back to the aerodrome.

"But the men were missing ... they encountered no meteorological conditions that might have forced them to land, there were no marks to indicate that the plane had been shot at. In the sand, around the plane, were seen the footprints of Day and Stewart. They were traced, side by side, for some 40 yards from the machine. Then, as suddenly as if they had come to the brink of a cliff, the marks (footprints) ended.

"Aeroplanes, armored cars, and mounted police searched. Rewards were offered. Tribal patrols searched unceasingly for four days. Nowhere beyond the point where the tracks in the sand ended abruptly, were other tracks found."

Fort then goes on to recount the many inexplicable mishaps and forced landings of French aviators during 1923 while flying in and out of Germany. The incidents were so frequent it was believed the Germans had perfected a secret ray that could pick planes out of the sky.

Not the Germans, Mr. Fort, but extraterrestrials, manipulating the controls of a plane and perhaps the mind of the pilot to make him land so they could kidnap the unsuspecting earthlings.

Perhaps if Charles Fort had lived long enough, he might have seen the striking similarity of the following case which began on July 15, 1951 (Special thanks to

UFO HOSTILITIES AND THE EVIL ALIEN AGENDA

Dave Swaim of the Pasadena Star News for an unfailing memory and for digging out the almost obscure details from the files concerning this incident.)

After renting a Cessna from the East Los Angeles Airport, Klaus W. Martens, 28, and his companion, June Walker, 22, flew toward Blythe, Calif., for a brief business appointment. He was a salesman for a Pasadena auto agency and she, a student nurse at Huntington Memorial Hospital. They had known each other for three months. Ms. Walker, clad in shorts, had told her roommate they planned to return that day.

They were never seen again. Because of thunderstorms in the mountains near Blythe, it was initially suspected that the plane had crashed. Civil Air Patrol, Air Force, and private planes searched in vain. After 10 days, 203 flights had been made in 165 planes.

Then, on July 30th, the downed Cessna was found southeast of Yuma, Arizona, at a deserted target range. The plane was undamaged and had enough gasoline for several minutes more flying time. The radio was in good working order and no airfield in the general area had received a distress call from Martens, an experienced WW II pilot.

It came to be called "Southwestern Arizona's Greatest Manhunt." The team included a 20-man ground Search and Rescue party from March Air Force Base, the Los Angeles County Sheriff's Aero Squadron, two PBY flying boats, volunteer pilots in private planes from the Tri-City Airport, as well as prospectors, 100 soldiers, and the Yuma Sheriff's posse.

The clues were as baffling as the situation. A note left in the cockpit said the couple was heading west. Yet they didn't take the compass out of the plane! And men familiar with the desert pointed out that only desolate, waterless country lay to the west!

Tim Beckley and "MIB Lady" Claudia Cunningham at the cemetery plot for Charles Fort, who once stated: "It may be that if beings from somewhere else would seize inhabitants of this Earth wantonly or out of curiosity, or as a matter of scientific research, the preference would be for an operation at sea, remote from observations by other humans of this Earth."

MISSING
UNRESOLVED CASE:
Glenn Lauder
Case number:
ML-40-CFR-763
DOB:
3/5/1969
Missing from:
Avondale, Arizona
Date Missing:
3/14/97
Unknown Circumstances
Anyone having information
should contact the Arizona Missing Database:
855-2FIND-THEM.
info@maricopamissing.com
UNRESOLVED CASE:
Mitch Adams
Case number:
ML-40-CFR-762
DOB:
2/4/1968
Missing from:
Avondale, Arizona
Date Missing:
3/14/97
Unknown Circumstances
Anyone having information
should contact the Arizona Missing Database:
855-2FIND-THEM.
info@maricopamissing.com
UNRESOLVED CASE:
Ryan Stone
Case number:
ML-40-CFR-761
DOB:
10/16/1970
Missing from:
Avondale, Arizona
Date Missing:
3/14/97
Unknown Circumstances
UNRESOLVED CASE:
Jacob Reynolds
Case number:
ML-40-CFR-764
DOB:
12/12/1969
Missing from:
Avondale, Arizona
Date Missing:
3/14/97
Unknown Circumstances
Anyone having information
should contact the Arizona Missing Database:
855-2FIND-THEM.
FROM A MOVIE ONE SHEET
FILM NOW STREAMING.
PHOENIX
INCIDENT
#DiscloseNow

UFO HOSTILITIES AND THE EVIL ALIEN AGENDA

There were no tracks around the plane indicating that the pilot, Martens, had walked around it to check for any trouble. Deputies in a jeep followed the couple's tracks for three and a half miles where they completely disappeared on top of a little ridge.

What made it look all the more suspicious was that their tracks crossed a road which, if it had been followed, would have led them into the town of Wellton, 30 miles to the north. And added to that were the mysterious marks of what appeared to be a second aircraft right where the couple's tracks ended!

Yuma County Sheriff Jim Washum believed the disappearance was staged but the facts didn't add up. Martens apparently wasn't running away from a marital situation coming in April; his former wife Nina had already obtained an interlocutory divorce. She discounted any possibility of a romance between Martens and June Walker. "If there had been any serious attachment between him and Miss Walker, or any other girl, I am confident I would have known about it – he would have told me," was the ex-wife's statement to the press.

A man planning to run away would certainly need some money, but Klaus Martens had not picked up his salary check at the automobile agency where he worked nor had he withdrawn any money from his small bank account.

While Yuma Sheriff Washum believed it all a hoax and discontinued the search, the March Air Force Base team continued the hunt through the sandy mesquite and cactus-covered desert.

No bodies were ever found. Sheriff Eugene Biscailuz

of Los Angeles, after hearing about the traces of what might have been another craft at the spot where the couple's footprints ended, said, "This leads to the belief the two were picked up at that point." But no one ever found out by whom!

In spring 1974, an aged Xerox copy of a photograph of Martens was given to the Certified Psychic Consultants without divulging any details of the case – only that the man had vanished. Sharon and Gary Travis, Southern Region Directors of the California group, gave the photo to four of their psychic readers to psychometrize (feel vibrations from an object). Many common denominators emerged from those readings:

· While the plane was in the air, Martens saw a bright light in the sky and then felt a pressure in his head, like a vice. This created a great deal of mental confusion for the pilot.

· A vapor began wafting out of the control panel in the cockpit that smelled like sulfur.

· There was a great deal of heat;

· Three lights were on something. There were eyes looking at him. (One psychic reader said it was kind of a cute thing and she had the feeling she wanted to take it home.)

And finally, a white light was seen, a light which the psychic team has come to associate with being symbolic of death. The readers scored well on the experiment, even though the case was 23 years old. The fact that Martens was in a plane was correct. The extreme heat probably indicated the soaring temperatures of the desert region at that time. The fact that a symbolic white light was seen is probably true, as the couple have long been presumed dead but ... did they die on this planet?

The dazzling bright light seen in the sky is certainly a characteristic description of a UFO, as well as the "three lights on something." The "eyes" looking at Martens is reminiscent of the famous Betty and Barney Hill UFO abduction case, when, as Barney looked through his binoculars into the eyes of one of the aliens on the craft, he was put under telepathic control.

Since nothing was found operationally wrong with the Cessna, the sulfuric vapor that filled the cockpit might well have been rigged by the alien craft to convince Martens that something was wrong with the controls and he must land immediately, even though he had enough fuel to get to his destination.

But the real key is the impression that one reader received as she looked at Martens' photo, the feeling that "it was a cute thing and she wanted to take it home." Rather than assuming she was looking through Marten's eyes at some cuddly or appealing extraterrestrial, in view of the resulting facts of the case, isn't it more likely that she was registering the alien's desires, wanting to take Klaus and June home – home to its planet?

UFO HOSTILITIES AND THE EVIL ALIEN AGENDA

Perhaps their mission wasn't quite finished and they didn't leave immediately for their home base, because one week after the couple disappeared, radar men at March Air Force Base tracked a UFO and pilots visually observed a silvery object circling above them.

On the day that little Tommy Bowman vanished, there was no doubting the UFO activity in the immediate area. It was on March 23, 1951, that the eight-year-old, visiting relatives with his parents in Altadena, Calif., simply disappeared around a turn in the trail as the family hiked in the nearby Arroyo Seco Canyon. Search and rescue efforts were extensive and thorough but Tommy's body was never found.

It is beyond our technological understanding how it is possible that persons can be kidnapped by an alien craft in the daylight without being observed unless they can put up some type of optical shield. But that same night at 9:55 p.m., a "light" was observed moving around this same general area. In the nearby town of Pasadena, which is directly to the south, Air Force Sgt. Dewey Crow and newsman Les Wagner watched the UFO maneuver slowly for more than an hour. White lights flashed from the object which was described as orange red in color. Air Defense radar was notified.

The craft appeared to be in no hurry as it moved silently over Los Angeles and the coastal towns. Police switchboards were flooded with hundreds of calls. An F-89 interceptor attempted to locate the object but the Air Force denied it was able to make contact, although at the same time witnesses on the ground could plainly see the craft.

Then, about midnight, the UFO was joined by three other saucers. Here is the official radar report: "At 2350 (11:50 p.m.) I was watching the radar scope when I noticed a target about 15 miles northwest and moving northwest. At first I thought it was a jet, then I noticed it was moving much faster than anything I had ever seen on the scope. About 40 miles northwest, it came to an abrupt stop and reversed course, all within a period of about three seconds. It then traveled back along its course for about 20 miles, reversing course again and disappeared off the scope.

"Five minutes later, two more targets appeared and disappeared off the scope in the same direction as the first and these we had time to clock. They traveled 20 miles in 30 seconds which figures out to 3,600 miles per hour. A minute or so later, a fourth target, appeared in the same area as the other three . . . and went off the scope to the northwest at 3,600 miles per hour ..."

About seven months prior to this incident, 11-year-old Brenda Howell and Don Baker, 13, vanished from the same general region in the San Gabriel Mountains, less than 15 miles from where Tommy Bowman mysteriously disappeared. The two youngsters had ridden their bikes into the canyon at six a.m. on the morning of August 6, 1956, and were never seen again.

Sheriff's deputies at the San Dimas Station searched all night and into the many fruitless days that followed. The children's bikes were found near Morris Dam in the canyon several days later.

Navy divers from the Ordnance Testing Station at the dam searched the lake, and the reservoir behind the dam was dragged but no bodies were recovered. To date, this unsolved case fills a large file at the Sheriff's station.

Could it be meaningful that a little over a week before the children vanished that the Air Defense Filter Center at nearby Pasadena trailed a "mysterious" object on radar which moved to the northeast (right over the vicinity of the dam)? The UFO was also seen by members of the Ground Observer Corps as a brilliant white light moving at variable speeds, alternately hovering and accelerating. It was clocked as going "faster than a conventional airplane." And, one month to the day that the children disappeared, a Western Airlines pilot reported white lights moving erratically over that same vicinity and there was visual confirmation of the UFOs from the ground.

Hovering lights which speed up, slowdown, and retrace their courses. Are they looking for something? Is there any real proof that aliens are behind our still unsolved and unexplained disappearances? That answer is a resounding yes. In fact, there are eyewitnesses to one kidnapping which occurred in the summer of 1954 in Long Beach, Calif.

It was a clear sparkling day and the party sitting on blankets and towels spread across the white sand of the beach were in a festive mood. But that day of celebration was to turn into a haunting nightmare which has never been satisfactorily resolved. The following is related by one of the relatives involved, Mrs. Alexandri Skolny of Alberta, Canada, in an interview with Ashly Pachal of the Edmonton Flying Saucer Society. She stated: "My cousins in Buffalo, N. Y., had adopted a boy and raised him. When he got married, he went on his honeymoon to Long Beach, Calif., where he visited my cousin and his uncle. They took him to the beach there. They went with a party of six. The rest of the family was seated near the shore and the young couple, who were just married, ran toward the water to swim."

Suddenly, Mrs. Skolny recalled, they all felt a strange breeze, a chilling kind of cold which gave them the shivers. Then it appeared without warning, a glowing green ball-shaped object estimated to be 12 feet in diameter, which zoomed out of the sky and swooped up the young couple, as well as a quantity of sand where they had been standing.

"Everyone was stunned by the amazing and horrifying sight. It was unbelievable! One moment the honeymooners had been running and laughing, and the next moment they had vanished, taken aboard that eerie green object which then shot straight up into the sky and out of sight. All that remained was an indentation in the sand where the couple had been, a space the exact width of the alien object.

"They (the couple) had no reason to disappear!" Mrs. Skolny said. "They had just been married. Their car was still there at the beach and locked. My relatives went to the authorities but all they did was laugh at them. All my family could do was go home and just say they disappeared. There hasn't been a sign of them ever since!"

Statistics from Missing Persons for the City of Los Angeles alone show that in 1973, the formal reports list Adult Missing-605; Juvenile Missing-6,099. The unofficial files, where no formal report has been made, are much higher: Adults-4,314; Juveniles-33,995. Sergeant Thomas of the Bureau advises there is a 98 percent clearance rate. But usually 2 percent of those reported missing are never found.

"When foul play is suspected, a body eventually turns up," the Sergeant stated. "But every once in a while, the persons have totally vanished and the department assumes them gone!"

Gone – like the circus convoy heading from London to Oxford, England, on Sunday Nov. 3, 1974, which completely vanished. The Toronto Sun reported: "The missing circus included the ringmaster, his wife and son, a troupe of French jugglers and trick cyclists, two elephants, two camels, six llamas, one donkey, two mules, six Shetland ponies, two Palomino horses, and a bison.

"After officials in Oxford reported the circus had not shown up, police scoured all main roads. But ... they failed to find even a solitary hoof print."

A spokesman for the police said, "Incredible as it may sound, they seem to have vanished into thin air!"

Gone – like the five officers and men standing on a rise overlooking the ocean at Iwo Jima during WW II. In the presence of seven startled men of the 24th Marines who were nearby, a dazzling white light enveloped the five and in an instant, they had vanished.

The witnesses knew it hadn't been a buzz bomb. There had been no sound or noise connected to the light. No hole was made in the ground and no body parts were found. The men were eventually listed as "Missing In Action."

Gone – like the pilot who had been training for nine weeks to qualify for the Navy's precision flying team, the Blue Angels. His was to be the fourth position with the highly-skilled outfit. On an airfield near Los Angeles, he went up into a clear blue sky with just a few small white clouds in the distance. The Blue Angel pilot who had trained him was on the ground with other personnel, proudly watching his pupil perform his first solo show. They were in constant radio contact.

As the plane flew into one small white cloud, radio contact was cut off completely. The pilot and plane never emerged from that cloud. Observers on the ground did not see or hear an explosion. No wreckage was ever found. Fighter jets were scrambled but they found nothing. The Blue Angel trainee had totally vanished. The land and air search, carried on for 15 weeks, was so intense that 13 small civilian aircraft, missed in other rescue efforts, were located. The Navy kept this disappearance out of the papers. There was no way to explain it.

The question of "how" these disappearances take place does not present as much of a problem as the "why." The enormous alien mother ships seen visually from the ground and as half-inch blips on radar are estimated at over 1,000 feet long – certainly large enough to scoop up a circus convoy, an F-89 jet, or 3,000 Chinese troops. But can they abduct adults, children, animals, and vehicles in broad daylight without being seen?

UFO researcher-writer John Keel suggests the invisibility factor, the technology of bending light waves around the craft so that the human eye cannot register their presence. He has cited instances of planes crashing into solid, unseen objects in midair as well as cars colliding with invisible "brick walls" on deserted country roads, impacts which have demolished the autos although no obstacles were visible.

As to the "why," it has been suggested that by making off with human beings, "they" can learn our language and eventually establish communication with us. That's certainly a strange way to establish interplanetary contact or make friends or open up trade negotiations or whatever "they" might have in mind!

Tell that to Belinda Cowden's mother, Mrs. Ruth Grayson Smith, who spends her every free moment at that lonely deserted campsite in Oregon, hoping and praying evidence will turn up to lead her to her daughter, son-in-law, and grandchildren. Tell her that because of the UFO activity in the state of Oregon during the time they vanished that they may have been kidnapped for an aliens' "Linguistic Research Program" with the people of planet Earth.

Tell it to Mrs. Viola Walker, mother of June Walker, who vanished in the desert with Klaus Martens. She knew her daughter didn't run away because she was more interested in finishing her nurse's training than anything in the world.

Tell it to all the parents and families of those missing loved ones who vanished from the face of the Earth; the people who still grieve but still hope...

The predicted confrontation with extraterrestrials in the near future draws closer. Shall we view them all as benevolent, angelic brothers of the universe?

Chapter 23

THE MISSING CHILDREN OF LETHAL LAKES
By George Andrews

The late researcher George Andrews was a prolific author on a variety of controversial topics – from drugs, to the sexual revolution, to the nature of UFOs.

"Perhaps the worldwide public appearance of UFOs and the appearance of LSD, both of which events took place shortly after the first atomic bombs devastated entire cities, are not isolated separated events, but on the contrary, closely linked. Both the UFO and the LSD experience are characterized by multi-colored luminous apparitions and otherworldly realities. LSD may have been deliberately planted among us to prepare us for the shock of confrontation with non-human intelligent beings."

Andrews saw the cosmos in terms of a duality of positive and negative. In his work he stated that, "Initially the people must be alerted to the danger represented by the predatory reptilian ETs, as a preliminary to ridding the planet of them. Most dangerous are the Reptilians in human form, camouflaged within the government. Most of the government has no idea of the extent to which they have been infiltrated. We are here to short-circuit the Reptilian strategy for transforming this planet into a slave colony, where humans would be used like we have been using cattle, pigs and chickens.

"We have only a slim chance of succeeding, but it is better than no chance at all. Once the State of Emergency is declared, under martial law it would no longer be possible to make the necessary changes. It would be the same horror scene as under the Nazis or the Aztecs, repeated all over again, with the death camps set up to feed the Reptilians.

"The thing to remember is that for the positively-oriented forces from elsewhere in the cosmos to intervene, they must have a platform for manifestation. The main thing that humans of Earth can do is to offer the necessary support, the positively-oriented energy that provides a platform for positively-oriented entities from elsewhere who are spontaneously in resonance with such energy patterns. The positively-oriented forces from elsewhere will not intervene unless a clear majority of us call out to them. They will not come unless we call them down. If evil is dominant in human hearts, they can't come through. So that is why the mission has to be. That is why we must get the message through. Those who align themselves with the positive energies bring in the positive forces. Those whose hearts are clouded over, who think only for themselves and are dominated by greed or other base emotions, call in the Reptilians. That is how it works. So long as the majority is of a base nature, the Reptilians will be dominant. We are here to drive out the Reptilians, because they are our enemies also.

"That is why the battle is for people's souls, why we have to reach everyone to get them to understand it is within their power to bring about this change. But they can only do it by aligning themselves with God, no matter how they perceive divinity or what names they use to describe it. No matter what religion they practice, they should sincerely practice it, or else invent their own. It is essential to live directly instead of vicariously through a TV screen, under the delusion that there's no more to life than watching TV, and being cynical about everything else. If enough people wake up in time to take appropriate action, Earth will not become a slave colony for the Reptilians. The positively-oriented ETs will drive them from the planet, and they will have to go elsewhere in the cosmos for the nourishment they seek.

"Even if the news media refuses to touch these topics, there is a telepathic effect that is independent of the news media. There is a psychic osmosis through which ideas whose time has come will spread, even if deliberately ignored by the media. Since it's a holographic universe, one person reaching supreme awareness can spread it from mind to mind telepathically, by-passing the media and the covert censorship it surreptitiously imposes."

George C. Andrews, © 1998

****

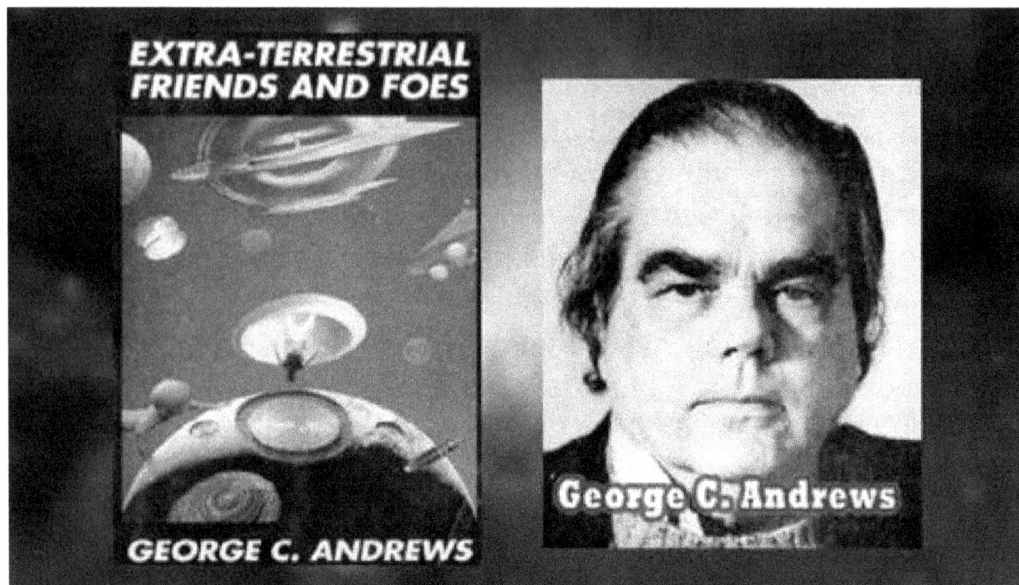

UFO HOSTILITIES AND THE EVIL ALIEN AGENDA

Lake Whitney is a beautiful vacation spot near the Dallas/Fort Worth area. Thousands of tourists go there every weekend. With only rare exceptions, they return home safely. Things go wrong occasionally at any large resort frequented by crowds of temporary visitors. However, when things go wrong at Lake Whitney, there are indications on some occasions of an unknown factor in the equation, a force in action that is not publicly recognized.

Local residents are reluctant to discuss the matter because the well-being of their community depends on the tourist trade. I do not wish to disturb the well-being of their community by frightening away tourists, but feel that for the well-being of all concerned (which includes residents, visitors and the whole wide world) the bizarre nature of some of these anomalous incidents should be investigated in order to bring this unknown factor into clear focus. It can then be dealt with in the light of conscious awareness instead of being repressed and continuing to lurk hidden in the depths, unrecognized and therefore able to literally get away with murder.

In order to give an idea of the kinds of incidents I am referring to, I now quote from an article by Glenn Guzzo, which appeared in the Fort Worth Star-Telegram of February 15th, 1976, entitled:

WHITNEY TRIANGLE? LAKE DEATHS STILL BAFFLING

Lake Whitney - The Texas Parks and Wildlife Department may have to conduct a séance to get any meaningful citizen input on how to keep motorists from driving off a boat ramp here and ending up on the bottom of the lake.

"It's not that the citizenry is unconcerned about three bodies discovered at the same location in a recent 20-month span. They just haven't been able to solve the mystery. Nor do they have answers to assorted drownings, airplane crashes and other unusual events that seem to be more in keeping with a science fiction novel than a casual recreation site in Bosque County.

"'There's something weird going on out there in that lake,' says a not-so-salty dog who has done his fishing in Lake Whitney for 15 years. 'But if you say I said that, I'm going to come looking for you.'

"Investigators from Warr Acres, Oklahoma, came looking at Lake Whitney in 1971, hoping to find traces – possibly the bodies – of two Warr Acres women who disappeared in 1963. Authorities thought they had a major break in the case with the discovery of a black and white 1963 Cadillac like the one Mrs. Margery Elston, 41, and her daughter Melinda, 18, were driving when they vanished.

"The empty Cadillac turned out to be the wrong car. But while they were at it, divers found two almost-identical cars nearby. During that same year, in the same area, divers found another dozen autos, all empty. Instead of solving a puzzle, the discoveries only led to more mysteries and the Oklahoma case remains unsolved. Citizens of the small towns that dot the lake's perimeter and the fisherman

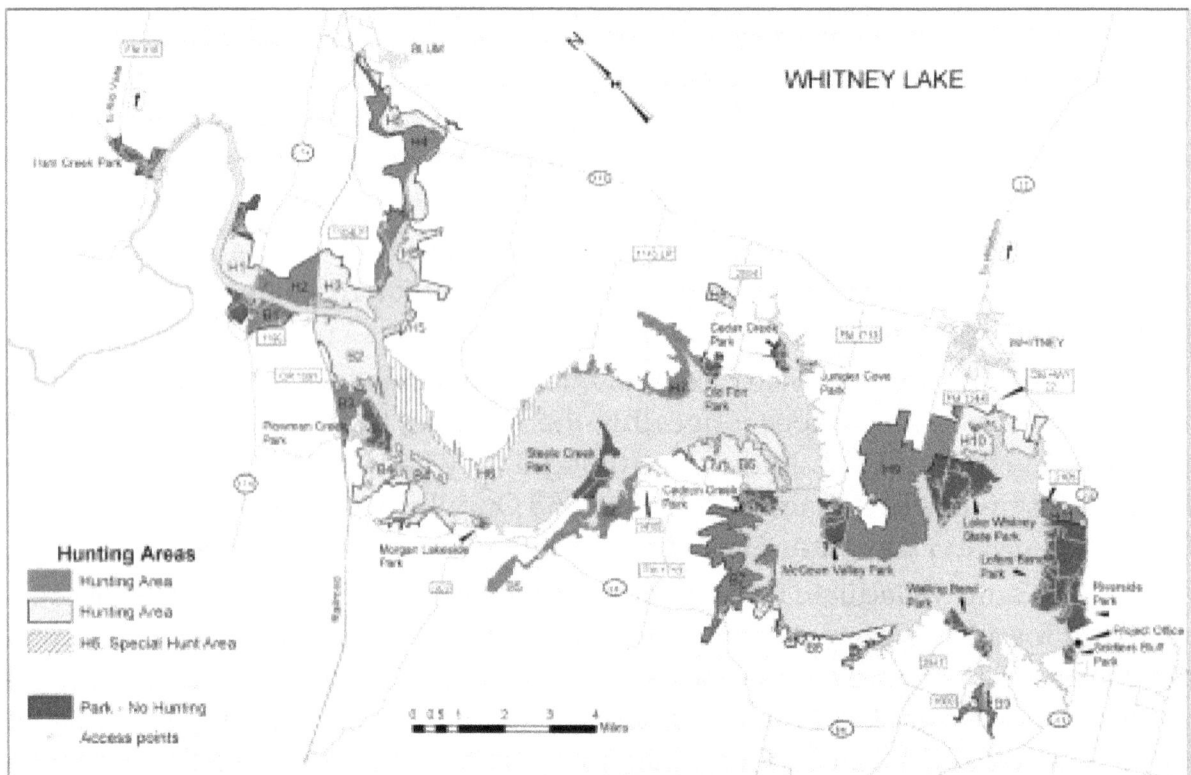

who doesn't want his name associated with superstitions aren't the only ones without explanations for numerous mysterious incidents at the lake. County officials are equally baffled, as are regional and state departments.

"Parks and Wildlife personnel in Austin say they want to hear public opinion before taking steps to remedy whatever hazards they can identify at the lake. They may have to do without it, unless they can contact the spirits of those who have died strangely on or near Lake Whitney. As expected, the lake has had its share of drownings. Small boats have sunk, fishermen have gone overboard, swimmers have gone down for the last time and drunks have turned an innocent dock into a pirate's plank.

"But there also have been unusual deaths caused by such occurrences as cars going off the end of a short road into 10 feet of water and airplanes falling into the lake. What makes the events so intriguing is that conditions that would be considered unusual in just a single accident have been repeated several times.

"When Robert and Dan Hutto of Cleburne dragged the area near the boat ramp in Lakeside Village January 3, they were looking for the body and car of their father, 47-year old Robert Alton Hutto. They found them. But they also found the body of Bobby Webb of Weatherford, whose sunken auto was near Hutto's. Hutto had been missing since New Year's Eve, but Webb, 46, had not been seen for 13 months. Webb, who had suffered a massive heart attack four months before he disappeared and was recovering slowly, was discovered nine feet below the surface.

"Hundreds, maybe thousands, of small boats had passed over Webb's car while launching and docking at the popular boat ramp without anyone noticing its presence.

"A mobile home park close to the water houses the population of Lakeside Village, listed in the 1970 census as 226. Several park residents and weekend visitors swear that the Webb and Hutto cars were pulled from Lake Whitney front first, as if they had entered the lake backwards.

None of the residents tries to explain it.

"Sheriff Baxter denies the reported position of the cars, but he confirms another village tale: the automatic transmissions of both cars were in park.

"'No one will ever know why they were in park,' Baxter said.

"Rex Broome, who manages the marina between the mobile home park and the water, suggested the drivers, discovering their dilemma too late, may have shifted into park as a last resort to keep from entering the lake.

"However, there were no skid marks on the boat ramp showing signs of sudden efforts to stop. And fishermen who look for bass, catfish and trout in that part of the lake say they like the rough surface of the boat ramp because it's not slippery.

UFO HOSTILITIES AND THE EVIL ALIEN AGENDA

"The Hutto and Webb bodies were only the second and third found recently off the boat ramp. On April 27, 1974, the body of Howard Wane Jackson, 33, of Waco, floated out of an open window of a car found 60 feet from the ramp. In that instance, too, there were no skid marks. Less than a mile from the boat ramp, easily within sight, is the spot where four people were killed July 20 when a plane crashed about 300 yards from the shoreline.

"Vernon K. Carter, 54, and his wife Lois of Fort Worth and their two sons, died in the crash. The Carter crash is similar to a two-fatality plane crash on the opposite side of the lake in Hill County five years before.

"Steve Bowman, 20, and his wife Sammie, 18, both University of Texas students, died Oct. 4, 1970, when their plane plunged into 20 feet of water 200 feet offshore and sank. In both accidents the pilots reportedly were flying at very low altitudes less than 100 feet above the water. When the pilots tried to climb suddenly, the engines stalled and the planes fell into the water.

"Coincidentally, both accidents were witnessed by people who knew the victims well.

Seven family members, including Bowman's parents, and a friend watched the UT couple crash after the plane had flown only 20 feet above the water.

"A close friend of the Carters told investigators he waved to the family, then watched as the plane wagged its wings, flew about a quarter-mile, then nose-dived 100 feet into the water.

National Transportation Safety Board records indicate the pilots erred in both accidents by flying too low. However, it never was established why 24-year-old Roland Carter, a flight engineer and co-pilot for Alaska Airlines, was flying his Cherokee 180 so low or exactly what caused the plane to crash.

"Residents of Lakeside Village say you have to be 'spooked' to believe a mysterious force drew the planes to their destruction. But the same residents and visitors talk about the highly unlikely string of events. 'Whenever you get four of five people together they're going to talk about it,' marina manager Broome said.

"But not everyone from the area likes to discuss the matter.

"'It's taboo,' said a fisherman who refused to give his name. 'You don't talk about things like that. It's a jinx.'"

Is it a jinx, or is the unknown factor in the equation to be defined in other terms? An aspect of the situation that Mr. Guzzo did not consider in his otherwise excellent article is that there have been many UFO sightings in the Lake Whitney area over the years, and sightings continue to be reported. There was intense UFO activity in the area of Calvert, Texas, about 60 miles southeast of Lake Whitney in 1973-4, the time period during which many of the incidents described in Mr. Guzzo's article occurred. There have been two UFO landings near Lake Whitney, which left physical evidence in the form of burn marks. This is how a woman, who

UFO HOSTILITIES AND THE EVIL ALIEN AGENDA

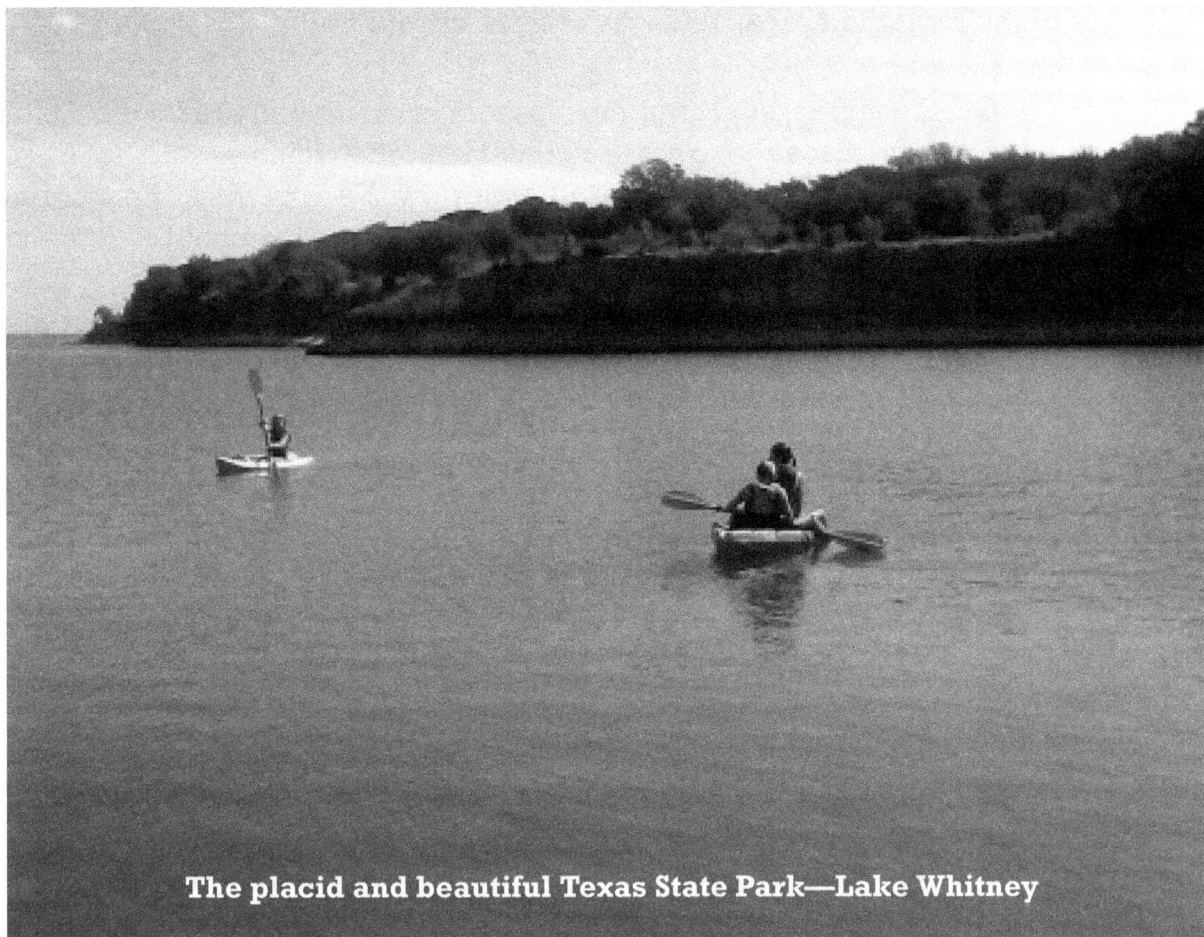

The placid and beautiful Texas State Park—Lake Whitney

wishes to remain anonymous, described the sinking of her boat in 1984:

"It all happened in less than three minutes. It was late in the afternoon on what had been a beautiful day. There were some storm clouds building far in the distance, about thirty miles to the south, on the rim of the horizon. Otherwise the sky was clear and blue, except for one small gray cloud that was almost directly above us. The lake was calm. There was not even a breeze. There were no waves at all. Then from one second to the next, out of nowhere came a wind of hurricane force. It did not come in gusts, it was like a wall of wind. Huge waves sprang up and pounded our boat, which was a 1980 model Baretta that was flotation-insulated to stay at least partially afloat, even if completely filled with water. The boat began to spin like a top, turning around and around in counterclockwise circles. Then suddenly the bow went up and the stem went down, and it sank dead straight vertically like a stone. Fortunately we were able to jump clear in time and to swim to shore. Later on, the insurance company sent a scuba diver down to locate the sunken boat, but he was not able to find it."

Several weeks after the sinking of the boat, a sweat lodge ceremony was performed on Serenity Point (a peninsula extending into Lake Whitney) for the specific purpose of making contact with extraterrestrials. When the participants

255

came out of the sweat lodge, a UFO was clearly visible in the sky and a photograph of it was taken.

Divers have disappeared. Diving instructors warn their students that Lake Whitney is dangerous, that although it is a contained lake which should not have an undertow, sometimes there is a powerful undertow. Here is what one student had to say: "When we went to do our scuba certification, the school took us to another lake. Our instructor told us a story about the last time he ever took a class to Lake Whitney. The diving spot they had used for a long time was about eighty feet deep and completely pitch black. There was no light at that depth. There was a tree he would tie equipment to at the bottom. He went down first to check things out. It was pitch black and he couldn't see his hands. Suddenly two small red lights about the size of nickels flashed on and blinked. He shot up a few feet and then started thinking: 'There must be some light reflecting on something down here. It must be my angle.' So he went back down. Pitch black again. Suddenly the two red lights came on again, and began to blink faster and faster. He shot to the surface: to hell with the bends! He never brought another class to Whitney."

The woman who described what the diving instructor had told her, but who wishes to remain anonymous, says she has also heard that similar types of strange events are occurring at a lake in Mexico. However, her informant was not able to specify which lake it was.

Lake Whitney may not be the only miniature Bermuda Triangle within the boundaries of the United States. Apparently there have been whole series of bizarre fish and bird kills, animal mutilations, and human disappearances in the vicinity of Lake Tahoe since early 1987.

According to this correspondent, who also demands anonymity, media coverage of the incidents is very different from the real story. Some sort of mysterious illness is spreading among animals, birds, and humans in the area. There have been scores of animal mutilations – which are never reported in the media as Lake Tahoe is heavily dependent on tourism and the mutilations have coincided with an epidemic of "missing persons." Media coverage attributes the massive fish kills to pollutants or salinity, but something else is responsible, which biologists involved in the investigations refuse to discuss. Quite a few horses have been among the animals mutilated. During the same time (January-April 1987), numerous animal mutilations occurred near Kingman, Arizona, which were systematically ignored by the news media.

Another disturbing reference to Lake Tahoe surfaced in an article by Hillary Johnson in the July 16, 1987, issue of Rolling Stone, entitled "Journey Into Fear: The Growing Nightmare of Epstein-Barr Virus." Some researchers consider chronic Epstein-Barr virus syndrome (CEBV) even more dangerous than AIDS. Where did the first outbreak of this disease occur? At Incline Village, which is on the shore of

UFO HOSTILITIES AND THE EVIL ALIEN AGENDA

Lake Tahoe, in 1984. Like AIDS, CEBV attacks the immune system, and there is no known cure.

Although CEBV is not now considered fatal, it destroys its victims both physically and mentally (middle-aged people develop symptoms similar to Alzheimer's), ruining careers and marriages as well as healthy normal functioning. CEBV is so difficult to detect and diagnose in the early stages that a doctor has compared it to the Stealth Bomber. There is disagreement among researchers as to whether the high levels of Epstein-Barr virus that accompany the disease are its cause, or merely a symptom. Dr. Robert Gallo of the National Institutes of Health, who discovered the AIDS virus, has discovered a virus called HBLV that many researchers think is more likely to be the real cause. Epstein-Barr virus has been well-known to the medical community for many years, but the HBLV virus, which is of astounding and unprecedented virulence, appears to be a newcomer on the scene. Hillary Johnson brings up this point in her interview with Dr. Paul Cheney, who treated the first wave of cases at Incline Village and alerted the medical community to the danger:

What's crucial, of course, is whether Gallo's virus is new to the planet, or the human race, or simply an old virus newly discovered. 'If I find out that the virus is 3,000 years old,' Cheney said, 'I'll do this.' He sighed with relief. 'After all, it's been around for that long, and we're still here. But if it's new, then no one knows what it can do. Can we handle it immunologically? Only time will tell. One thing – it's difficult to imagine an old virus could suddenly cause an outbreak.'

If Gallo's virus is new to this planet and to the human race, just like the AIDS virus that it resembles in so many ways, including the time frame of its inception, then we must at least consider the possibility that these two previously unknown viruses came to this planet from elsewhere.

Yet another indication of intervention from elsewhere is to be found in the following brief article, released by the Associated Press on May 28, 1987:

DEATH BOLT STRIKES TOP ATTORNEY

Bossier City, La. – Prominent attorney Graves Thompson stood on his new boat, raised his hands, looked skyward and declared, "Here l am." Then a lightning bolt struck him dead.

There was no thunder, lightning or sign of lightning before the killer bolt struck.

Thomas, 40, died of electrocution on Memorial Day.

It was not immediately determined why Thomas gestured toward the sky just before he was killed.

"He just got the boat last week, and he's been out with it just about every night after work," said Lisa Hester, the lawyer's secretary.

Thomas came from a family of prominent Louisiana judges and lawyers, and his cases regularly made headlines.

All of this brings to mind the bizarre story of the "exploding lake" in the small African nation of Cameroon in August 1986, which was never satisfactorily explained and in which there was very heavy loss of life. Tom Adams of Project Stigma summarized the little that is publicly known about this mysterious incident in the 1987 issue of Crux (P.O. Box 1094, Paris, TX 75460):

World scientists disagree still on the cause of the death of over 1,700 people in the African nation of Cameroon in August 1986. The deaths of people and animals were attributed to a cloud of toxic gases emanating from the volcanic Lake Nios (Why is the matter in dispute?). A moderate-length Associated Press article on the incident appeared in the first edition of the Kansas City Star of August 26, 1986. The second edition that day included the identical story but with the omission of this one paragraph:

"The French news agency Agence France-Presse quoted the Rev. Fred Tern Horn, a Catholic missionary, as saying, 'It was as if a neutron bomb had exploded, destroying nothing but killing all life. In the first village we went through, we saw men, women and animals stretched out dead on the ground, sometimes in front of their huts or on their beds, sometimes on the path."

Could it be that this paragraph was omitted because it hints that these massive unexplained deaths may have had an artificial rather than a natural cause?

Approximately three months before this massacre, on May 13, 1986, there was an apparition of the Virgin Mary in a jungle area of Cameroon, not far from the little village of Nsimalen. (Liberation, Paris, France, July 31, 1986.) The nearest town of any size is Yaounde.

The apparition was visible at the top of a tree continuously for several weeks and was accompanied by other types of paranormal luminous phenomena, such as the sun appearing to dance and stars appearing to come down from the sky. Thousands of pilgrims from all over the nation flocked to see it, and did so, returning to their homes with vivid descriptions of what they had seen. Many healings were reported. What the relationship may be between this unusually long-sustained apparition, witnessed by such large crowds of people, and the large-scale tragedy that occurred less than three months later is difficult to determine. .

In the special issue of Critique devoted to the subject of Evil (P.O. Box 11368, Santa Rosa, CA 95406), a horrifying hypothesis was suggested: that the massacre at Lake Nios had indeed been a neutron bomb test, prearranged with the Cameroon government by the U.S. government through the intermediary of Israel. Until we know more about the matter, let us keep our minds open and continue to investigate all of the possibilities.*

* A remarkable example of synchronicity occurred just as this book was about to go to press. I was invited to a gathering of friends in the south of France.

UFO HOSTILITIES AND THE EVIL ALIEN AGENDA

Among the guests was Lt. Col. Roger Vanni of the French Army. During our conversation, it developed that he had been stationed for many years in Cameroon. I asked him what he knew about the tragedy of Lake Nios, and was staggered when he replied that he had been among the first to arrive on the scene afterwards, and had been in charge of burying the cadavers. It was he who wrote the report, before the team of international scientists took over.

In addition to his military career he has also served with the Paris Fire Department. So he has extensive professional experience in identifying causes of death of cadavers found under a wide variety of circumstances. The cadavers he found and buried near Lake Nios displayed symptoms characteristic of asphyxiation by carbon dioxide. Other symptoms indicated the presence of SO_4. He not only discussed the matter at length with me, but later provided me with a whole stack of reports that describe down to the minutest details everything known or reported by witnesses about the tragedy. The evidence that the cause was in fact an eruption of volcanic gas is precisely detailed, massive and irrefutable.

Although its relevance may be only marginal, it seems appropriate at this point to quote the hypothesis proposed a quarter of a century ago by William Burroughs in his "Nova Express" (Evergreen Press, New York, 1964):

"Let me explain how we make an arrest: Nova Criminals are not three-dimensional organisms (though they are quite definite organisms as we shall see) but they need three-dimensional human agents to operate. The point at which the criminal controller intersects a three-dimensional human agent is known as 'a coordinate point.' And if there is one thing that carries over from one human host to another and establishes identity of the controller it is habit: idiosyncrasies, vices, food preferences ... a gesture, a certain smile, a special look, that is to say the style of the controller. A chain smoker will always operate through chain smokers, an addict through addicts.

"Now a single controller can operate through thousands of human agents, but he must have a line of coordinate points. Some move on junk lines through addicts of the earth, others move on lines of certain sexual practices and so forth. It is only when we can block the controller out of all coordinate points available to him and flush him out from host cover that we can make a definitive arrest. Otherwise the criminal escapes to other coordinates. Virus defined as the three-dimensional coordinate point of a controller. Transparent sheets with virus perforations like punch cards passed through the host on the soft machine feeling for a point of intersection. The virus attack is primarily directed against affective animal life. Virus of rage hate fear ugliness swirling round you waiting for a point of intersection and once in immediately perpetrates in your name some ugly noxious or disgusting act sharply photographed and recorded becomes now part of the virus sheets constantly presented and represented before your mind screen to produce more virus word and image around and around it's all around you the invisible hail of bring down word and image. What does virus do wherever it can dissolve a hole and find traction? It starts eating. And what does it do with what it eats? It makes exact copies of itself that start eating to make more copies that start eating and so forth to the virus power the fear hate virus slowly replaces the host with virus copies. Program empty body. The classic case presented to first year students is the Oxygen Impasse: Life Form A arrives on alien planet from crippled space craft. Life Form A breathes 'oxygen.' There is no 'oxygen' in the atmosphere of alien planet but by invading and occupying Life Form B native to alien planet they can convert the 'oxygen' they need from the blood stream of Life Form B. The occupying Life Form A directs all the behavior and energies of Host Life Form B into channels calculated to elicit the highest yield of 'oxygen.'

"Health and interest of the host is disregarded. Development of the host to space stage is arrested since such development would deprive the invaders by necessity of their 'oxygen' supply. For many years Life Form A remains invisible to Life Form B by a simple operation scanning out areas of perception where another life form can be seen. However an emergency a shocking emergency quite unlooked-for has arisen. Life Form B sees Life Form A and brings action in The Biologic Courts alleging unspeakable indignities, mental and physical cruelty, deterioration of mind body and soul over thousands of years, demanding summary removal of the alien parasite. To which Life Form A replies at the First Hearing: 'It was a question of food supply – of absolute need. Everything followed from that.'

"Iron claws of pain and pleasure squeezing a planet to keep the host in body prison working our 'oxygen' plants. Knowing that if he ever saw even for an instant who we are and what we are doing (Switched our way is doomed in a few seconds) And now he sees us planning to use the host as a diving suit back to our medium where of course Life Form B would be destroyed by alien conditions. Alternative proposed by the aroused partisans fumbling closer and closer to the switch that could lock us out of Life Form B and cut our 'oxygen' lines. So what else could we do under the circumstances? The life form we invaded was totally alien and detestable to us. We do not have what they call 'emotions' – soft spots in the host marked for invasion and manipulation. The Oxygen Impasse is a basic statement in the algebra of absolute need. 'Oxygen' interchangeable factor representing primary biologic need of a given life form."

The research accomplished by Ted Holiday during the final years of his life contains some extremely valuable insights, which are eloquently expressed in The Goblin Universe (Llewellyn, 1986). Colin Wilson's brilliant introduction sets the stage for an unforgettable journey through a multi-dimensional hall of mirrors.

UFO HOSTILITIES AND THE EVIL ALIEN AGENDA

Ted Holiday was no ordinary researcher. He was involved in a quest on the heroic scale, and his life was at stake. His investigations led him into areas of forbidden knowledge, guarded by entities at whose nature he could only guess. Holiday's book basically records the cycle of events that culminated in his own death.

The first evening he spent on the shore of Loch Ness, he had a feeling of foreboding "hard to define and impossible to explain," that "Loch Ness was better left alone." Yet he would not, could not, leave it alone. He was drawn back to it again and again, as if by a powerful magnetism, haunted to the point of obsession by the mystery of its depths.

Strange signals from the remote and recent past clustered around his activities, weaving a web of synchronicities in which the Elizabethan Magus, Dr. John Dee, Aleister Crowley, and the Fairy Folk made their presence felt. UFO activity erupted around him, as he penetrated more deeply into the labyrinth. Although I am not always in agreement with the theories he proposes to explain UFO phenomena, the case histories he presents are fascinating and provide a welcome supplement to my own work.

If neither UFOs nor dragon-like lake monsters are psychic projections, but are in fact as physically real as meteors (which eminent scientists of the not-too-distant past indignantly denied existed), some unacknowledged hidden variables are indeed at work, which will disrupt the neat and tidy academically approved version of reality each time they surface.

Ted Holiday accumulated strong evidence that creatures of the "phantom menagerie" – such as dragon-like lake monsters, Bigfoot, and certain anomalous large cats and dogs – do not follow the rules obeyed by normal terrestrial animals. They appear in areas where the available food supply is not sufficient to sustain animals of that size. They leave behind no excrement or skeletal remains. They are able to materialize and dematerialize, just as UFOs do. They are capable of provoking amnesia, hypnotic somnolence, or sudden blind panic, which may be so intense as to result in suicide or sudden death.

We tend to automatically assume that the triggering of such a state of panic would be an act of diabolical malevolence, but we should at least consider the possibility that from the point of view of the anomalous creature, it may simply be self-defense. It is not so long ago that human warriors, such as the Japanese Samurai and the American Indians, used war cries before going into battle that were intended to magically strike their adversaries with panic or paralysis. The anomalous animals associated with UFOs may use some sort of ultra-sound to exert a similar effect.

Some American Indian shamans believe that the same entity which manifests as Bigfoot can also manifest as an aquatic monster or a panther. In Scandinavian mythology, Loki was a shape-shifting trickster, a malevolent renegade among the gods. Case histories of UFO abductees indicate the existence of a renegade category among extra-terrestrials. Certain types of E.T.s may use such anomalous animals to explore or act as sentinels in terrestrial or aquatic environments. Certain types of renegade humans may invoke their materialization in order to communicate with them and obtain power from them, as is indicated by the grimoires of black magic.

Perhaps it was this deeply entrenched, ancient, and tremendously potent negative force that Ted Holiday confronted when he collaborated with a Christian priest in an attempt to exorcise Loch Ness. The experienced exorcist was not harmed, but the semi-skeptical Holiday was vulnerable to the backlash, as he lacked the single-pointed concentration necessary for self-protection under such circumstances, no matter what the religious framework.

The day after the exorcism, Holiday was discussing with friends whether he should investigate the scene of a UFO landing near the Loch. He was advised not to go and accepted that advice. At the moment he decided not to go, there was a sound like a tornado, followed by several loud thuds, and "a pyramid-shaped column of blackish smoke about eight feet high revolving in a frenzy. Part of it was involved in a rosebush, which looked as it were being ripped out of the ground." A beam of white light focused on Holiday's forehead, and the phenomena ceased as suddenly as it had started.

The next morning, he woke early and went for a walk before breakfast, during which he met a man dressed entirely in black, including gloves, helmet, and mask, who disappeared inexplicably.

One year later, Ted Holiday had his first heart attack on the precise spot where he had encountered this man in black. Upon recovering from it, he completed the manuscript of "The Goblin Universe." Five years after that first heart attack, he had his second heart attack, which killed him.

Tim Dinsdale, a down-to-earth 62-year-old engineer, has picked up the trail Ted Holiday left. In an interview with Alison Leigh-Jones of Titbits (London, England, March 1987), Mr. Dinsdale said, "There's much more to the place than the monster. It's a very peculiar region indeed." Although he was unwilling to discuss the many peculiarities of the area in full, he did refer to the place as "magical" and was willing to share some of his information. Near the Loch is a rock formation known as the Rock of Curses, where witches' covens have met nocturnally since prehistoric times.

The late Aleister Crowley believed that energy emanating from a nearby mountain, named Mealfuorvonie, was helpful to his progress in black magic, and that the whole area of the Loch was propitious for his activities. Mr. Dinsdale mentioned unexplained nocturnal lights and human disappearances. If anything, the exorcism of the Loch mentioned earlier appears to have made matters worse. Similar paranormal phenomena occur at the nearby Loch Morar, which is the deepest lake in Britain. Mr. Dinsdale discreetly continues his investigations. He would not be surprised if the Loch Ness monster turned out to be somehow linked with UFOs.

More interesting material comes from an article entitled "Aliens and You," by Rev. Lynn Johnson and published in The Mind Science Journal, Box 1302, Mill Valley, CA 94941.

As everyone knows, in recent years there has been an unprecedented epidemic of children mysteriously disappearing. The standard explanation is human kidnappers or sexual deviates, and this is certainly valid for many of the cases. However, some law enforcement officials and criminal psychologists have pointed out that the number of such cases is too great to be attributed entirely to kidnappers and sexual deviates. Edmond Cunningham, Ph.D., is one of our most esteemed criminal psychologists, and he recently made this statement: 'The numbers are far too great for that many kidnappers to be that cunning. Human behavior is such that the conscience either eventually overtakes the criminal and he starts making major mistakes in his actions, thereby revealing his identity, or he simply kills the kidnap victim and a body usually shows up. The vast majority of the children whose pictures are currently being displayed on milk cartons, grocery bags, posters and TV have vanished without any ransom being demanded or any corpse being found.'

I don't know of any evidence directly linking UFOs to the current epidemic of missing children, and one must be careful not to jump to conclusions in such matters. However, neither should we ignore the possibility that there may be some sort of UFO involvement. All we can do at this point is to keep an open mind and our attention alert, carefully examining each case history for clues as it comes in.

Two newspaper reports that have arrived from Australia provide support for this frightening hypothesis. The first is from the Sunday Mail of Adelaide, S.A., Australia, on March 29,1987:

"Perth: Weird lights and strange phenomena have surrounded the area where two young boys have gone missing near the Gibson Desert. Two men have reported lights, which looked like a big bus or a convoy of trucks in an area where there are no roads, and which seemed to follow them near where teenagers Simon Amos, 17, and James Annetta, 16, went missing."

The article goes on to describe unusual lights, loud noises and circular burn marks in shrubbery reported by four local "level-headed businessmen," who were reluctant to discuss their findings. The large semi-trailer truck in which two of them were driving was followed by a UFO for approximately 50 miles. They had a camera with them, and got a clear picture of the saucer-shaped object. All the reports concerned the area in which the two youths had disappeared.

The second newspaper article appeared in the Sunday Territorian of Darwin, N.T., Australia, on April 12, 1987:

"A 23-year-old Casuarina man has brushed aside fears of being labelled a 'nut case' to give his account of the night he was followed by an unidentified flying object near Kununnurra.

Damien Monck came forward with his account after newspaper reports last month that three men travelling from Lake Gregory station to Kununnurra saw mysterious lights in the area where two teenagers disappeared in December.

"Mr. Monck said he saw a strange light about 90 km. from the area one month before this sighting.

"He said he was with two passengers driving along the main Darwin to Broome highway and was 70 km. from Kununnurra on a clear November night when a bright light followed his car. He said it travelled with the car for about 40 km. and appeared to be coming from a round object travelling at the same speed as his car but at least 4 km. away. It was first spotted by seven-year-old passenger James Newton, who pointed it out to his mother Karen. Mr. Monck ignored Ms. Newton's pleas to turn back and look for the light after it disappeared over a hill 30 km. outside Kununnurra."

It seems appropriate at this point to quote a passage from UFO Contact From Reticulum by Wendelle C. Stevens and William J. Herrmann, which was privately published by Lt. Col. Stevens in 1981. Mr. Herrmann has just been abducted for the second time and has established friendly relations with the short humanoid captain of the spacecraft, who is showing him around the ship:

"I looked at the instruments and the checker-board lit up and grew bright. .. the occupants in the room moved back and forth in an organized pattern ... Again the leader spoke.

'Our velocity is now decreased from 2,000 to 60. You will notice the people below have pulled over and stopped their cars to observe. Our visor scans will bring the facial expressions with five foot scanning.'

"The screen showed a Pinto station-wagon in front and a Buick Electra directly behind it. Both cars were pulled off the road. A lady was standing at the driver's side of the Pinto, and a man was standing on the driver's side of the Buick. The screen then moved up to a close-up picture. I felt less than five feet away! The looks on their faces will be with me for a long time. The leader spoke, 'Do you recall such wonder expressed? We will move forward. Observe the woman get into her car. The man will do likewise.'

"Even though inside the object, I couldn't detect motion. Sure enough, the woman ran around the passenger side and jumped in, locking her door. The man calmly opened his door and got in, locking the door. At the same time the leader spoke, 'A two-foot scanning will show her holding the lock down ...The man will just sit and watch. A useless gesture on their part if we wished to direct observe, but this is not our purpose ...we will now continue on our way.'"

The following is from "Intruders" by Budd Hopkins, published by Random House in 1987:

"Lucille F. wrote that this alien society seemed to be 'millions of years old, of outstanding technology and intellect but not much individuality or warmth.' She had the sense that 'the society was dying, that children were being born and living to a certain age, perhaps preadolescence and then dying.' There was 'a desperate need to survive, to continue their race. It is a culture without touching, feeling, nurturing ... basically intellectual. Something has gone wrong genetically. Whatever their bodies are now, they have evolved from something else. My impression is that they wanted to somehow share their history and achievement and their present difficulties in survival... I saw a child about four feet tall, gray, totally their race, waving its arms ... it was in pain and dying. I was told that this is what is happening now.'.... In this unexpected scenario the UFO occupants – despite their obvious technological superiority – are desperate for both human genetic material and the ability to feel human emotions – particularly maternal emotions. Unlikely though it may seem, it is possible that the very survival of these extra-terrestrials depends upon their success in absorbing chemical and psychological properties received from human abductees.

Don Ecker is the MUFON state section director for Idaho. The following quotation is taken from his article entitled "Report on Human Mutilations," which was published in the July/August 1989 issue of "UFO Magazine."

"According to a recent report just received from Westchester County, NY, researchers have discovered that in a small area of the county, which has been the site of numerous UFO overflights and reported human abductions, over 3,000 missing children reports have surfaced. After extensive investigation by local police departments, these children have not yet been found at centers for young runaways or in red light districts. Researchers and law enforcement officials are baffled."

Le Monde Diplomatique (5 rue Antoine-Bourdelle, 75501 Paris) is a French monthly of the highest quality, featuring sophisticated analysis of political events worldwide, and is remarkably free from bias. In its issue of August 1992, it carried an article by Maite Pinero on abductions of children, not by aliens, but by humans motivated by financial greed, who then sell the kidneys and other body parts of these children that they have either kidnapped or "adopted."

There is an enormous traffic in such organs, particularly in impoverished regions of South America, much of it flowing in the direction of the United States. The article covers a whole series of scandals concerning this traffic that have erupted throughout South America since 1985, and documents a number of occasions when direct or indirect pressure from the U.S. government was used to silence local protests, routinely attributing the incidents to Soviet or Cuban propaganda. One of the recent cases, in February 1992, involved an American lawyer, Patrick Gagel, who was responsible for having exported three thousand children in thirty months from Peru for "adoption," which then disappeared without a trace. Yet when the Peruvian police arrested Gagel, pressure from on high obliged them to release him. Who applied that pressure from on high? The U.S. Embassy? Gagel's operation was but one of many, merely the tip of the iceberg, according to Defense of Children International. Throughout South America, each time local police have made an arrest, the Gagel scenario has been repeated over and over again. Pressure from on high has been exerted, on a number of occasions openly by the U.S. government, and those arrested have been released. The number of children involved is far greater than the number of sick people wealthy enough to pay black market prices for both the transplanted organ and a clandestine transplant operation. Is this type of abominable activity related to the secret treaty with the Grays?

SUGGESTED READING:
EXTRATERRESTRIALS FRIENDS OR FOES
EXTRATERRESTRIALS AMONG US

Chapter 24

MAN KIDNAPPED BY GLOBES
AS REPORTED IN THE PAGES OF THE APRO JOURNAL IN 1962
Report by Olavo T. Fontes, M.D.

The tragic case I am going to relate here was investigated to the limits that any investigation can reach when every piece of evidence is analyzed and evaluated, when every clue is exhaustively followed, and when every conventional explanation is explored and dismissed for lack of proof. In a case of this kind, however, we could only establish a definite conclusion if that conclusion could be negative. We could be sure, in other words, if the body of the victim was found; if facts or motivations in the past life and personality of the victim or witness were demonstrated, showing that they couldn't be trusted, were psychotic, or had reasons to simulate the whole thing; or if additional evidence was not uncovered connecting the facts in the case with the sighting of unconventional aerial objects. As the body was not found, facts or motivations of such a kind were lacking and there was definite evidence concerning the sighting of UAOs – the case must be accepted as possible despite the lack of absolute proof. On the other hand, we must recognize the witness obviously cannot present more than was presented: his report about the mysterious disappearance of his father. You can believe it or not. The absolute truth, that only the witness himself can be sure about it.

The readers may not like this report. For one reason, it will show them that such things as UAOs from other planets could be interested in kidnapping humans, in order to become better acquainted with them. In fact, personal experiences such as the one related here will cause you to wonder about missing people – about how many of the individuals who are yearly reported as missing might have chanced to come upon a UAO in some lonely spot and were captured as specimens.

A noise of running steps, a shadow seeming to float into the dark room, two ball-shaped objects emitting light and discharging a strange yellow smoke, a human being disappearing into that yellow mist before the eyes of his terror-stricken son – such is the fantastic story told by a 12-year-old boy, Raimundo de

UFO HOSTILITIES AND THE EVIL ALIEN AGENDA

The picturesque city of Diamantina

Aleluia Mafra. His father, Rivalino Mafra da Silva, is missing. Raimundo states that he was kidnapped by two strange objects, on the morning of August 20, 1962, just in front of his house, in a place called Duas Pontes, District of Diamantina, State of Minas Gerais.

The sixteen thousand citizens of Diamantina are divided in their opinions about the happening: some believe there was merely a murder with a missing body; others think that Rivalino ran away for some unknown reason and his son is telling a tale to cover him; still others think the whole thing was "the work of the Devil"; many others are certain that the child is telling the truth. On the other hand, the situation is quite different in Biribiri, Mendanha and Rio Vermelho, small villages in the vicinity of the area where the strange events took place. The residents at those places – those who still remain there – are living with panic in their hearts. They do not dare to go outside at night. They do not risk walking alone through the fields.

Their doors and windows are closed after 8 p.m. Their streets are empty and silent at night. Many families are leaving toward Diamantina. You cannot laugh at them – they are too frightened, haunted by the pitiful cries of a child: "They've got my father! I want my father back! Help me!"

They are under the terror of the almost unreal, of something alien and unexplainable, something so different from common sense that your mind is repelled by it. This may sound to you like the science fiction stories you customarily read in science fiction magazines. Perhaps this is best for your sanity – if you don't believe in UAOs from other planets.

RAIMUNDO ALELUIA MAFRA'S REPORT

Rivalino disappeared on August 20. That same day the police were called and the investigation started. Rumors began to spread and reporters were alerted on August 24. The case hit the headlines on August 26. On the evening of August 25, the boy Raimundo was interviewed by the press about the circumstances related with the disappearance of his father.

In spite of his undernourished aspect and obvious anxiety, the youngster was able to give a clear and detailed account of the tragic event. He falters only when forced by direct questions about his father: then he begins to cry. He is only a small boy, who has never attended school and doesn't even know the alphabet. He lives in a small house in a lonely spot, about 28 kilometers from Diamantina. He helps (or helped) his father as the oldest son, taking care of the two small brothers and doing all the house work. His mother died about one year ago. In his ignorance, living in a deserted place outside the civilized world, he has never heard about flying saucers, "space beings" from other planets, comics, or even radio and television.

His report was given in the presence of Lieutenant Wilson Lisboa, police chief at Diamantina. For the twentieth time, according to the information of that authority, he tells about the incredible drama lived by his father before his startled eyes. He says that things started in the night of August 19. The whole family was in bed – himself, his father and his two brothers (Fatimod, 6-years-old, and Dirceu, 2-years-old). He cannot tell the time, because there is no clock in the house, but he was awakened by the sound of steps and got the impression that people were walking hurriedly through the room. He called his father, who lit a small candle.

Then, under the flickering light, they saw a strange silhouette, more like a shadow, floating in the room without touching the floor.

A globe-shaped UFO Sighted above the city.

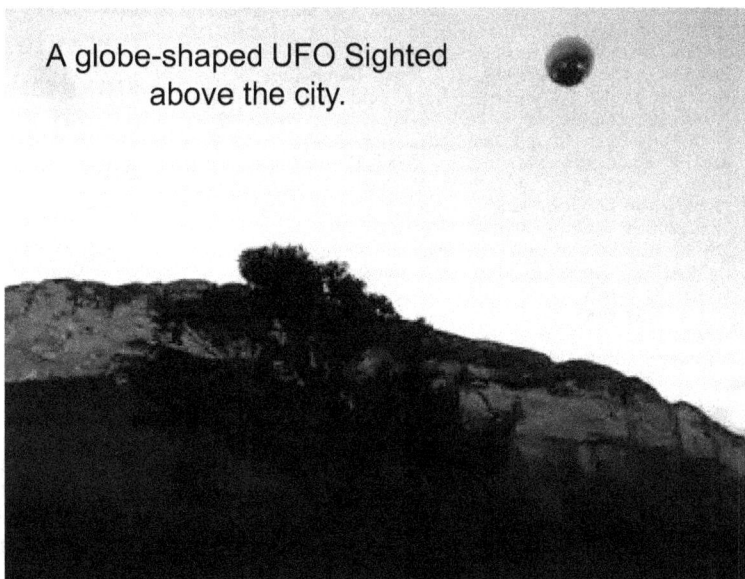

"It was a weird shadow, not looking like ours because it was half the size of a man and not shaped like a human being. We remained in the bed, quiet, and the shadow looked at us; then it moved to the place

Goodness gracious, great balls of fire

Was it a bird? Was it a plane?

No, the bright, speeding object which flashed across the sky south of Hamilton yesterday was probably a fireball, says a leading New Zealand astronomer.

Dr Wayne Orchiston, the director of the Carter Observatory, in Wellington, believes Waikato people who spotted yesterday's UFO actually saw a small rock — the size of a thumbnail — from space, disintegrating in the Earth's atmosphere.

Hamilton's chief air traffic controller Graeme Opie was among several Waikato people who saw the UFO.

Mr Opie saw the object, which had a bright head and a long sparkly orange coloured tail, from the Hamilton airport control tower about 1.20pm.

Dr Orchiston, who has a special interest in meteorites, said the phenomenon was probably a fireball. Fireballs ranged in size from thumbnail to fist size and were much smaller than a meteorite which reached the Earth intact.

Fireballs usually disintegrated in a flash of heat and light at between 25km-50km above the Earth.

He said the observatory received a report of a fireball every couple of months.

Mr Opie said the UFO looked like a meteorite, but did not behave like one. It shot across the sky maintaining a level flight path. It travelled from east to west at a height of about 2500m and roughly 16km south of his position.

Initially Mr Opie thought he may have seen a jet aircraft but he rang radar operators in Auckland and they had no targets on their scopes south of Hamilton.

But yesterday Dr Bob Valkenburg, of the Te Kuiti-based Unexplained Phenomenon Research Society, said there had been a lot of UFO activity in the area recently and he expected it to increase.

"We cannot be alone on this tiny piece of dust in an infinite universe."

Several other people rang the *Waikato Times* to report the incident. Te Kawa couple Gail and Keith Stanley also saw the object as they were driving south along SH3. Mrs Stanley said it caught her attention because it was so bright. She said it had a rainbow-coloured tail.

Two Cambridge schoolboys said they saw what looked like a comet flying through the sky.

Brad Hooker and James Gill, both 9, were playing cricket during lunch at St Peter's Catholic Primary School when they spotted the object.

□ Meanwhile, a woman rang the *Times* yesterday to recount an experience at Horotiu on Sunday. She said a triangular shaped craft hovered only about 5m over her vehicle and followed her towards SH1.

She was reluctant to tell her friends in case they thought she was going "nuts".

where my brothers were sleeping and looked at them for a long time, without touching their bodies. Afterwards, it left our room, crossed the other room and disappeared near the outer door. Again we heard steps of someone running and a voice said: 'This one looks like Rivalino.' My father then yelled: 'Who goes there?' There was no answer. Father left the bed and went to the other room, when the voice asked again if he really was Rivalino. My father answered it was right, that Rivalino was his name, and there was no answer. We came back to bed and heard clearly their talk outside, saying they were going to kill father. My father started to pray aloud and the voices outside said there was no help for him. They talked no more.

"We passed the night awake. In the morning, still afraid, I had the courage to go outside to get my father's horse in the field. But then I sighted two balls floating in mid-air side by side, about two meters from the ground, one meter from each other and a few meters away from our door. They were big. One of them was all black, had a kind of irregular antenna-like protuberance and a small tail. The other was black and white, with the same outlines, with the antenna and everything. They both emitted a humming sound and appeared to give off fire through an opening that flickered like a fire fly, switching the light on and off rapidly. I was frozen by fear. I called father to see those strange flying objects. He came out of the house, still praying and asking about what those things could be, his eyes locked on them. He warned me to stay away and walked toward the objects. He stopped at a distance of two meters. At that moment the two big balls merged into each other. There was only one now, bigger in size, raising dust from the ground and discharging a yellow smoke which darkened the sky. With strange noises, that big ball crept slowly toward my father. I saw him enveloped by the yellow smoke and he disappeared inside it. I ran after him into the yellow cloud, which had an acrid smell. I saw nothing, only that yellow mist around me. I yelled for my father but there was no answer. Everything was silent again.

"Then the yellow smoke dissolved. The balls were gone. My father was gone. The ground below was clean as if the dust had been removed by a big broom. I was confused and desperate. I walked in circles around the house looking for father, but I found no tracks, footprints or marks. Was this the work of the Devil? My father had disappeared in mid-air. I have searched the plains, fields and thickets with no results. I have watched the flight of vultures, looking for clues to locate his body, but I saw nothing. Five days have passed and nothing was found. Is my father dead, taken by the globes? I want my father back."

(Unquote-Belo Horizonte DIARIO DE MINAS, August 26. Rio de Janeiro CORREIO DA MANHA, September 4. Lieutenant Lisboa's report, transcribed verbatim).

THE POLICE INVESTIGATION

Lieutenant Wilson Lisboa, Police chief at Diamantina, was called the same day of the strange disappearance. He put the Raimundo boy under cross-examination but failed to make him change his incredible story. He then decided to make a complete investigation. Policemen were ordered to look for Rivalino's body, to find him at any cost. The search was started inside Rivalino's house. No clue was found. The surrounding country was covered by trained men. At the spot Rivalino had been kidnapped by the objects, according to the boy's report, the ground was clean of dust in an area about five meters in diameter, but no tracks or marks were visible. About fifty meters away, a few drops of blood were found. Lieutenant Lisboa collected samples and the analysis identified them as human blood. The search for the missing body spread through the whole district of Diamantina. It took ten days. Police dogs were sent from Belo Horizonte, bloodhounds trained by the military police to follow tracks and find missing people. They found nothing. A complete investigation of Rivalino's past life, possible love affairs, enemies, friends, relatives, etc., was undertaken. No clues were found.

Another investigation was made concerning the possible sighting of UAOs over the region by other witnesses. The following reports are of interest: (1) The vicar at Diamantina's cathedral, priest Jose Avila Garcia, contacted Lieutenant Lisboa to inform him that, by a strange coincidence, a friend of his had reported an unusual fact the night before Rivalino's disappearance. That friend, Antonio Rocha, was fishing at the Manso River, close to Duas Pontes, when he sighted two

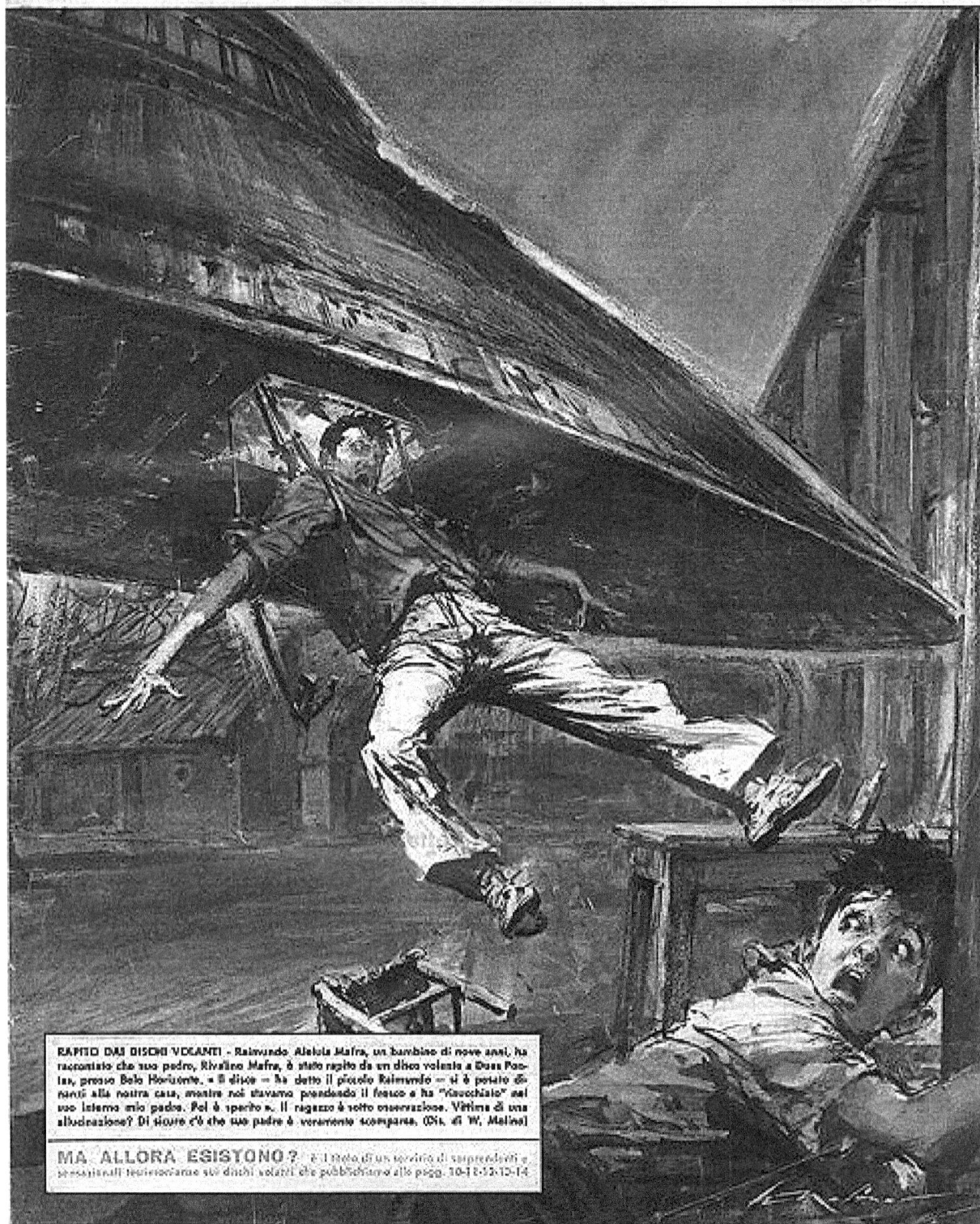

DOMENICA DEL CORRIERE

Anno 64 - N. 39 - L. 50   Settimanale del CORRIERE DELLA SERA   30 settembre 1962

RAPITO DAI DISCHI VOLANTI - Raimundo Aleixo Mafra, un bambino di nove anni, ha raccontato che suo padre, Rivalino Mafra, è stato rapito da un disco volante a Duas Pontas, presso Belo Horizonte. « Il disco — ha detto il piccolo Raimundo — si è posato di fronte alla nostra casa, mentre noi stavamo prendendo il fresco e ha "risucchiato" nel suo interno mio padre. Poi è sparito ». Il ragazzo è sotto osservazione. Vittima di una allucinazione? Di sicuro c'è che suo padre è veramente scomparso. (Dis. di W. Molino)

MA ALLORA ESISTONO? è lo stato di un terrore di sorprendenti e sensazionali testimonianze sui dischi volanti che pubblichiamo alle pagg. 10-11-12-13-14

272

ball-shaped objects hovering over Rivalino's house. Priest Garcia believed that the boy had dreamed and his father had been murdered, but it was his duty to report all facts to the police – even those opposed to his own opinion.

Mr. Antonio Rocha, who worked at the Mail Department in Diamantina, was called and confirmed his story. On the evening of August 19, he was fishing at a place near Rivalino's house. "At 4:00 p.m.," he said to Lieutenant Lisboa, "I sighted two strange ball-shaped objects in the sky. They were flying in circles over Rivalino's house. They came very low and were gone a few minutes later. I don't know anything about Rivalino's disappearance, but from the report given by his son Raimundo, I have the impression he saw the same objects I sighted."

(2) Rivalino Mafra da Silva was a diamond prospector. Lieutenant Lisboa interviewed other prospectors in the district who did their mining work with Rivalino. They told him a startling story. Rivalino had informed them that, on August 17, when going back home, he had seen two strange persons digging a hole in the earth at a spot near his house. When approached, the creatures ran away into the bushes. They were approximately three feet tall. A few moments later, he sighted a strange object which took off from behind the bushes and disappeared into the sky at high speed. According to Rivalino, this object was shaped like a hat and surrounded by a red glow. They didn't believe him, of course.

(3) Doctor Giovani Pereira, a physician living at Diamantina, went to the police to report the sighting of a disc-shaped object over his own house two months before. He had had a night call from a patient and was driving back in his car. When closing the car's door to go inside his home, he suddenly sighted a brilliant object, shaped like a disc, hovering low over his house. He stopped and watched it for several minutes. Then it moved away at high speed after crossing over the sleeping town. He said he had kept the sighting secret because he knew nobody would believe him.

(4) On the morning of August 24, a UAO crossed over the town of Gouveia, about 42 kilometers to the south of Diamantina. The sighting was witnessed by more than fifty people, including the local police chief, Lieutenant Walter Costa Coelho. According to the observers, the object was white colored, shaped like a soccer ball and encircled by a kind of fluorescent glow – remaining in sight for about two minutes. It was traveling to the north (toward Montes Claros), then changed course to the northwest.

A few minutes later, the same (or similar) object was spotted in the sky by more than one hundred citizens at Brasilia de Minas, a small town about 120 kilometers to the northwest of Montes Claros. Again it looked like a soccer ball and was surrounded by a white glow. According to the local priest, this UAO hovered for a long time over the town's church before disappearing to the west at high speed.

UFO HOSTILITIES AND THE EVIL ALIEN AGENDA

This sighting was printed in the papers on August 28. The news had a deleterious effect on the residents of small towns and villages in the same area of Duas Pontes. The coincidence was too much for them: a ball-shaped object, sighted over the same region, just four days after Rivalino's disappearance. The consequence was panic and hysteria.

The last step in Lieutenant Lisboa's investigation was to check and recheck the sanity of Raimundo and the reliability of his report. Raimundo was intensively questioned by the police and, at police request, a psychiatric examination was made by Dr. Juan Antunes de Oliveira. He was found to be sane. Dr. Oliveira then decided to make a last experiment. It was a cruel test, but justified due to the circumstances in the case. In the presence of witnesses, the boy Raimundo Mafra was taken to a room where there was a human body covered by a cloth. "Raimundo," said Dr. Oliveira, "this is the body of your father. He is dead. You lied when you told us that he had been kidnapped. Tell now the whole truth: what really happened on the morning of August 20?" The boy began to cry, in a state of great emotion, but continued to affirm that his story was not a lie, that his father could be dead but had been taken by the two objects. "Perhaps they brought him back dead." was his conclusion. This ended the experiment. Dr. Oliveira was interviewed by the press. He said: "I don't wish to discuss the facts in the case. They are beyond my competence. But I can tell you that the boy is normal and he is telling what he thinks to be the truth."

An attempt was also made to cross-examine the kid applying the technique known as hypo-analysis. The attempt failed because Raimundo was not receptive to hypnosis.

So Lieutenant Lisboa came to the end of his investigation and found himself in a very peculiar and difficult position. He was certain that Rivalino had been murdered, but the corpse had vanished. He had thought that the boy was lying or out of his mind, but failed to get proof showing he couldn't be trusted, was psychotic, or had reasons for simulation. He didn't belief in flying discs or balls, yet the evidence he had collected pointed in that direction. At this point, he decided to do two things: (1) to continue the search for Rivalino's body; (2) to make a written report to the Secretary of Public Security of the State and send Raimundo Fafra to Belo Horizonte, the State capital.

RAIMUNDO MAFRO GOES TO BELO HORIZONTE

On August 30, the boy arrived at Belo Horizonte. He had a companion, Mr. Antonio de Carvalho Cruz, the Commissioner of the State Child's Department at Diamantina, who had the mission of taking the boy to the proper authorities in the capital. At that moment, general curiosity had been aroused about the kid, and he was interviewed by the whole press and even appeared on a television program. Then Colonel Mauro Gouveia, Secretary of Public Security in the State of Minas

Gerais, took charge of the case. Raimundo was questioned, cross-examined, photographed and again submitted to medical and psychiatric examination. Three days later, he was taken into custody by military authorities. An Air Force plane took him to Rio de Janeiro, where he disappeared behind the protection of a tight security ring. No one knows where he is now.

EPILOGUE

A month after the mysterious disappearance of Rivalino Mafra da Silva, the police at Diamantina decided to stop their investigation and to close the case. The body is still missing and every effort to find new clues has met with complete failure. Lieutenant Lisboa and his policemen are depressed. One of them, policeman Clemente, said to the press: "Nobody expects to find a satisfactory explanation with respect to Mr. Rivalino's disappearance." (Belo Horizonte ULTIMA HORA, September 22, 1962).

At this point, it would be preferable merely to present the evidence and to allow the reader to draw his own conclusion; but I find it necessary to call attention to a very important thing: Reports indicate space creatures have been investigating the Earth closely for many years. They appear to follow a very methodical plan, step by step. Available evidence indicates they have already taken specimens of terrestrial flora, water, rocks and soil. An investigation of the fauna was apparently lacking, as far as the collection of specimens is concerned. However, would it not be logical that they eventually turn their attention to collecting specimens of fauna? It would be surprising, indeed, if they overlooked the most interesting example of terrestrial fauna – man himself.

The case of Rivalino Mafra da Silva appears to be the first one – in the whole UAO history – where vanishing people and UAOs are definitely connected by direct evidence. Therefore, in spite of some incredible details we cannot explain (i.e., the happenings at night inside and outside Rivalino's house), I am forced to conclude he was kidnapped by two ball-shaped UAOs – in the presence of his own son. There is no other alternative. And, as they always repeat their moves, it is reasonable to expect that other similar cases will happen soon.

Elijah being summoned to "heaven" by an extraterresrtrial

Human Mutilations
And Death At The
"Hands" Of Outsiders

Chapter 25

ARE UFOs RESPONSIBLE
FOR MYSTERIOUS HUMAN MUTILATIONS?
By Tim R. Swartz

On August 4, 2002, 39-year-old Todd J. Sees left his house just after 5:00 AM to take a quick trip up Montour Ridge near Northumberland, Pennsylvania. He was driving an ATV (all-terrain vehicle) to assess the local deer population before hunting season began. He told his wife that he would be back by noon. When Todd failed to return, his 18-year-oid son went out to search for him. At the top of the ridge, he found the abandoned ATV, but there was no sign of Todd.

When the police were called in to the search, a massive volunteer search team was formed. With the assistance of search dogs, the search party scoured over six miles of the ridge, the surrounding area around the Sees' residence and even a pond that was less than 200 yards away from the home was searched using divers. The search lasted for approximately 36 hours, and with the exception of one of his boots found 50 feet up in the branches of a tree, no trace of Todd was found.

On the morning of August 6, a family member noticed something white in the bushes around the pond. For almost a half an hour, police and rescue workers hacked away at brush and small trees to finally reach what turned out to be the body of Todd Sees.

Sees, clad only in his underwear, was emaciated to the point of actually being shrunken, and had a dark burn patch on his left temple. The police report stated that the corpse had a "look of horror on its face." Reportedly, the FBI became involved in the case, refusing to allow Sees' wife to identify his body. To further complicate matters, the funeral home said that Sees' body was so decayed that it would only allow a closed casket funeral. This was especially odd considering that the volunteers who found Sees' body did not report any unusual deterioration.

The police eventually released a statement that Todd Sees had died accidently from "cocaine toxicity." This result did not sit well with Sees' family and friends, who argued that Todd was not a cocaine user. Investigators Butch Witkowski and Lon Strickler have conducted exhaustive research into this unusual case. Lon's website, Phantoms and Monsters, details several UFO reports in the Montour Ridge area that had been submitted to Peter Davenport, President of the National UFO Reporting Center.

"The craft was seen on Montour Ridge on August 04th, 2002 at 5:30 in the morning by a farmer and also 3 fishermen on the Susquehanna River. It was at the very top of the mountain on the western end, just hovering above the power lines. The farmer said it appeared out of nowhere, he could see some sparks flying from the tower and dropping to the ground. The whole incident lasted about 10 minutes, then it (the object) got very bright and took off very low heading west, then it stopped and went straight up, it was gone in seconds. The horses on the farm were very upset and nervous for about 12 hours.

"Farmer saw an object above Montour Ridge at the power lines, on top of the mountain; it was round and very still over the lines. Suddenly it moved what looked like a few hundred feet to the east. It stopped and a beam of blue and white light shined to the ground. What was seen next was very unnerving: he saw what appeared to be a man suspended in the light, he was being pulled up headfirst, he was moving his arms slowly in the light. What looked like a man was pulled up into the bottom of the craft. A few seconds later and it started shuddering, then went west very fast stopped briefly, then went straight up and out of sight."

This unsettling incident is just one of many similar cases that involved seemingly deadly encounters between people and UFOs. There could actually be many more that have been covered up because the reality of such encounters is just too disturbing to contemplate.

Researchers of the UFO phenomenon have openly speculated that if UFOs do represent extraterrestrial visitors, why do they operate in such secrecy? UFO contactees and some experiencers say that the extraterrestrials are friendly and are here to help mankind. However, there is compelling evidence that there is a sinister, hidden agenda being perpetrated by non-human intelligence against planet Earth.

BLOOD IS THE LIFE

Over the years, strange attacks on animals and humans have been recorded and attributed to predators, other humans and even supernatural entities such as vampires. What makes these incidents stand out is the general lack of blood found on the bodies. Primitive man believed that blood was sacred, the source of life in all creatures. When you lost your blood, you lost your life. So it made sense that the life force must be contained in blood. The Old Testament is a good example of ancient beliefs regarding blood. Leviticus 17:14 states, that "the life of every living creature is its blood." The verse goes on to say that it is forbidden for anyone to eat blood because it is the source of all life.

Because of these early beliefs, man has always had a superstitious horror when dealing with unusual attacks that involve the loss of blood. Throughout history, there have been numerous reports of strange attacks and mutilations that seem to go beyond normal animal predators. In 1874 near Cavan, Ireland, for several months something killed as many as thirty sheep a night, biting their throats and draining the blood. In 1905 at Great Badminton, Avon, sheep were again the target for attacks. A police sergeant in Gloucestershire was quoted in the London Daily Mail, "I have seen two of the carcasses myself and can say definitely that it is impossible for it to be the work of a dog. Dogs are not vampires, and do not suck the blood of a sheep, and leave the flesh almost untouched."

In a single night in March of 1906, near the town of Guildford, Great Britain, fifty-one sheep were killed when their blood was drained from bite wounds to the throats. Local residents formed posses to hunt down whatever was killing their livestock, but nothing was ever caught, and the killings remain a mystery. Events of this kind have probably occurred regularly throughout history. The cases that have received media attention are those involving a large number of deaths, but there are probably hundreds of smaller attacks that have gone unnoticed over the years.

These strange livestock attacks are eerily similar to the attacks by the so-called Chupacabra, which means "goat sucker." Confining itself chiefly to the southern hemisphere, the Chupacabra has been blamed for numerous attacks on small animals and even some people. The animals have had their throats bitten and their blood sucked out by the creature that reportedly stands on two legs, has large black or red eyes and is about four feet tall. Unlike past killings, the Chupacabra has been seen by shocked eyewitnesses whose descriptions seem to describe an animal that superficially resembles the "Grays" of flying saucer lore. As in past cases, attempts to track down the Chupacabra have met with failure. If history is any indication, the Chupacabra will never be caught, and the strange events will remain a mystery. It is as if the mystery mutilators appear out of thin air, do their damage, and then, just as quickly, disappear again.

WHAT IS KILLING OUR CATTLE AND WHY?

Before we go further into the human mutilation mystery, it is crucial to examine, albeit briefly, cattle mutilations and how there may be similarities with human cases. The mutilation of cattle seems to involve a different set of circumstances than past vampire-like attacks on livestock. While cattle mutilations almost always involve the complete draining of blood, physical mutilation of the flesh is so apparent that seasoned ranchers are shocked by the unusual nature of

Evidence at human mutilation sites prove that the techniques are identical to those of animal mutilations.

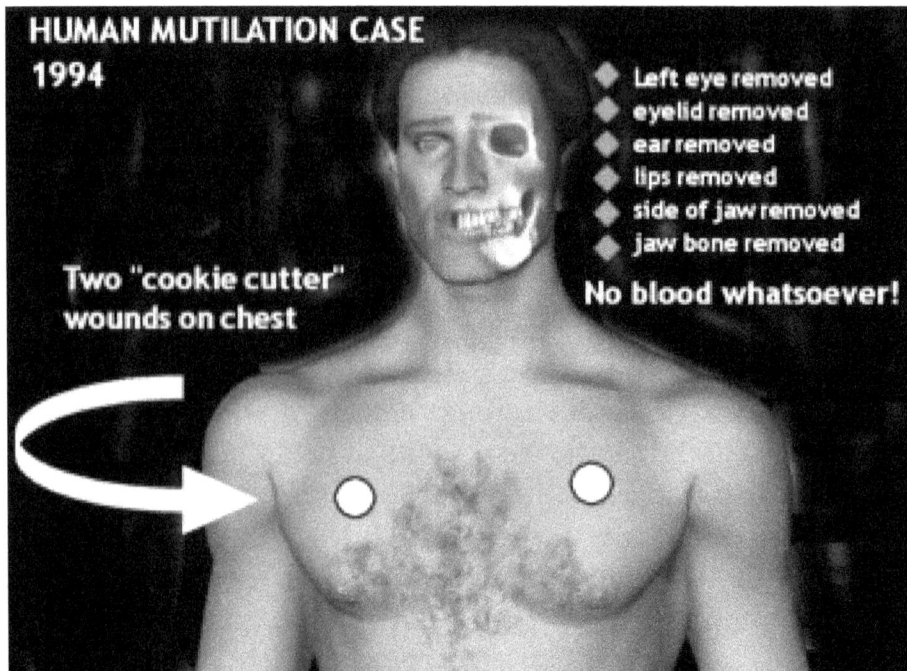

HUMAN MUTILATION CASE
1994

◆ Left eye removed
◆ eyelid removed
◆ ear removed
◆ lips removed
◆ side of jaw removed
◆ jaw bone removed

Two "cookie cutter" wounds on chest

No blood whatsoever!

This chart shows areas of mutilation performed upon one man.

the deaths.

No one really knows when the first unusual cattle mutilations began. Records show that in the middle of 1963, a series of livestock attacks occurred in Haskell County, Texas. In a typical case, an Angus bull was found with its throat slashed and a saucer-sized wound in its stomach. The attacks were attributed to a wild beast of some sort, a "vanishing varmint." As the attacks continued through the Haskell County area, the unknown attacker assumed mythic proportions and a new name was created, "The Haskell Rascal." Whatever was responsible for the mutilations was never caught, and the attacks slowly stopped.

Throughout the following decade there would be similar reports of attacks on livestock. In 1973 the modern cattle mutilation wave can be said to have begun in earnest. It is interesting to note that a huge UFO flap was occurring across the country in 1973, with many sightings taking place in the same areas that cattle mutilations were happening. In November of 1974, rumors began to connect the sighting of UFOs with mutilated cows that were being found in large numbers in various Minnesota counties. Dozens of UFOs were reported in Minnesota and dozens of cattle were found dead and mutilated. Although the sightings and mutilations were never correlated, many felt that the number of sightings was added proof that the UFOs were somehow involved.

In 1975, an unprecedented onslaught of strange deaths spread across the western two-thirds of the United States. Mutilation reports peaked in that year,

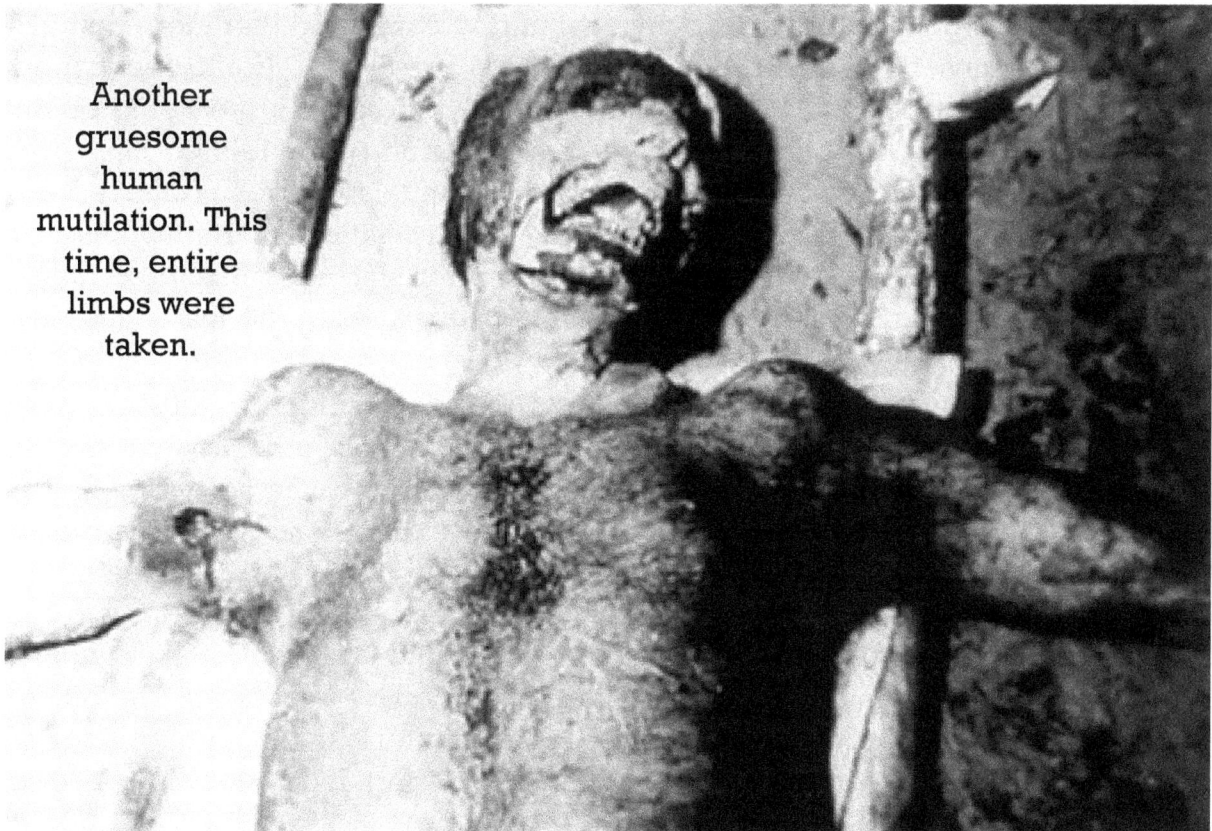

Another gruesome human mutilation. This time, entire limbs were taken.

accompanied by accounts of UFOs and unidentified helicopters.

In an attempt to explain the strange nature of the killings, wild stories and rumors swirled among investigators. One that stood out was that the aliens reportedly absorb nutrients through their skin. They harvest the blood of cattle which is then "painted" on their skin, allowing absorption of the required nutrients. Supposedly human blood is preferred by the aliens, but cattle blood can be altered to serve the same purpose.

If the stories are true, some would ask instead of animals, why aren't the aliens catching and mutilating humans? The truth could be that human mutilations and deaths are occurring on a regular basis, but that the stories are too horrible to contemplate. If murderous, UFO-related human mutilations have taken place, they have either gone unrecognized for what they really are, or have been adeptly covered up by official intervention.

A GRISLY COVER-UP

Thousands of people worldwide disappear every year, never to be seen again. An entire series of books, "Missing 411" goes into great detail examining bizarre missing people cases, starting at first in U.S. National Parks, but also in more populated and even urban areas. Even though author David Paulides avoids stating any conclusions about these cases, the extremely mysterious nature of many of these disappearances seems to go far beyond the more mundane explanations such as homicides or other common situations.

Some disappearances are so unusual and unexplained that more disturbing scenarios must be examined. In 1956 at the White Sands Missile Test Range, an Air Force major reported that he had witnessed a disk-shaped flying object kidnap Sgt. Jonathan P. Louette. After three days Louette's mutilated body was found in the dessert near the test range. Louette's genitals had been removed and his rectum cored out with surgical precision. Like many cattle mutilations, Louette's eyes had been removed and all of his blood was missing. The Air Force filed a report stating that Sgt. Louette had died of exposure after being lost in the dessert.

The late Leonard H. Stringfield, a former Air Force intelligence officer, wrote in his self-published book, UFO Crash/Retrievals, Status Report No. 6, about the testimony given by a "high ranking Army officer" whom Stringfield says he has known for several years and who is allegedly a "straight shooter." The officer claimed that while he was in Cambodia during the Vietnam War, his Special Operations group was involved in a fire fight with aliens, whom the soldiers came across sorting human body parts and sealing them into large bins. Subsequently the unit was held for several days and interrogated under hypnosis. The officer claimed that he and his men were given cover memories which only began to surface years later. The implications here are staggering. If this story is true, then

UFO HOSTILITIES AND THE EVIL ALIEN AGENDA

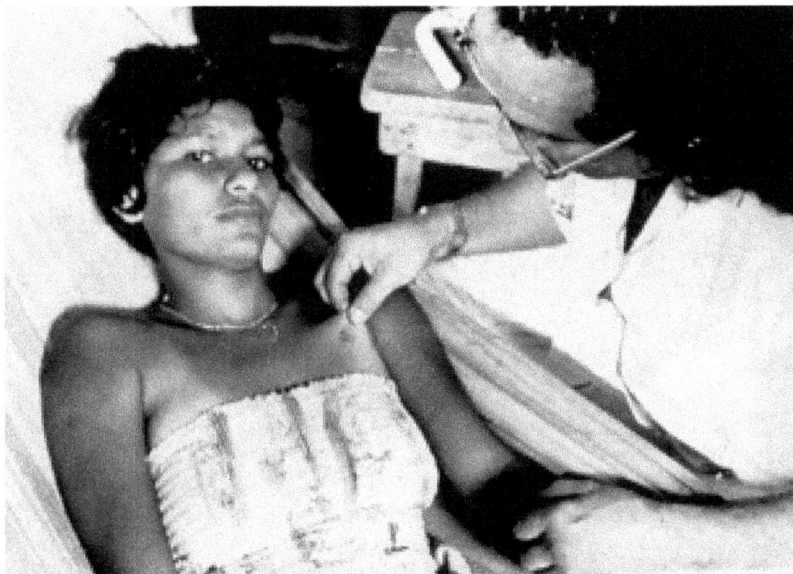

This young lady was reportedly bitten by a Chupa

the possibility exists that military and government officials are aware of the alien's interests in the physiological makeup of the human body.

Don Ecker, at one time, took a special interest in human mutilation cases. In his article "The Human Mutilation Factor," Ecker lists several cases of "overt or covert hostility on the part of UFOs." In 1989 the mysterious death of a man a decade earlier came to the attention of the MUFON State Director of Idaho, Don Mason. According to the report, in 1979, two hunters in the Bliss and Jerome area of Idaho stumbled across the almost nude body of a man that had been hideously mutilated. The body's sexual organs had been removed, its lips were sliced off, and the blood had been drained. Although the body was found in very rugged country, its bare feet were not marked, and no other tracks, animal or human were evident. After the police were notified, an intensive search was mounted and the man's possessions were recovered miles from where the body was found. No one knows how the body ended up where it was found, or even more importantly, what happened to him. It should be noted that this area, over the years, had had many unexplained UFO reports and cattle mutilations.

Ecker also uncovered a conspiracy of silence that occurred in Westchester county New York, in 1988, concerning bizarre late night break-ins at several morgues. Fresh human bodies had undergone mutilations involving partial removal of the face and total removal of the eyes, stomach, thyroid gland and genitals. An assistant medical examiner, who had broken the silence concerning the case, stated that checks were immediately run on the employees who were on duty at the morgues. No links connecting morgue employees with the crimes were found. While there is no evidence that UFOs were responsible for the bizarre incidents, once again we see human bodies being mutilated in the same ways that cattle and other animals are being mutilated.

Ecker admits that he had been warned by a "prominent UFOlogist," that there was a lid "screwed down tighter than you would believe in regards to human mutes." Despite his efforts to unscrew this lid, Ecker was stonewalled at almost every step into his investigations.

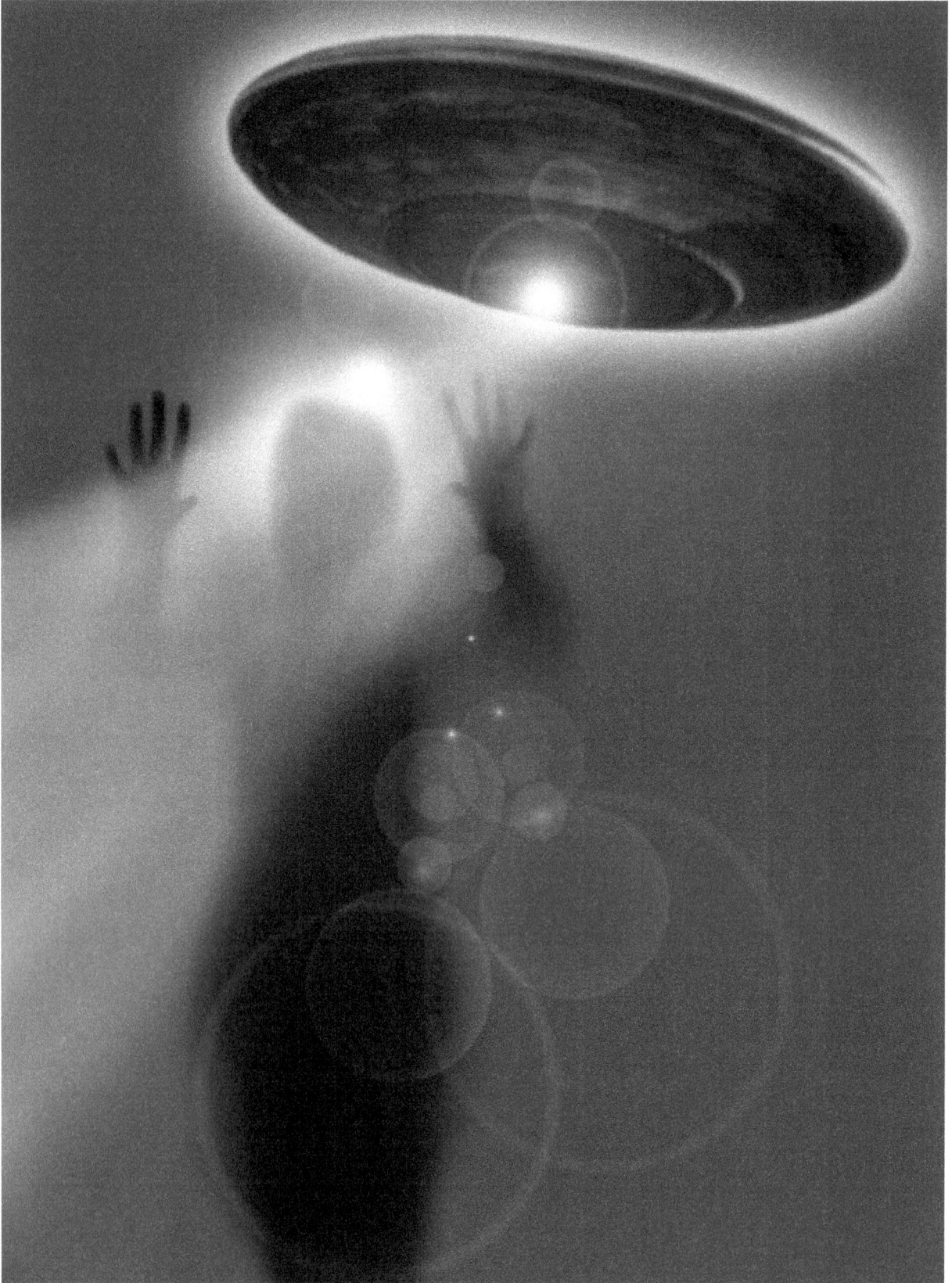

Contactees and abductees claim that the saucer beings can create false thoughts and "screen images" to conceal what really occurred during the event.

UFO HOSTILITIES AND THE EVIL ALIEN AGENDA

THE GUARAPIRANGA RESERVOIR INCIDENT

Another interesting case that has received little publicity in the United States is the Brazilian Guarapiranga reservoir case. Brazilian ufologist Encarnacion Zapata Garcia and Dr. Rubens Goes uncovered a series of sensational photographs obtained from police files. The photos are of a dead man whose injuries are similar to the wounds of countless UFO-related animal mutilation cases. The body had been found near Guarapiranga reservoir on September 29, 1988. The name of the man has been withheld from the media and UFO investigators at the request of his relatives. After studying the photos, Encarnacion Garcia was impressed with how similar the wounds of the body were to those found on the carcasses of so many mutilated animals. The initial police report noted that the body, although extremely mutilated, showed no signs of struggle or the application of bondage of any kind.

The body appeared to be in good condition. Rigor mortis had not set in and it was estimated that the victim had been killed approximately 48 to 72 hours previously. There were no signs of animal predation or decay which might be expected. Strangely, there was no odor to the body. Bleeding from the wounds had been minimal. In fact, it was noted that there was a general lack of blood found in the body or on the ground around the body. Police photos show that the flesh and lips had been removed from around the mouth, as is common in cattle and other animal mutilations. An autopsy report stated that "the eyes and ears were also removed and the mouth cavity was emptied." Removal of these body parts, including the tongue as here, is common enough in animal mutilation cases.

The "surgery" appeared to have been done by someone familiar with surgical procedures. The lack of profuse bleeding suggested the use of a laser-like instrument producing heat, thus immediately cauterizing the edge of the wounds. The autopsy report states that, "The axillary (arm-pit) regions on both sides showed soft spots where organs had been removed. Incisions were made on the face, internal thorax, abdomen, legs, arms, and chest. Shoulders and arms have perforations of one to one and a half inches in diameter where tissue and muscles were extracted. The edges of the perforations were uniform and so was their size. The chest had shrunk due to the removal of internal organs."

The autopsy report continues, "You also find the removal of the belly button leaving a 1.5 inch hole in the abdomen and a depressed abdominal cavity showing the removal of the intestines."

The report also noted the victim's scrotum had been removed, and that the anal orifice had been extracted with a large incision about three to six inches in diameter.

It is significant that the police and medical examiners were convinced the holes found in the head, arms, stomach, anus and legs were not produced by bullet wounds. What is most disturbing about the anal incision and the extraction of anal and digestive tract tissue is that it is a carbon copy of the surgery seen in so many UFO-related animal mutilation cases.

While no evidence linking the Guarapiranga reservoir mutilation case with UFOs has been found, Brazilian ufologists and police have hinted that there may be at least a dozen or more cases similar to this one.

A human mutilation disturbingly like the Guarapiranga reservoir case allegedly occurred in the U.S. and was reported by New Zealand researcher Karen Lyster on her website www.karenlyster.com/humanmute.html<http://www.karenlyster.com/humanmute.html>. Lyster writes that in 1994, she was approached by a physician who told her that he had recently come back from a trip to the United States where during that time he had stayed with another physician friend who worked for the U.S. Military.

Lyster was told that the U.S. physician was notified late one night that there had been an accident and a body was brought in to him on base so that he could conduct a preliminary examination. The doctor detailed on how the body was brought in with horrific injuries the likes he had never seen before. He said the body had been mutilated...an eye was missing, an ear. Photos had been taken of the victim and these photos accompanied the body. Two of these photos the physician had managed to secure for himself.

The physician talking to Lyster then produced the two photos that his friend had given him, asking that he "hold onto" them in case anything happened to him. Lyster said that the first photo was the full frontal of a white male and the second photo was of the same male turned onto his stomach.

The body had been photographed lying in a grassy field. It was difficult to determine how old the victim was as the face had been badly disfigured. However, due to the physical appearance of the body, she estimates that he was possibly 20 to 35 years old.

The body's left eye had been surgically removed, including the eyelid. The lips had been removed in a very clean and precise cut, so had the left ear. The left-hand side of the jaw had been removed, taking part of the jaw bone and teeth with it. The incision along the jaw line had extended past the removed segment of jaw bone to include a further strip of tissue that stopped about two and a half inches from the base of the chin. No blood was seen around the wounds or on the ground around the corpse.

The torso had two very clean, round areas that had been removed from the chest area - these would have measured approx two inches in diameter. The sexual organs had been completely removed, and again, no blood was visible.

The second photo, of the body on its stomach, showed that the entire rectum had been completed cored out, leaving a large cavity. Lyster remembers saying "God, I hope this poor bastard was dead before they did this."

UFO HOSTILITIES AND THE EVIL ALIEN AGENDA

Of course that leaves the question of just who "they" are.

SPECIAL NATO UFO TASK FORCE

It has long been rumored that there is a special "UFO Task Force," probably international in nature, whose job is to be quickly dispatched to possible UFO hotspots/landing areas. The 60s television show by Gerry Anderson, "UFO," dealt with this very premise. In the mid-1990s, reporter Derrick Gough came in possession of documents and information via an anonymous UK military source who worked on secret projects for NATO and the US military. Gough's informant revealed that The United Kingdom, under the authorization of Margaret Thatcher and in collaboration with President Ronald Reagan, created a specialized task force that serves as a branch of NATO. This task force is referred to in documentation as "Group 5 8 Security."

Group 5 8 Security's main function is to serve as a "deploy and secure" operation. The group is dispatched to a location to stop civilians, or the local police, from wandering into the area until a military presence with scientists and doctors arrive. At first glance this special task force sounds like the Navy Seals, designed to secure places that may have a suspected terrorist presence. However, Gough's whistleblower said that the true nature of Group 5 8 Security was to handle sites where human beings had been discovered surgically mutilated like cattle.

The military informant told Gough that he was part of a five-man-team who would be dispatched to anywhere in the world within 45 minutes notice of an incident occurring. During his time with the group, he visited many European locations, Russia and the U.S. His first incident, however, occurred at the Brecon Beacons in South Wales. He said that the team quickly found two mutilated corpses, a man and a woman, lying side by side. They both had their genitals removed and the man was skinned. The woman had her breasts cut off and later investigation revealed that twelve inches of the inside of her anus had also been removed.

Gough's informant was told that the victims had been killed by "the enemy," but no one had any idea as to why this had happened. He was also told that the mutilated individuals would be placed on a missing persons list indefinitely, with their families having no idea as to their ultimate fate. He mentioned how at every scene the bodies were always laid out in perfect lines. At one scene the victim's arms and legs had been removed and placed adjacent to the bodies, though in some areas the right and left arms/legs were switched. At one location they discovered the remains of more than 30 mutilated individuals.

The military source described to Gough other incidents where it had appeared that UFOs had crashed. Amongst the wreckage was discovered not only human and animal body parts, but also the corpses of "extraterrestrials." He describes them as looking nothing like the typical depictions of aliens in popular culture, and how he and his companions called them "demons" due to how they looked. He said that they appear very physically strong, saying that they could

UFO HOSTILITIES AND THE EVIL ALIEN AGENDA

"rip you apart."

During his time working for Group 5 8 Security he found out how NATO knew in advance where an incident site would be. Apparently, NATO has numerous sites around Europe where they monitor for UFO activity. If possible, they then shoot these craft down using laser-based weapons. Two such weapons are located in the UK, one is in Germany and the others are in the United States.

Ufologist Richard D. Hall, after hearing Derrick Gough's story, took it upon himself to look further into the bizarre case. Hall was able to meet with Gough's informant and the man confirmed that he was the military source, offered his official NATO identification and retold all the events in perfect fashion, erasing all doubt from Hall's mind.

After this, Hall produced a two-hour documentary called "UFOs and NATO: The Human Mutilation Coverup" which is available on YouTube. The film offers dates and times, the recording of the original meeting with Derek Gough, discussions with other experts, photos of human mutilation victims and more.

According to Hall, the Group 5 8 Security is still active, meaning that whatever is causing these gruesome murders and mutilations is also still active.

Richard Hall's documentary can be seen here:

www.youtube.com/watch?v=45hL8e-QuXQ<http://www.youtube.com/watch?v=45hL8e-QuXQ>

ATTACK OF THE CHUPA-CHUPA

Possibly one of the reasons that NATO initiated its secret UFO task force was the "Chupa-Chupa" incidents in Brazil. The July 12, 1977 edition of the "JORNAL DA BAHIA" reported that, "A fantastic story of a flying object emitting a strong light and sucking blood from people circulated from mouth to mouth among the population of the counties of Braganca, Vizeu and Augusto Correa in Para, where many people fear leaving their homes during the night so they won't get caught by the vampire-like light from the strange object which, according to our information, already has caused the death of two men. No one knows how the story started, but the truth is that it reached Bele'm and grabbed headlines in the local newspapers."

Months later, on October 8, the newspaper "O LIBERAL" launched the first in a series of reports, about the Chupa-Chupa (suck-suck) phenomenon. "Sucking animal attacks men and women in the village of Vigia: A strange phenomenon has been occurring for several weeks in the village of Vigia, more exactly in the Vila Santo Antonio do Imbituba about seven kilometers from highway PA-140, with the appearance of an object which focuses a white light over people, immobilizing them for around an hour, and sucks the breasts of the women leaving them bleeding. The object, known by the locals as 'Bicho Voador' (Flying Animal), or 'Bicho Sugador' (Sucking Animal), has the shape of a rounded ship and attacks

Messages and prophecies received during contact and abduction appear to be geared toward resolving the mystery of humanity's place in the cosmos.

people in isolation.

"One of the victims, among many in the area, was 46-year-old Mrs. Rosita Ferreira, a resident of Ramal do Triunfo. Ferreira a few days ago was sucked by the light on the left breast and passed out. Increasingly it looked like she was dealing with a nightmare, feeling as if there were some claws trying to hold her. She was attacked around 3:30 in the morning. Another victim was the lady known as 'Chiquita,' who was also sucked by the strange object with her breast becoming bloody, but without leaving any marks."

THE CASE OF THE MELTING MAN

The Chupa-Chupa attacks bear an uncanny resemblance to a 1946 incident that happened in the tiny Brazilian village of Aracariguama. The newly revealed details of this story are from Pablo Villarubia Mauso, translated by Scott Corrales, Institute of Hispanic Ufology.

On March 4, 44-year-old farmer Joao Prestes Filho had spent the day fishing while his wife and children were attending a carnival in a nearby town. That evening when he returned to his empty house, he found that he was locked out and had to enter through a window. Suddenly, something resembling fire or a "fiery torch" entered the room in which he was standing. Joao felt his body burning and ran into the street to his sister María's house, near the Aracariguama church.

Prestes's second cousin, Vergílio Francisco Alves, remembers clearly what had happened to his cousin.

"When I got to María's house, I found the sheriff speaking with Joao. He was in bed and having problems using his tongue. His skin, which was fair, was toasted, reddish, as if he'd been roasted. His hands and face had the worst burns. The hands were twisted. His hair didn't burn, nor did his feet or clothing. He was only burned from the waist up."

Joao's brother, Roque, who was a deputy policeman at Santana de Parnaíba, soon arrived and helped take him to the Santana de Parnaíba hospital. Roque said that his brother was in "great pain" from his burns and was having difficulty talking. A few hours after arriving at the hospital, Joao Filho died of cardiac arrest. Even as he was dying, Joao kept repeating that the light had attacked him and that it was "otherworldly." Subsequent reports stated that by the time he had reached the hospital, his flesh was melting to the point of falling off his bones. However, those relatives who were with him when he passed away say that those stories weren't true.

However, an interesting side story to this strange case came from Prestes's second cousin, Vergílio Francisco Alves.

"When Joao was a tropero (cattle driver), he was still young and lived with his father in Aracariguama. One day at sundown, as he led the donkeys over a hill, he saw a fire that fell from the sky - a fireball. He was near a chapel that had a cross, and he could feel the fireball passing him, almost knocking him down. Joao would tell me that at that spot you could sometimes see ten or twelve balls emerging from the sky. Some of them were red, others Moon-colored. Sometimes five or six of them would fall to the ground and explode. People would call them the boitata lights."

According to the Portuguese colonial chronicles and the stories of Canarian priest José de Anchieta from the 16th century, the word "boitata" is of native origin and means mysterious lights that would pursue and even kill the native Indians.

Compared to reports of mysterious animal attacks and mutilations, reports involving humans are somewhat rare. The probable reason is that many such incidents involving people are not recognized for what they are. The possibility is that a massive cover-up by officials world-wide exists to hide the fact that something is preying on humans. If we consider that extraterrestrials are visiting Earth, the likely reason for such visitations is scientific exploration. Consider that with billions of galaxies and the likelihood that there are multitudes of different kinds of life scattered across the universe, the Earth is just another source of specimens for extraterrestrial scientists to gather and study. While man's ego would like to think that we are special in the universe, the hard reality could be that we are just curiosities to be collected, studied, and possibly exploited, and then finally pickled in a jar someplace with the notation: HUMAN, MOSTLY HARMLESS.

A German Hanebau saucer
paces a B-17 and a B-25

Chapter 26

MORE FRIGHTENING CASE HISTORIES
By Sean Casteel

One of the better known investigators of the negative, hostile aspects of the UFO phenomenon is Bill English, son of an Arizona state legislator and himself a former captain in the Green Berets. While in the army, English was assigned to a British Air Force "listening post" north of London as an information analyst. In the course of his duties, he was asked to prepare an analysis of the elusive GRUDGE 13 report.

GRUDGE was the precursor to PROJECT BLUE BOOK, the U.S. military's first repository of reports by the USAF's investigation into UFOs. Reports one through twelve of GRUDGE/BLUE BOOK were generally innocuous and contained no classified or truly sensitive material.

But report number 13 was something altogether different. Many years after he laid eyes on it, English dictated two audio cassettes outlining what he remembered from the GRUDGE 13 report. The report included some bizarre case histories, one of which was called "Darlington Farm Case," out of Ohio, which took place in October 1953.

A man, his wife and 13-year-old son were sitting at the dinner table when the lights began to grow dim. Dogs and animals raised a ruckus outside and the boy got up to see what was going on. He called his parents to come look at the strange light in the sky.

As the parents went out onto the porch, one of the family's dogs broke loose from its leash and began to run. The boy chased it into an open field, and, as the parents watched, the light came down from the sky. They described it as a round ball of fire; it began to hover over the boy and dog. The mother and father heard the boy scream for help, whereupon the father grabbed his shotgun by the front door and ran into the field, with the mother following. When the father got to the field, he saw his son being carried away by what looked like little men into a huge fiery-looking object. As it took off, the father fired several rounds at the object, to no avail.

UFO HOSTILITIES AND THE EVIL ALIEN AGENDA

The dog was found with its head crushed, but there was no sign of the boy and no footprints left by the little men who carried him off. The parents called the Darlington police, who officially reported that the boy had simply gotten lost in the forest which bordered the farm. Within 48 hours, however, the Air Force ordered that the parents be "relocated" to a secret site for the sake of national security. The parents were in a severe state of shock and required a great deal of psychotherapy and "deprogramming."

According to the report, as recalled by English, there are at least four relocation sites across the United States, which include extensive medical facilities to deal with all medical emergencies, to include radiation poisoning.

Another case history in the GRUDGE 13 report gets even more frightening. A case was witnessed by Air Force personnel in which an Air Force Sgt. E-6 by the name of Jonathan P. Lovette was observed being taken captive aboard what appeared to be a UFO at the White Sands Missile Test Range in New Mexico. The abduction took place in March of 1956 at about 3 am local time and was witnessed by Major William Cunningham of the USAF Missile Command near Holloman Air Force Base.

Cunningham and Lovette were out in a field downrange from the launch sites looking for debris from a missile test when Lovette went over the ridge of a small sand dune and was out of sight for a time. Major Cunningham next heard Sgt. Lovette scream in terror and agony. Thinking the sergeant had been bitten by a snake, Cunningham ran over the crest of the dune and saw Lovette being dragged into what the major described as a silvery disc-like object hovering about 15 to 20 feet in the air above.

A long, snake-like object was wrapped around the sergeant's legs, dragging him into the craft. Cunningham stood frozen in shock as the UFO rose quickly into the sky. The major used his jeep radio to report the incident to Missile Control, who confirmed seeing the object on radar. Search parties were sent into the desert to find Sgt. Lovette while Major Cunningham was admitted to the White Sands dispensary for observation.

After three days of searching, Lovette's nude body was found approximately ten miles downrange. The body had been mutilated; the tongue had been removed from the lower portion of the jaw. An incision had been made just under the tip of the chin and extended all the way back to the esophagus and larynx. He had been emasculated and his eyes had been removed. Also, his anus had been removed, and the report commented that the procedures had all been done with practiced surgical skill.

Also noted was that a number of dead predatory-type birds were found in the area around Lovette's body. The birds had apparently died trying to consume the sergeant's corpse. The report included several grisly black and white photos.

UFO HOSTILITIES AND THE EVIL ALIEN AGENDA

English continued to stand by his story of reading the infamous Report Number 13 in spite of years of ridicule and denial from doubters and naysayers.

* * * *

Jacques Vallee is one of the best known, most respected researchers in the UFO community. In his 1990 book, "Confrontations," he recounted several UFO sightings that had resulted in the deaths of the people who had witnessed them.

Even though the following case has been covered extensively elsewhere, it is worth repeating in order to have the preeminent Vallee offer up his personal take on what took place.

The event happened in August 1966 in Rio de Janeiro in Brazil. A local teenager came upon the corpses of two men while looking for his kite. The scene was puzzling. Both men lay on their backs with no signs of having put up a struggle. Both were dressed in neat raincoats and suits, and both had lead masks near to their heads. A medical examination concluded that they had died from a heart attack, which by itself would not be extraordinary. However, the fact that two men, both dressed the same and with the same makeshift protective headgear, would suffer a heart attack in the same place at the same time – it doesn't sound likely, does it?

When Vallee studied the case notes, he surmised that the two men had been witness to a UFO. He even went so far as to say that the two men were at least hopeful, if not expecting, to witness such a craft due to the lead masks that he theorized they intended to use to protect them from any harmful rays. They were identified as Jose Viana and Manuel Pereira da Cruz – both local television transmitter engineers.

Upon hearing of the deaths of the two young men, others came forward to tell of strange sightings around the area where the bodies had been found. Usually reported was an oblong-shaped object that appeared to have an orange glow and emitted strange beams to the ground. Was this the craft that the two television transmitter engineers had hoped to see?

Vallee visited the site of Viana and da Cruz's death as part of his research and was quickly overwhelmed with other strange stories from the region – some dating back 20 years prior.

One story was said to have happened in 1946. A man named Prestes Filho was returning to his village of Aracriguama a little past 7 pm when from a fishing trip when an unusual object appeared overhead, shooting out a strange beam of light as it did so. The light "hit" Filho, who nevertheless was able to reach his front door and tell his sister what happened.

But soon after, his condition grew frighteningly worse. It was reported to Vallee that Filho's "flesh had literally detached itself from the bones. It was as if he had been boiled in hot water for a very long time, so that his skin and underlying tissue fell!" By the time Filho made it to the hospital, he was dead. Even after he died, his skin continued to come away from his bones.

Three years after the original incident that had led Vallee to visit Brazil, an incident happened in the small town of Anolaima in Columbia, in July 1969. Just after 8 pm, two children saw a strange bright object in a nearby field. They grabbed their flashlights and began to signal in the object's direction.

The object reacted by moving rapidly closer to them. The other family members, in response to the children's cries of terror, came out to see the same bright craft. A total of 13 family members witnessed the object overhead. They all saw it saw it take off at tremendous speed in the direction of where it had initially landed.

The children's father, one Arcesio Bermudez, grabbed a flashlight out of his child's hands and ran off in the direction the brightly lit craft had gone. He soon returned, obviously unsettled, and said he had witnessed a "strange creature" in a clear part at the top of the craft. After he had shone his flashlight on it and made himself visible to the UFO occupant, the object lit up again and disappeared completely.

In the ensuing days, however, Bermudez became more and more ill. Dark blue spots were seen on his skin and he was constantly nauseated. He died one week after the sighting. His death was officially declared to have been caused by gastroenteritis.

The reader may already be aware that Jacques Vallee was an associate of Dr. J. Allen Hynek, the legendary UFO investigator who created the Close Encounter numbered hierarchy of experiences. A UFO experience that results in the witness' death is called a Close Encounter of the Sixth Kind. While this type of encounter doesn't occur frequently, it is given a numbered place in the cataloging of Close Encounters overall.

Jacques Vallee and John Keel

Bill English

Chapter 27

AN ODOR-FILLED SHIP AND AN ALIEN LADY
One Man's Alien Confrontation Inside a UFO
By Nomar Slevik

Editor's Note: Once again, the experiencer in this case found himself in a room onboard the UFO which he says "smelled of cleaning solutions," and in which aliens seemed to stalk him. Researcher Slevik is confident this is an authentic case worthy of consideration and not the subject of a hoax. The outlined incident also involves an interaction with a female alien. Researcher Slevik has his hands full in coming up with an adequate explanation for this encounter. TGB

****

Bangor's origins read classically like many other inland Maine towns. There's a river running through the city, Native American tribes were already in Penobscot county and settled before Europeans arrived, and there are countless stories involving all aspects of the paranormal. Bangor is also home to probably the most famous of modern writers, Stephen King! His home, on West Broadway Street, was built in 1858 along with numerous other mansions as the city enjoyed exceptional wealth during this time due to the bustling lumber industry. While some have been maintained, others burned during the great Bangor fire of 1911. (The fire started small but quickly spread due to unusually high winds at the end of April that year.) Many of the remaining mansions were converted into apartment houses. One of these old buildings is where our next encounter begins. The report comes from Ken Pfeifer's website, *World UFO Photo and News.* A pseudonym has been used.

Kyle visited with his brother one evening in October of 2010. After dinner, drinks, and brotherly conversation, Kyle left the apartment at around two o'clock in the morning. As he walked to his vehicle, he heard a female voice call his name. He described it as "rhythmic and all encompassing." He looked around and thought he would find his brother's girlfriend playing a trick on him, but no one appeared

to be on the street. The voice continued, and was being perceived by Kyle as a "...soothing, almost hypnotizing female voice." His eyes gradually became heavy and his walking slowed to a standstill. Eventually, his eyes closed completely and the voice continued in its tranquil manner, "...we're here to help, it's ok, we're not going to harm you, we're going to make you feel better." After a moment, he fainted.

He started to wake; his eyes stirred under their lids as he lay motionless. They popped open and in an instant ached as a bright light from above blinded him. His hands reflexively shielded his eyes and he lifted his head to spy his whereabouts. He was in a room that smelled of cleaning solutions and noted that it felt sterile. As he looked around the room he saw two tall, "shadowy figures" opposite him. The beings frightened him, and he looked on in disbelief as they turned to face him. He described them as, "...two grey aliens. They had long fingers, very tall, had a little mouth that moved in an up and down motion like a fish, but nothing was coming out." He clenched his eyes shut, his thoughts manic. He was losing control. He lay there terrified but had an urge to understand the situation that he was in. He started to control his breathing to calm himself. He wanted to know what was happening to him and tried to speak.

No sound was heard as he mouthed the words he was trying to say. He then realized that his words were interpreted through thought – communications had turned telepathic. They said they did not intend to hurt him. Kyle elaborated, "I asked the alien if they aren't going to harm me then what are they going to do to me? They replied back with *we are going to help you and fix you*."

In an instant, the two aliens were on either side of Kyle. They stood him up, and he felt like an escorted prisoner as they walked him to another room. They walked through a corridor and he could see inside another room that was to the left and observed more beings, "...a huddled circle formation of at least seven to eight greys..." When they arrived in the second room, he was placed against a board while still in an upright position in the center of the room. He said that this area also smelled of chemicals and that the aliens walked around him, skulking as they looked him over.

One of his captors stepped close to him. Eye to eye, they stared at each other for a moment, then the entity carefully placed one of its hands over Kyle's nose. He froze with fear as it inserted a black tube into his nostrils. His breathing stopped for a moment; he gasped and screamed in reaction. He then heard, "...we're here to help, it's ok, we're not going to harm you, we're going to make you feel better." The voice he had heard previously suddenly came back. It reminded him of an airport's intercom announcement, and he could tell it was artificial and an effort to calm him. It didn't.

They laid him down on the board and it seemed to hover above the floor. Despite the hypnotic voice continuing, Kyle thrashed about on the table. The aliens struggled to hold him in place, but a leg broke free and he kicked one of his captors! It was with such force that it stumbled back, hit the wall, and slid down. The other alien appeared frightened of Kyle and backed away. As more entities rushed into the room, the young man looked around and felt overwhelmed and outnumbered. He tried to fight them off as they grabbed him, but he felt weak and dizzy. He fainted once more.

As he came to, he smelt asphalt. He woke up on the side of an unfamiliar road and clumsily stood. He looked around, confused by his surroundings. It was daylight, but he felt that only an hour had passed. He checked his watch and found that it was nearly six o'clock in the morning. He wasn't sure which town he was in, and looked for a familiar landmark. Just then, a police officer approached him. To Kyle, he seemed to come out of thin air. Kyle asked him what town he was in and what street he was on. The officer explained in an oddly angry tone that he was in Bangor and pointed him towards a street sign. From Kyle's perspective, the officer was gone in an instant, and he was confused by the officer's hostile demeanor. He was far from his brother's apartment but knew a friend's house nearby. When he arrived, he explained the previous night's events. He felt that he was believed, although he knew his friend could see the state that he was in and was trying to be comforting more than anything.

For weeks after his encounter, Kyle was cautious about who he would share his experience with. His brother and friends did provide him with a support system, but he needed to speak with someone or anyone with otherworldly experience. After conducting research on the alien abduction phenomenon, Kyle finally reached out to other Mainers about his experience and spent time at the "Experiencers Speak" conferences. These meetings are held annually in southern Maine and are organized by Audrey Hewins, founder of Starborn Support, which is a non-profit organization that helps alien abduction victims cope with their encounters.

Kyle is still on his journey, and his hope is to find out what truly happened to him and others like him. Despite his first, horrid encounter, he did share other abduction experiences that were more pleasant in nature. Through the other abductions, he has found a bit of peace from the aliens. He now feels that they are giving him a message of compassion and love. He explained, "Be happy and as loving as possible. This is the message I am getting."

It is interesting to note that this is similar to the experience recorded in the Travis Walton case from Arizona. Walton has stated in the last few years that he now believes his experience, while traumatic, was intended to save his life. A theme such as this could open a new paradigm to the abduction experience in popular culture. Films such as *Extraterrestrial, The Fourth Kind* and others have shown the malevolent nature of alien abductions. We are living in tumultuous times; perhaps a connection of compassion in their shared perceptions is exactly what we need to believe right now.

Chapter 28

MYSTERIOUS FUMES!
POISONOUS GAS WARFARE FROM OUTER SPACE?
By Tim R. Swartz

"**OFFICE** Workers Sickened by Mysterious Gas." "Fumes From Meteorite Crater Have Sickened 600." "Fumes From Corpse Cause Evacuation." News reports from all over the world have shown a drastic increase in cases in which people are sickened and harmed by mysterious gases. Many reports have come from seemingly common, normal buildings and schools. Dubbed "sick buildings" by the press, explanations on the causes range from chemicals used during construction, to mass hysteria.

All across the U.S. recent suicides using household chemicals and the possibility of terrorist attacks using poisonous gas have focused attention on the frightening scenario of gas attacks on large population centers. However, little attention has been paid to the evidence that mysterious fumes have already been responsible for sickening hundreds, maybe thousands of innocent people. Reports of unexplained gas attacks go back a number of years; some seem to indicate a disturbing connection with the frequency of attacks and an increase of UFO sightings.

One of the more familiar cases of reported gas attacks was the 1944 series of incidents in Mattoon, Illinois. However, there have been other, similar cases, such as the gas attacks of Botetourt County, Virginia in 1933. These episodes have been completely written off as classic cases of mass hysteria. However, there were elements to these cases, most notably in the form of physical evidence, that have been repeated constantly in the mysterious fumes and gas attack reports that continue to this very day.

Cold War Fears and Flying Saucers

Starting in July 1950, reports of UFOs associated with mysterious fumes began to trickle in. The July 1 edition of *The Cincinnati Post* ran the front page headline, "Saucers whirl over city."

"Flying saucers were reported over Cincinnati at widely separated points.

UFO HOSTILITIES AND THE EVIL ALIEN AGENDA

At least three reports were received around noon. The first saucer was sighted around 11 a.m. by Mrs. Katherine Willis at 25 Murray Road, St. Bernard, and her daughter, Beverly Ann. A few moments later, Jack Earls of 4713 Paxton Road, reported seeing saucers over Mt. Lookout and Lunken Airport. Control tower operators at the airport said they saw no saucers, and nothing unusual.

"Saucers also were reported at the same time by a Mt. Washington resident. 'Beverly saw it first and watched it for about two minutes,' said Mrs. Willis. 'It was white, way up in the sky, and I could tell that it was spinning,' said Mrs. Willis. 'I couldn't tell how high it was. It looked like it was going as fast or faster than the airplanes.' In the past two days, flying saucers have been reported by officials in Cairo, Ill., and Louisville, Ky."

The reports of UFOs coincided with reported low-flying aircraft leaving a noxious exhaust, causing an outbreak of mystery fumes throughout the Cincinnati area. The press at the time did not make the obvious correlation. The *Cincinnati Enquirer* reported on July 9th a similar incident in Illinois.

The *Cincinnati Enquirer's* headline reflected the Cold War fears that were on the minds of everyone during the decade. "Only a horrible nightmare. Fears of Russian gas bombing arise when foul odor from passing truck spreads through seven towns near Moline, Ill." The article goes on to report that a foul smell choked seven towns, sending some residents into hysterics and raising fears of a Russian gas bombing.

The noxious odor crept through Moline, East Moline, Silvis and Rock Island, Ill., and then spread across the border into Muscatine, Bettendorf and Davenport, Iowa. No injuries were reported aside from upset stomachs. Some residents, in hysterics, called police. One man insisted to Silvis police that "the Russians are flying over and gassing us."

The evil-smelling fumes routed citizens from the beds and from taverns, almost forced the closing of two farm machinery factories, jammed police switchboards and kept firemen on a near-emergency basis. An official of the Iowa-Illinois Gas and Electric Co. said the odor probably resulted from a leak in a tank of Pentalarm being hauled through the area on a truck headed west. He said the truck was seen passing through Moline shortly before midnight. Pentalarm is an odorant used to inject a smell into natural gas, normally odorless, to permit detection of leaks. The official said the odor is not injurious but can cause nausea.

Several taverns in Silvis and East Moline lost their patrons in a hurry when the smell entered. Some 20 persons jammed the East Moline police station. Police at Muscatine, Iowa, said anxious citizens jammed the switchboard with calls. Some became hysterical and left town. Others just closed all their doors and windows and tried to go back to sleep.

In Moline, a reporter said, "The police are being run ragged, calls are com-

The Norwood Searchlight Incident of 1949, as it has been called, is an outstanding series of 10 visual sightings of a strange aerial object that took place in or near the Norwood, Ohio area from August 19th of 1949 to March 10th of 1950.

1949

Cincinnati, Ohio, 1949. During the Jitney Carnival at the St. Peter and Paul Church in Norwood, Cincinnati, Ohio, Reverend Gregory Miller, purchased from army surplus, an 8 million candle power searchlight. Sergeant Donald R. Berger of ROTC of the University of Cincinnati would operate it. During the height of the festivities, at 08:15 pm, Sgt. Berger's sweeping searchlight suddenly flashed across a stationary, circular object in the sky. Reverend Miller and later others joined in and observed. This was only the beginning. Davidson took ten "still" photographs of the large disc-shaped object that flew in and out of the searchlight beam.

Cincinnati, Ohio, 1949
Norwood Police Sgt. Leo Davidson

ing in from all over the town, and the squad cars are going all over the place. People are heading for the high ground away from the river." The smell was strongest in the lowland areas along the Mississippi.

The utility company, flooded with calls, dispatched more than 30 repairmen to find what was first believed to be a leak in a gas main. The men hunted for three hours but found no leaks. Authorities at the Rock Island arsenal said, "Everything is in order here," in response to queries on whether the smell might have originated there. The smell lingered in the Illinois area for three hours, then hit the Iowa cities. It disappeared at daylight.

Subsequent attempts to locate the mysterious leaking truck failed. Some residents said the gas was nothing like the smell of

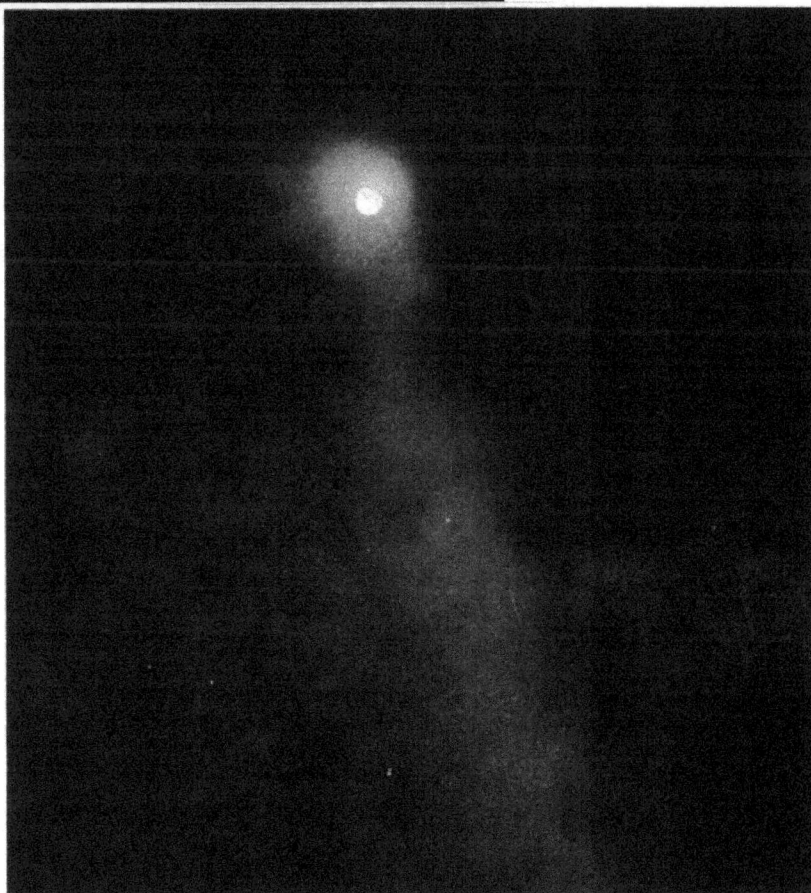

UFO HOSTILITIES AND THE EVIL ALIEN AGENDA

Pentalarm. The gas they smelled was incapacitating, and felt like, "the air was being sucked out of the lungs." On July 7th and 8th, newspaper and police switchboards were flooded with reports of UFOs flying overhead. The Moline/Davenport area also had an unusually high amount of UFO reports during the week of the 8th.

THE RETURN

Sometimes the mysterious fumes inexplicitly return years later to the same location. *The Cincinnati Times-Star* reported on October 24, 1949 that mystery fumes that gave off an offensive odor in south Reading, Ohio, damaged paint on more than 100 homes and made at least three persons ill. The article also reported that silverware in the affected area was strangely tarnished by the gas.

An 18 to 20 block area was affected by the fumes, whose odor was described as "like rotten eggs" and like "a burning brake band." Police Chief William Martin and Safety Director Harry Veddern said there was no evidence of the source of the odor. Martin added that he would have, "a chemical analysis made of the rain drippings from an affected house."

Officials at the Carlisle Chemical Works, which has a plant in southeast Reading, told city officials they did not believe their plant was responsible and offered their facilities to assist in the investigation. The odor was first reported to police at 3:00AM by patrolman William Appenfelder, who was cruising in south Reading. He told the dispatcher "the smell is so strong I can hardly breathe."

Newsmen said the paint on houses in this area apparently had dissolved, and was washed away by the rain. They noted that spots not struck by the rain appeared unaffected. Most serious damage was at the home of Mr. and Mrs. Charles Ringo, 219 Reading Road, where the south side of the house was left entirely bare of paint. Another resident, Mrs. George Bradford, 266 Burkhardt Avenue, said the side of her house looked as though "it was plastered by a mud shower."

In the same neighborhood forty-seven years later, on April 18, 1996, The *Cincinnati Enquirer* reported that a mysterious odor caused the evacuation of a local plant. About 600 Employees at the Hoechst Marion Roussel Inc. complex were ordered out shortly after noon when an odor was detected there and at the nearby Standard Textile Co. plant.

Toni Sweeney, spokesperson for Hoescht, said workers reported smelling propane or sulfur, and the building was evacuated as a precaution. Employees were kept outside for about an hour, she said. After firefighters arrived, they and members of Hoechst's Environmental Safety and Health Team toured the facility and found no source for the odor. Tests found no levels of dangerous chemicals or gases.

Mrs. Sweeney said the smell probably originated outside the plant, particularly because it was also reported at Standard Textile, east of Hoechst. "It may have been brought in through the ventilation system," she said.

Another strange gas incident happened at 11:30 PM, October 13, 1996, according to a WLW radio news update. The odor was detected in the Lockland region. The complaint originated from an area around the business, Office Depot. Eleven people were treated at the scene during the early morning hours. Lockland Fire Dept. Captain Bill Welchans said that there is no explanation for the fumes. All industries initially thought responsible were ruled out, and Welchans stated that there were no industrial operations within the vicinity of the odors.

SICK BUILDINGS

The phenomenon of "sick buildings" has some people worried that a new, though subtle form of terrorism is taking place. An alarming increase of UFO sightings in the same areas as reported sick buildings has lead some investigators to suggest that there could be a connection. More mundane explanations such as chemicals used during construction, to outgassing from new carpets have taken the brunt of the blame when mysterious fumes are reported. However, air checks usually can find no trace of any potentially hazardous chemical or gas to account for the strange symptoms reported by stricken individuals. Other, less credible attempts at an explanation generally leans towards cases of mass hysteria, or workers suffering from the "blue flu."

In Margate, Florida on April 7, 1997, a grocery store was forced to close because of a strange gas that made at least 30 ill. Officials say about 100 people were in the store when workers and customers began complaining of sore throats and watery eyes. Hazardous materials investigators say noxious fumes that caused the apparent respiratory problems among store customers had dissipated by the time they arrived, making it difficult to trace their origin.

Of the people who complained about symptoms, 11 were taken to area hospitals, but all were released following treatment for what were described as "minor" problems. Some customers said the fumes smelled like chlorine, while others described them as smelling like pepper.

Tamarac, Fla., Fire & Rescue Battalion Chief Dennis Peloso said the varied descriptions made it difficult to determine the source of the smell. "We've checked for chemicals, gas, Freon. We looked at the refrigeration system. Nothing," said Peloso, who called the negative tests "a little weird."

Schools also seem to be a favorite target of "sick building syndrome." On January 14, 2008, St Helens, Oregon High School was evacuated after students and faculty fell ill after noticed a strong odor like "rotten eggs." Nearby, people at the local Safeway store, and at the bank, also became ill. Several people had to be treated for nausea, dizziness, burning of skin and eyes, and respiratory complaints.

UFO HOSTILITIES AND THE EVIL ALIEN AGENDA

After some initial speculation that it was a natural gas leak, it was finally determined that it was, instead, a mystery. There was no leak anywhere on the school grounds, and chemical sniffers detected no natural gas. Local officials promised to look into the matter, but no investigation was ever conducted.

Mysterious fumes circulating through a classroom in Jamaica triggered a school evacuation and sent five children to the hospital gasping for breath, on January 25, 1997. Traces of the fumes were detected inside Room 315 of PS 37. Children were complaining of headaches, chest pains, watery eyes and breathing difficulties, Fire Department spokesman Luis Basso said.

Taking emergency precautions, the Fire Department evacuated the building immediately. Basso said 25 people from Room 315 - 23 children and two adults were treated at the scene for inhaling the mysterious fumes. Five of those were taken to a nearby hospital for further treatment.

Superintendent Celestine Miller of Community School District 29 said she did not know the exact source of the children's discomfort. However, the investigation by environmental authorities and the city Health Department was unable to trace any hazardous materials released into the school environment.

So far no explanation has satisfactorily answered the questions concerning the causes of "sick building syndrome." If the culprit is a mélange of "common" chemicals, then air tests should have found the suspected contaminants. The same goes for deliberate gas attacks. Air tests should be able to determine what the fumes are. There is also no good reason for the fumes to suddenly appear, and then disappear just as quickly. Chemicals in the carpet or in the structure of the building would leak slowly and evenly into the atmosphere. Making it easy for modern air check systems to find and fix the cause of the problem. Easy answers however, are not forthcoming.

UNKNOWN "FUMES" SICKEN WAL-MART CUSTOMERS AND EMPLOYEES

A "mysterious illness" that caused nausea, vomiting, and upper respiratory issues caused the evacuation and shut down of an entire Wal-Mart shopping center in Pennsylvania on February 12, 2013, and officials still have no idea what caused the strange event.

Firefighters from Center Township were called to the store at around 7:00 PM after both employees and customers began vomiting and having trouble breathing. Wal-Mart employees believed the cause of the odd event was emanating from the grocery section, roping it off from customers before the sickness began overtaking the rest of the store.

Center Twp. Fire Chief Bill Brucker mentioned feeling his airways close up when called to investigate, but couldn't explain it.

"I wasn't back there long before I started to feel something in my throat," Brucker told the Beaver County Times. "I couldn't pinpoint a smell or a cause, though."

At least two customers were affected so badly that they were rushed to the hospital in ambulances, with at least another three driven by car.

When the firefighter's own air quality tests didn't turn up anything odd, they called in the county's hazmat team who also came up empty handed, though Brucker wondered if that might have anything to do with the fact that the store's air conditioning had been reversed to filter the air by the time the team had arrived.

Not too long afterwards the store re-opened without any further incidents of unexplained sicknesses. Wal-Mart did not issue a statement, and no medical test results from those who fell ill were ever released.

"I don't know if we'll ever find out the cause," Brucker said.

Around the same time as the Wal-Mart incident, several faculty members from a school in Shawnee, Oklahoma started to experience strange symptoms while in the building, but reasonable explanations for the sickness were never pinpointed.

"I think it raises alarms when you have more than one experiencing problems," Shawnee Superintendent Dr. Marc Moore told Oklahoma News 9.

The issues came to a head at a January 11 meeting where teachers all began to complain of a "different kind" of headache, dizziness, and a "fogginess" clouding their minds, leaving them a tired mess when they get home from work.

The first thing the faculty did was call in the air quality testers from Oklahoma Natural Gas to check for leaks, but when they came up empty handed, they moved on to installing carbon monoxide detectors. Those, too, came up clean.

"If it is something here, we're doing everything we can to find it," said Taylor.

The school employees believed the source of the strange sickness emanated from the kitchen, however, just as mysteriously as it appeared, the cause quickly vanished.

The mysterious fumes made another appearance in February, 2013 when residents of the Autumn West apartment complex in Bangor, Maine began complaining of a "God awful" smell that they couldn't quite put their fingers on. Some thought it smelled a bit like burning rubber, others like a kind of musty, dirty smoke. One resident, Laurie Baker, even began to notice strange symptoms that coincided with the appearance of the odors; a sunburn-like skin rash appeared all over her body and her dog, Bella, began to experience respiratory problems.

The strange smells and their accompanying side effects eventually got so unbearable that many of the renters had to be evacuated to a local hotel.

Both the police and local fire department were called out to the location on separate occasions, and while the air tests showed elevated levels of ethanol or other chemicals often found in "clandestine laboratories" (read: meth labs), each investigation came up empty handed for a source of the mystery fumes. The housing authority and an air quality testing company have also conducted their own tests which have yielded similar dead ends.

"I wish we could have found something," housing authority director Mike Myatt told The Bangor Daily News. "We're hoping that it's resolved quickly and that we can get these people back in their homes."

MYSTERY REMAINS OVER FUMES THAT SICKENED 60 PEOPLE

In Springfield, Massachusetts, over 60 people were sickened and 10,000 workers and students were forced to evacuate a two-block area when mysterious fumes blew through the western Massachusetts city on Thursday, May 29, 1986.

People fell to their knees complaining of dizziness, shortness of breath and nausea. Eight people were hospitalized for the night, including a young man who suffered seizures.

Officials were uncertain what tests to perform on blood and urine samples taken from those who were stricken, said Dr. John Santoro, chief of emergency medical services at Baystate Medical Center.

"You can't put them in a magic machine and get an answer. We need some clues from the chemists working on what the possible toxin was," he said.

Officials examined sewer lines for substances that could have reacted after being dumped down different drains, investigated an old supermarket being converted into offices, sampled air and looked into an old refrigeration system inside the store for possible ammonia leaks. None of these has yielded any clues, they said.

The fainting spells began at about 10:00AM Thursday in the first floor of the old grocery. Two employees of Kavanagh Furniture Co., which rents part of the building, fainted.

Authorities said they could find no connection between Thursday's events and a report two weeks ago from city workers who said they smelled a strange odor while planting trees in front of the store.

"It's possible that we may never identify the source," said Robert Terenzi, a member of a state environmental emergency response team.

Georgia McDonald's Toxic Fumes a Deadly Mystery

Fire officials in Pooler, Georgia were stumped about what toxic chemical or chemical mixture knocked two women unconscious and sickened eight others at the local McDonalds restaurant on September 7, 2011. One of the women, Anne Felton, 80, of Ponte Vedra, Fla., died after going into cardiac arrest. Firefighters administered oxygen to Carol Barry, 56, of Jacksonville, Fla., before she was admitted to a Savannah hospital, Pooler Fire Chief G. Wade Simmons said.

"Every one of the 10 people that had some sort of symptoms...had been or were in that restroom," Simmons said.

No one anywhere else in the restaurant was affected.

Among other confounding aspects of the case was how quickly the gas disappeared. "It was there, and then it was gone in the next hour to hour and a half we were doing things at the scene," Simmons said.

By the time a Savannah hazardous materials analyzed air samples from the restroom, they found nothing detectable.

That left law enforcement officials and toxicologists to speculate about what the victims might have inhaled, and how it ended up in the women's room. "We've heard everything from terrorist attacks to carbon monoxide to sewer gas to God knows else," Simmons said.

Much of the speculation centered on the possibility that the women were sickened by a noxious combination of cleaning chemicals. Labels on toilet bowl cleaners, drain openers, window and glass sprays and scouring powders usually caution against using more than one product at a time.

Simmons said that based on employees' routines at the Pooler McDonald's, workers would have cleaned the women's room early in the day, before serving up Egg McMuffins to the morning breakfast crowd. But the initial report of someone choking didn't get called in until just before noon, further deepening the mystery of why people suddenly became ill so much later. None of the products on the cleaning cart had spilled, he said, and the cart wasn't even near the bathroom when patrons began developing symptoms.

"Cleaning chemicals are common culprits in bathrooms," said Dr. Kelly Johnson-Arbor, a medical toxicology specialist at Hartford Hospital in Connecticut.

"Perhaps the people in the bathroom mixed together bleach and ammonia," which would produce chloramine gas, an irritant. "It doesn't usually cause people to die, but if it's in a high enough concentration and/or the person had underlying cardiopulmonary disease (such as asthma), it could certainly be potentially fatal."

Despite all of the possible explanations, no cause for the gas was ever determined.

TOXIC PEOPLE

The case of Gloria Ramirez remains one of the most baffling events in modern medicine. On February 19, 1994, 31 year old Gloria Ramirez, who had been recently diagnosed with cervical cancer, fell ill at her home in Riverside, California. Her family comforted her as she began to vomit and her condition worsened.

UFO HOSTILITIES AND THE EVIL ALIEN AGENDA

In the evening, they called for an ambulance.

Ramirez was brought into Riverside General Hospital's emergency room at 8:14 p.m., suffering with severe cardiac distress. Her blood pressure was dangerously low because her heart was beating too rapidly, a condition highly unusual in someone so young. The ER staff administered drug treatment and pumped air into Ramirez's lungs. When her heartbeat suddenly began to falter, they tore open her shirt to use defibrillation paddles. Nurse Sally Balderas recalls seeing an odd, greasy film on Ramirez's skin, like a puddle of oil on pavement. Some witnesses recalled a scent similar to ammonia that suddenly began to fill the air.

Registered nurse Susan Kane drew a blood sample from Ramirez, and she immediately noted a strong odor coming from the syringe. She handed the syringe to respiratory therapist Maureen Welch, who sniffed it and reported an ammonia smell. Welch in turn gave the syringe to medical student Julie Gorchynski, who observed specks in the blood which have been described as "white crystals" and "manila-colored particles."

Within seconds of drawing the blood, Kane collapsed. Trying to maintain consciousness, she complained of burning pains on her face and was taken away on a gurney. Soon after, Gorchynski reported lightheadedness and then passed out. She had severe difficulty breathing and was the most severely affected member of the ER staff. Dr. Mark Thomas next felt ill, although he was able to remain standing. Welch was the third to faint, thrashing her limbs involuntarily on the ER floor. Nurse Sally Balderas experienced a burning on her skin and began to vomit. Balderas spent ten days in hospitalization, while Gorchynski, remained in intensive care for two weeks. Out of 37 ER staffers present, 23 reported an affliction of some kind. The total number of complainants would later reach thirty-two.

Dr. Humberto Ochoa, emergency room director, ordered an evacuation of all patients from the ER as he continued the effort to save Ramirez. He and three other staffers remained well enough to continue the struggle for the better part of an hour. In fact, Ochoa never felt ill at all, and could smell nothing unusual. Unable to revive her, Ochoa pronounced Ramirez dead at 8:50 PM. Her body was sealed in an airtight bag and left for the Riverside County hazardous materials team and the coroner's office to investigate. However, it would yield no easy answers.

From the start, the truth of the Ramirez incident has been clouded in confusion. In the instant media frenzy, some news reports blamed Ramirez's chemotherapy as a possible catalyst for the toxicity, when in reality she had never undergone chemotherapy. The first investigation into the Ramirez case came from the Riverside County hazardous materials (HAZMAT) team, which arrived at the scene two hours after Ramirez's death. The HAZMAT team tested the air in the ER and in other parts of the hospital for poisonous chemicals, and found nothing.

UFO HOSTILITIES AND THE EVIL ALIEN AGENDA

An autopsy was also inconclusive. Led by Riverside coroner Scotty Hill, a team of pathologists conducted the examination inside airtight contamination suits. They collected blood and tissue samples, as well as air that had been sealed in Ramirez's body bag. Apparently mystified by the case, Hill did not announce the autopsy results until April 29, 1994, over two months after the fact. The official cause of Ramirez's death: cardiac dysrhythmia as a result of kidney failure, which had been brought on by her cervical cancer. Hill had found no identifiable toxic substance that might have played a role in her death or in the illnesses of those present in the Riverside General ER.

As for how toxic fumes may have originated from Ramirez's body, no reasonable explanations have been suggested. The most complex theory, albeit a highly controversial one, comes from the Lawrence Livermore National Laboratory in a California. Livermore investigators found that Ramirez's body contained a high concentration of the harmless compound dimethyl sulfone. Its presence may be explained; they felt, by the use of DMSO, a common folk remedy for ailments such as cancer pain.

With two oxygen atoms added, dimethyl sulfone becomes dimethyl sulfate, an enormously deadly chemical. Livermore offered the theory that such a reaction had occurred within Ramirez's body, resulting in the mystery fumes.

Most scientific authorities call the Livermore hypothesis impossible. Dimethyl sulfate causes eyes to tear, does not take effect immediately upon exposure, and would most likely cause death, none of which was true of the alleged Ramirez toxin. Furthermore, proper conditions for dimethyl sulfone to gain two oxygen atoms do not exist in the human body. To add to the confusion, the Ramirez family denied that she had ever used DMSO.

The official explanation issued by Riverside County Department of Health, and the one most widely accepted, is that the incident was the result of "mass hysteria." Given the lack of a physical explanation and the inconsistent reactions of those present, stress and anxiety were judged the true source of the spontaneous afflictions. This mass hysteria could have been triggered by an incidental odor in the environment, such as cleaning chemicals or smelling salts.

Two of the stricken ER staffers, Dr. Julie Gorchynski and Maureen Welch, forcefully objected to this conclusion. Dr. Gorchynski in fact suffered from a degenerative condition to her knees after being contaminated by the mysterious fumes. She also suffered from breathing difficulties, muscle spasms and other symptoms. "I had chemical burns in my throat and nose," Gorchynski told reporters. "My lungs are working at half capacity, biopsies show my knees are dead, there has been a drop of my enzyme levels and crystals in my blood as well, and it's all medically documented. You don't get these kinds of symptoms from mass hysteria."

UFO HOSTILITIES AND THE EVIL ALIEN AGENDA

Possibly in deference to Gorchynski's $6 million lawsuit pending against Riverside County, the health department later revised its opinion to state that Gorchynski, Welch and Sally Balderas were in fact not casualties of mass hysteria. Ultimately, there have been no conclusive answers to the Ramirez case, and for the time being at least, it will continue to remain a baffling medical mystery.

A week after the Riverside incident there was another outbreak of mysterious fumes at the Mercy Hospital in Bakersfield, California. The emergency room was evacuated after doctors inserted a breathing tube in the trachea of a 44-year old woman suffering with shortness of breath. As at Riverside, emergency room personnel noticed a gaseous cloud rising from the patient. They complained that a potent chemical odor originating with the patient's blood left them with burning eyes, nausea and headaches. Fortunately, no serious injuries resulted from the exposure.

What could be the cause of the mystery fumes? Can we blame modern industry for their continued use of dangerous chemicals? Should we point a finger at the strange reports of UFOs that seem to coincide with an increase of people sickened by strange unknown gases? Or is the human mind to blame, with doctors and other health professionals assuring the panicked populace that the shortness of breath, vomiting, paralysis, unconsciousness, and long term physical damage is nothing more than mass hysteria?

The mystery continues.

Ultra-Terrestrials
Is There An Alien
Cult Of Evil?

Chapter 29

AN ALIEN INVASION – THE HYBRIDS ARE HERE!
By Sean Casteel

If you corner those who have made a study of the abduction phenomena – or what publisher Tim Beckley calls the Alien Invasion Hypothesis – you will come to understand that there is a certain degree of uneasiness among the handful of researchers who have come face-to-face with what many consider to be otherworldly parasites.

Here several of the top abduction investigators and experiencers reveal their findings – and it should be noted that two of them are no longer with us to update their discoveries, though Dr. Jacobs and Whitley Strieber continue to be passionate in spreading what they have learned about an often traumatizing topic.

****

The late Budd Hopkins still stands tall among the many researchers of alien abduction. It was Hopkins who was often credited with popularizing the idea of alien abductions as involving genetic experimentation. In his book "Intruders" (1987) Hopkins told the story of a young woman named Kathie Davis (a pseudonym) who was impregnated by aliens while still a teenager. She was later "presented" with the alien/human hybrid child she was the "mother" of onboard a UFO.

Hopkins likened Kathie's experiences, and the phenomenon in general, to rape, and felt that no good could come from these encounters, which he described as "severe" and "nightmarish."

While both men and women reported to Hopkins abductions by aliens that included sexual encounters, allegedly for some form of extraterrestrial eugenics, women in particular seemed to a part of a "highly technological colonization scheme."

When I interviewed Hopkins in the years after the publication of "Intruders," I asked him what he felt about the good versus evil aspects of alien abduction. He told me that it wasn't an easy question to answer. He said one doesn't ever

UFO HOSTILITIES AND THE EVIL ALIEN AGENDA

Intruders author, Budd Hopkins
with Peter Robbins

make that same judgment about people, and that same moral ambiguity should logically extend to the UFO occupants as well.

For example, Hopkins explained, if some person you know says or does something mean and hurts your feelings, you say well, "I know that person and normally he's not that way." But you don't say, "Now I realize that he's a demon." Also, if someone takes you out to dinner or gives you a present or does something nice, you don't say "Now I realize that he's a god." Those kinds of blanket judgments also don't work when projected on to the alien abduction phenomenon.

"Communion" author Whitley Strieber, whose understanding of his own experiences began after he contacted Budd Hopkins and asked for his help, was also morally ambivalent about the "Visitors," his name for the alien abductors. In one interview I did with Strieber, he told me the aliens were indistinguishable from demons and that it was impossible to overstate how "awful" the abduction experience is. Stieber would later claim to have been raped by the Visitors and subjected to many forms of humiliation by them.

And yet Strieber will at other times say he is learning deep spiritual truths from the Visitors, and that he is beginning to converse with them through meditation and other similar techniques. The abduction experience seems to bring with it an emotional rollercoaster that leaves experiencers baffled and trapped between darkness and light.

* * * *

THE SAD TRUTH, ACCORDING TO DR. DAVID JACOBS
By Sean Casteel

Retired Temple University professor David Jacobs, Ph.D., the author of the highly regarded books **The UFO Controversy In America** and **Secret Life**, has spent more than 50 years researching UFOs and alien abduction. But it was only recently that he came to feel he had solved the mystery to his own satisfaction.

UFO HOSTILITIES AND THE EVIL ALIEN AGENDA

The solutions he arrived at are the subject of his third and fourth books: **The Threat: What The Aliens Really Want And How They Plan To Get It** and **Walking Among Us: The Alien Plan to Control Humanity**. Finding what he believes to be the answers was not a happy event for Jacobs. He told us recently that he now approaches the subject with an attitude of dread and deep concern about the future of humanity and the planet we call home.

Dr. Jacobs with Visitors

****

Casteel: What do the aliens really want?

Jacobs: Well, you know, the ultimate question I think to ask for the UFO phenomenon is "Just what the hell do you think they're here for?" That's the question that I've tried to address in this book—what is this all about? What is happening here? Why is this happening? Why are people saying that these events are happening? So what I've done then is try to answer those questions as best I can by using as much information as I can from eleven years of fairly intensive research into abductions.

And what I've been able to find is that this is a program. They're not here just because they're examining people, or studying people, or experimenting on people. I don't know, Sean, if you remember I gave a talk about that in Los Angeles when I saw you. So they're not here to sort of "examine" us in some way. They're here on a mission. They're here with a goal in mind. They've got a program, and it's a program with a beginning, a middle and an end. It's a program that is goal-directed and I think we're entering into sort of the end-phase of this program. I think that we're moving towards the end of this.

And the program ultimately is not abducting people. Abductions, you have to remember, are a means to an end. They're abducting people for a purpose, for a reason. The physical act of abducting people, which is the abduction phenomenon, really is only part of the program. So what I've done is kind of divided it into component parts and fleshed it out a lot more. So what we have here is an abduction program, a breeding program, which accounts for all the reproductive activity that we see, and a hybridization program, which is why people see hybrids all

321

the time—as babies, as toddlers, as adolescents, and then as adults.

And then, finally, I think all this is leading to an integration program in which ultimately these hybrids, who look very human, will be integrating into this society. And who will eventually, I assume, be in control here because they do have superior technology and superior physiological abilities that we do not have. We would therefore be sort of second-class citizens, I think.

Now, I find this to be very disturbing. And the interesting thing is that I don't really see other scenarios. I know that people feel it's positive and it's wonderful, and all the rest of that. And they're here to help us. But in the cases that I've investigated, very carefully, very thoroughly, for a very long time, I have not had people discuss that. When people discuss the future, generally speaking, they are discussing this integration program that they're confronting, and we're all confronting. I've been involved with UFO research for about 32 years now, since about 1965, and I have never been downcast or depressed about the phenomenon. I have never been pessimistic about it. I've always been filled with wonder and awe and amazement at it. I've been enthusiastic and optimistic about it.

But I must say that now that I've learned as much as I have learned, and I think I've learned an awful lot, I am very, very unsettled and upset by what I see. I don't like what I see. I wish I didn't see this. I wish I hadn't uncovered this. I despair of it. It's thrown me into a tremendous sense of concern about the future and unease. I just don't like it very much. I wish I did. I don't want to be this way. I don't want to be the bearer of bad news. I could not have ever imagined that I would come to this position. What I'm seeing now, what I've found with the phenomenon, I could never have imagined.

Now, though, I am persuaded by the evidence. I think that we are looking at a very serious business happening in front of us. As you know, the UFO and abduction phenomena is very, very widespread. And people have seen tens of thousands, hundreds of thousands, maybe millions of UFOs around the world for a long time now, at least through the 20th Century, and certainly since 1947, and before that as well. It means that the amount of time and energy put into this program is really quite enormous. This means that it has a tremendous amount of importance to these beings.

And there's another aspect to it also that is disconcerting. It's a secret phenomenon. They don't want us to know what they're doing. They don't want us to interfere. This is a consciously-arrived-at and successful secrecy program to prevent us from knowing. Gosh, that makes me very uneasy, Sean.

So anyway, I've become depressed about the whole thing.

Casteel: So the reason for all this negative feeling and depression is because you feel that you and I and people who are natural human beings will somehow be subject to a higher form of oppression?

UFO HOSTILITIES AND THE EVIL ALIEN AGENDA

Jacobs: Of authority, right. I do think that something like that is going to happen. The way I look at is, I have one scenario which I like. All the rest of them I don't like. The one scenario which I do like is that one day, they will come to the abductees and say, "Our program is done now. We have accomplished our goal. We've taken what we need. Thank you so much for your help. Thank you so much for your cooperation. We'll be leaving now. You'll never know we were even there. People will wonder forever whether they were abducted or not. Now, goodbye and good night." That's my favorite scenario. I love that scenario. But in fact, we never hear that. We always hear a scenario about the future in which these beings say they're going to be here with us.

And everything is going to be wonderful. Everything is going to be great. It's going to be just delightful. We're going to like it, they're going to like it, everybody's going to like it. That's the future according to them, but when I take a look at their society, and when I take a look at a future in which they would be in control because of their superior technology and physiological abilities, I see a very, very different society than the kind that we live in now—a society that's far more restricted and far more controlled. The whole concept of individual freedom in this kind of society would be under serious question. I don't like that. I don't want it. I would rather have human beings make their own mistakes and fix their own problems and do things by themselves. I think we're perfectly capable of doing it. I think we can all live together into a happy future. I think that's within the realm of possibility.

Casteel: So they paint a Utopian picture of what's going to happen?

Jacobs: Well, they paint a picture of what they consider to be good for themselves. And they live in a controlled society. They live in a society where everybody knows his or her job. They live in a society where everything is controlled. The ability for people to act independently is very, very circumscribed in this kind of society that they live in. I just don't like it. I'm filled with apprehension over this. Now the key thing here is they are here for a reason. They are not studying or experimenting on us, and they're keeping their activities secret from us so we won't find out.

Casteel: That's what you were saying at the lecture, that they were way beyond the experimental stage.

Jacobs: Right. In fact, that was the title of the lecture "Is this an experiment or a program?" You don't have to be a rocket scientist to see what's going on. This doesn't look like an experiment to me, you know. It's been worldwide with millions of people for 50 years. Day in, day out, 24 hours a day. What kind of an experiment is that? And there are a lot of other reasons why it's not an experiment.

So it's disconcerting. And I never used to think this until I began to put it together—until I began to come to these conclusions and realize I think this is what it is. I think that in the book, basically, I've advanced hypotheses which might very well essentially be what this UFO phenomenon is all about. This is not the final aspect of studying this phenomenon, but I do think that I've fleshed out what the goals and purposes are. We're not exactly sure of all the "whys." Why they would want to do it in the first place? What's the point? We don't really know that. But I think this is certainly a hypothetical answer to the UFO puzzle. I think pretty much we've answered it.

That's what we're looking at. And therefore I think we're looking at a very difficult future.

Casteel: So the "how-they-plan-to-get-it" part would be through the breeding and hybridization programs?

Jacobs: Yeah, but how it's going to be played out, I don't really know. There's a lot of different scenarios. There's the Disaster Scenario that abductees keep talking about over and over again. We've had this for years and years. I don't quite know how that's going to happen. Whether there's going to be a disaster or not. There's a scenario where they just sort of naturally and nicely integrate into this society and we never even know it's happening. I guess there's things in the middle. We don't really know how what the aliens and abductees call "The Change," is going to take place. We don't really know that quite yet.

Casteel: But they're given visions like the world on fire or natural disasters, that kind of thing?

Jacobs: Yeah, well, it's all sorts of disaster scenarios, which includes atomic war. It includes asteroids hitting the earth. It includes floods, plagues, famine, whatever. It's sort of a generalized disaster and you just fill in the blanks as to what kind of disaster it will be. It's really non-specific, although people report more atomic war or the earth cracking in half or being destroyed by a comet or something like that more than other things, I guess.

I don't think the specifics are all that important, but the idea of a disaster is the most important thing. But I don't know whether that's true or not. We really don't know that yet because it might be a very different scenario. But I'm certainly going to stick by my guns and say that this is an integration program. However it's worked out, they will be integrating into this society and that's what this is leading to.

As I say about all my books, there's no possibility, Sean, that I have avoided error. I'm going to be wrong somewhere, somehow, in God knows how many things. But I think that this hypothesis that I'm presenting here is supportable by the evidence. And that's what makes it more disturbing. Everything I've written in this book is evidence-driven. That's why it's such a difficult book to deal with.

Casteel: Well, it's like "Invasion of the Body Snatchers," where you end up mouthing the words, "It's better this way. We have no pain now."

UFO HOSTILITIES AND THE EVIL ALIEN AGENDA

Jacobs: You are exactly right. I've thought about that, too. And of course one of the things I've been criticized for is because it has such a science-fiction quality to it. And then people say, "Therefore it is science-fiction." People have picked it up in the culture and that's why I'm hearing this. What they do is they make the mistake by finding similarities and saying that the similarities are in fact equalities. Which, of course, they are not. The fact is though is that it does have what they call "Programmatic Content." It does have content whereby we can see the inner workings of what they're doing and what's happening. And there are parallels in science-fiction, and certainly one of them is "Invasion of the Body Snatchers." That is in fact one of the parallels we see. But there's a lot of other parallels in science-fiction also. And if you look hard enough, you're going to be able to find a bit and a piece of it here and there and everywhere. But I don't think that this is science-fiction.

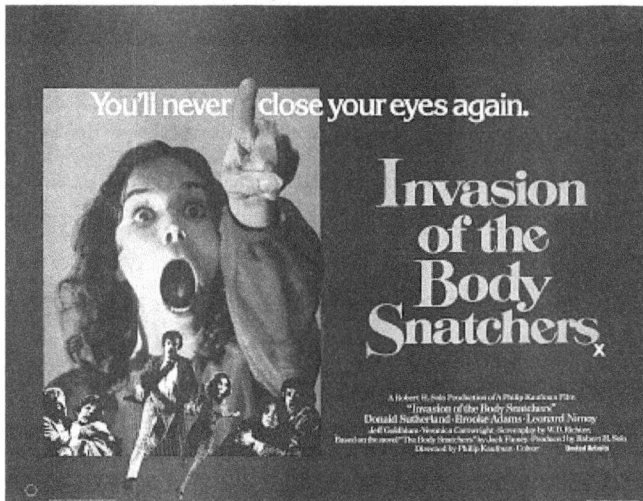

You've got to remember that most abductees are really not much into science-fiction. Most people I work with say they don't follow science-fiction. They haven't seen those movies. They don't know about that kind of stuff. It's not something where it just pervades the society. As people try to tell me, "Well, it's just sort of everywhere." Well, it isn't exactly like that. You've got to remember that the abduction phenomenon, while similar to science-fiction here and there, is really very different in almost all of its areas. It really is a different kind of situation, as you know.

And of course you have situations where people see other people being abducted and people are physically missing from their normal environment. There's a strong physical component to it that's very tough to explain.

Casteel: One thing I thought was interesting was the way you went over the varying degrees of hybrids. Varying percentages of human versus alien combinations.

Jacobs: Yeah, I tried to put forward a new concept of hybridization. One that makes more sense and one that's more in line with the evidence as it's presented by the abductees.

Casteel: Like subtle degrees between the various kinds?

Jacobs: Right. And you know, it does make sense that way. It answers a lot of things. It makes sense primarily because abductees have been saying this. So I

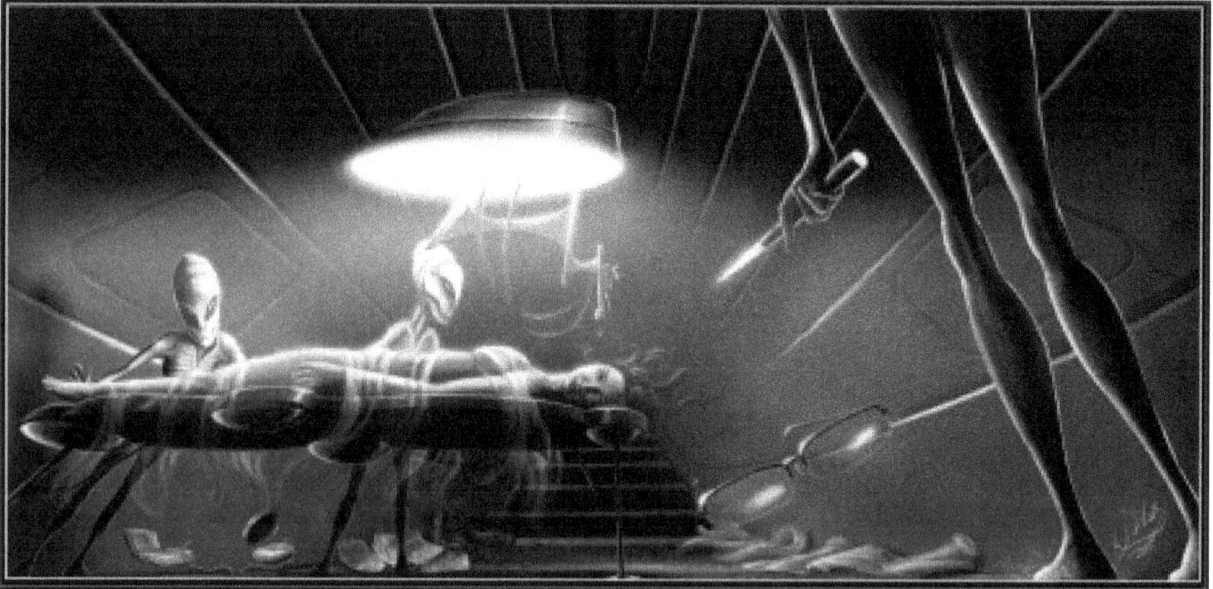

was able to divide it into sort of like late stage, middle stage, and early stage hybrids. But one of the interesting things about this phenomenon, Sean, is that you know I talked about toddlers and what kind of toys they played with and young children and what they play with. And I even had some sort of "widget" that the adolescents tinkered with. Remember? There was that one sequence where he had a box and he had to press certain things and if he pressed them the right way there was a flash. Remember that?

Casteel: Yeah, like an alien Nintendo or something?

Jacobs: Right. But in terms of toys, we basically know quite a lot about alien toys, about kids' toys, and all that. And the hybrids and what they do. I spent a lot of time on hybrid children. If this were psychological, I couldn't do that. We wouldn't be able to describe certain toys that they used that other people have described as well. We're learning so much about this phenomenon, it's just extraordinary. And yet everything we learn points in the direction of the integration program.

Remember I talked about this one woman who was involved in this sort of learning situation in which she was standing in front of a class of hybrids. And a picture of a dog came down.

Casteel: Right, and she was supposed to explain all the earthly things.

Jacobs: And she was asked "What's a dog for?" And she said, "Well, you know, it's a companion" and all that. But see, that points to integration into the society. Everything points to it. This concern, this interest in earth things. I don't think it's just sort of "interest." I think they're interested in it for a reason. The interesting thing here is they're not saying, "How do you elect a president?" "What do you do Saturday night?" and things like that. They're not interested in politics, economics, culture, society. That they're not concerned with. And we almost never, never get questions about that. But if this were psychological, we would be getting questions like that. But they're interested in physiology, they're interested in anatomy, they're interested in the natural world, the environment, animals, things like that. Things that are not man-made necessarily. As if in the future, it's not going to matter what we have built. The only thing that's going to matter is what they do. That's one way of looking at it.

But it's disconcerting. You want them to ask questions about Clinton, you know, and things like that. Something where you can see they really are learning about society, but in fact, if they integrate into this society, there's going to be an overlay of their society. And ours isn't going to matter a whole lot. That's my interpretation. Now I might be awfully wrong about that, but it certainly is my interpretation of it.

Casteel: Well, given the inevitability of it all, you just kind of want to hang on to some kind of shred of hope that it'll be a good thing somehow.

Jacobs: Well, the one hope I have right now is not much of a hope. But the hope that I have right now is the fact that it's still secret. That is to say that as long as it's still secret, they must assume that they are still vulnerable and therefore there's a way that we can affect the program. That's not a whole lot to hang on to, for me. But you know I despair that the scientific community is going to realize the import of what's happening in front of them. I think that even if they do, there's so much water under the bridge and we're so far down the line with this that indeed it may make no difference. Maybe 30 or 40 years ago it might have made a difference, but I'm looking to the future where I just don't see the scientific community getting interested in this subject. It hasn't happened in the past, and failing some sort of sudden event, some sort of sudden revelation, some sort of incredible thing, "Clinton Exposed As Alien Himself," or something like that, I just don't see them becoming interested in it. They've had half a century of the ability to have that interest and have not utilized it.

I don't like what I'm seeing here. I've spent my entire adult life studying this subject intensely. Year after year. I have a professional degree with a Ph.D. in the subject and I teach the only course in the country on the subject, the only regularly scheduled, full credit course in the country, which I've taught for 19 years. I've written three books on the subject and many articles. And I've never really felt the despair I feel now that I think that we've broken it open and we're looking at it and examining it. And it's just not what I expected. It's not what anybody expected. I just wish it was not that way. I just don't like it.

However, this is one of those situations where you can despair of it—you can feel Oh, my God, this is awful—but you have to lead your life as though it's not happening. It's the only way you can get along. It's the way I get along. And I've got two kids, you know. So I look at them and I look at the book and I don't know what kind of future they're going to have. That's true. I really don't. That wasn't just words. This is of great concern to me. I really don't know what kind of future they're going to have.

Casteel: Well, again, I guess the one ray of hope is the possibility that it won't be a cruel form of oppression to live under them.

Jacobs: Well, that certainly might be possible. I don't think it's going to be a cruel form of oppression. I just think it's going to be very different and not to our liking. I don't see an oppressive situation necessarily. I don't think that we're going to be whipped like a slave in a galley on a Viking ship or something like that. But at the same time, I do not see the freedom of movement and action and activity that we have now. Individual freedom and freedom of thought and all that to be the same in the future as it is now.

You've got to remember these beings are telepathic beings. They tap right into your thoughts. I don't want anybody tapping into my thoughts. When I was down in Brazil earlier I gave a paper on what it's like to live in a telepathic society based on abductees' testimony about the society that the aliens live in and the kind of telepathy that the abductees experience. And Sean, you don't want that society. You want to be private. You want to keep your thoughts private. You want to have individual expression, individual thoughts. You want to be able to do what you want to do without anybody knowing. And in their society, that's not necessarily true. It's a different kind of society.

****

So, is alien integration into Earth's society already a given? Will we lose many of the freedoms we currently enjoy to the superior capabilities of the aliens and the hybrid offspring that are also a part of us?

The old truism "Only Time Will Tell" seems operative here, as it does with so much of the UFO mystery. And in the words of the late rock singer Tom Petty, "The Waiting Is The Hardest Part."

* * * * *

MORE FROM DAVID JACOBS
ON ALIEN DECEPTION AND MANIPULATION
By Sean Casteel

It has long been theorized that we face a greater threat from alien/human hybrid creatures than we do from the diminutive gray aliens. The grays are at least a separate species with clearly recognizable, although humanoid, alien features. But the alien/human hybrids that look essentially human can pass among us unseen and with impunity as they pursue whatever their secret agenda turns out to be.

UFO HOSTILITIES AND THE EVIL ALIEN AGENDA

Longtime abduction researcher Dr. David Jacobs has made a thorough study of alien abduction as an element of the colonization program of the UFO occupants.

In his book "The Threat: What the Aliens Really Want and How They Plan to Get It," Jacobs writes, "The production of a hybrid species appears to be the means to the aliens' goal. So far, researchers have been unable to uncover any other purpose for the UFO and abduction phenomena, and the Breeding Program. Why are aliens producing hybrids? This has long been one of the fundamental mysteries of UFO and abduction research."

Researchers have posited for years, according to Jacobs, that the aliens are a dying race and must pass on their genes to hybrids in order to maintain their existence. They either cannot reproduce or cannot reproduce in enough numbers to sustain their species' viability. Although dismissed by many UFO researches, Jacobs says the evidence suggests that there may be some merit to this theory.

Jacobs makes reference to an abductee he worked with named Allison Reed, who endured a four-and-a-half long abduction event in which an alien escort showed her a "museum" type room with strange, life-size holograms of several beings that represented earlier attempts at hybridization. Each of the holograms had a different flaw that made them unable to reproduce. The human race, says Allison, is not the first race the aliens have tried to breed with. But we as humans are the most compatible gene-sharing species at present.

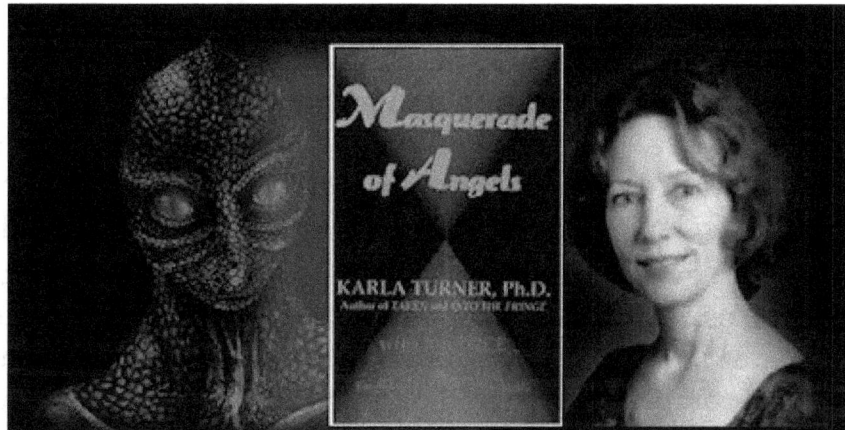

THE DIFFERENT STAGES OF HYBRIDIZATION

Over the many years spent working with abductees who report experiences with the Breeding Program, Jacobs has come to believe the process works in stages.

"It is clear from abduction reports," Jacobs writes, "that hybridization starts in vitro with the joining of human sperm, eggs, and alien genetic material. The result of this union, which is 'grown' partially in a human female host and partially in a gestation device, is a hybrid being who is a cross between alien and human. Many of these hybrids look almost alien. They have large black eyes with no whites; small, thin bodies; thin arms; thin legs; thin, nonexistent, or sparse hair; a tiny mouth; nonexistent or tiny ears; and pointed chins. They have no genitals. Some look so much like aliens that abductees often mistake them for 'pure' aliens."

There is a second stage in which are joined a human egg and sperm with the genetic from a first stage hybrid. The resulting offspring is a cross between a hybrid 1 and a human, called a hybrid 2. But these beings still look quite alien. They have an oddly-shaped head with a pointed chin, high cheekbones and only a small amount of white in their eyes; their hair is still quite sparse but there is more of it; their bodies are thin but larger. But they are still likely to be unable to reproduce.

"When mature," Jacobs writes, "these early-stage hybrids often help aliens with the abduction procedures and are an integral part of the alien workforce. Abductees see them taking care of hybrid babies and toddlers and executing other important tasks."

The breeding process gradually works to a point Jacobs calls "late-stage hybrids," who have normal-looking eyes with perhaps only a slightly enlarged pupil. Their skin color is humanlike but sometimes a little too even. They often have short-cropped hair but some have curly or long hair. Some do not have eyebrows or eyelashes, and most do not have body hair or pubic hair. Their frames are sometimes thin, sometimes muscular, but never overweight. They are often blond and have blue eyes, but a broad range of hair and eye color is reported.

An even later stage in the hybridization program produces human looking creatures who also possess the aliens' extraordinary mental abilities. They can engage in staring procedures, mind scan, visualizations and so on. They have nearly complete command over the abductees, who report having a little more physical and mental control during hybrid abduction activity, but not enough self-control to effectively resist abductions.

This later stage hybrid can also reproduce with humans via "normal" intercourse and bypassing the standard egg and sperm harvesting phase of abductions. The resulting hybrid is barely distinguishable from "normal" human beings. The evidence points to the development of an increasingly human-looking and human-behaving hybrid armed with the aliens' ability to manipulate humans.

JOINING THE ALIEN WORK FORCE

Once the hybrids are born, Jacobs says, the aliens funnel them into specific types of service. By the time the young hybrids have reached adolescence, the aliens have given them new tasks and responsibilities within the abduction program. The hybrids also begin to interact with abductees more on a sexual and social level.

After the hybrids reach adulthood, their responsibilities increase and they are more involved in the abduction routine. Although they still function in an "assistant" or subordinate capacity, some adult hybrids conduct the full range of physical, mental and reproductive procedures. They work alongside the gray aliens and become partners working toward a common goal.

Some abductees prefer being with the hybrids rather than with the grays. For them, there is comfort in the familiarity of the human form. Others find the late stage hybrids to be frightening and prefer the more predictable gray aliens, who act according to a well-defined system which, over time, many abductees have come to find comfortable.

"For the most part, the hybrids act like the grays," Jacobs writes. "Task-oriented, efficient and clinical. But their presence injects a note of emotionality and unpredictability. Their very humanness almost makes them party to a crime involving the kidnapping of men and women. Many women feel more emotionally vulnerable around late-stage hybrids."

The aforementioned Allison Reed told Jacobs: "It sounds crazy, but I feel more comfortable with the little gray guys than being left alone with these people-looking hybrids. They don't have that compassion, I don't feel it. I don't know if they're anything like human beings. Maybe that's why I'm scared, because human beings can be so cruel. Whereas the gray guys, they do their job and they don't want to hurt you. But they don't want to, you know, give you kisses and love you either. They're just kind of neutral in a way. But human beings can be so cruel."

OUTRIGHT ABUSIVE HYBRID ALIENS

Jacobs writes about other women who experienced abusive treatment from hybrid aliens.

For example, a woman called "Beverly" was visited in her bedroom by three hybrids she had encountered before. They pulled her out of her bed and subjected her to a night of sexual intimidation and terror.

First, they made Beverly remember a conversation with a trusted confidant during her adolescence who had warned her not to give her body away unless she was sure because except for her heart it was her most precious possession. Then the hybrids told her they could take her body whenever they wanted and that she was always vulnerable and never safe. One hybrid raped her, and she was forced to perform fellatio upon another. They punched her, twisted her skin, and hurt her without leaving marks. They pushed an unlit candle into her vagina. Then they told her she had caused her children to be abducted.

Further experiences of similar sexual abuse occurred. During yet another event, the hybrids sat Beverly in a chair, stood around her, and filled her mind with horrendous images. She saw a graveyard with the bodies of people she loves, including her children, who had been hacked to death and were covered with blood. She also saw a crucifixion scene with loved ones, again including her children, hanging on crosses. Then the hybrids put images of religious figures in her mind and assaulted her.

UFO HOSTILITIES AND THE EVIL ALIEN AGENDA

"What are the reasons for this sadistic Independent Hybrid Activity?" Jacobs asks. "It seems possible that some women are selected for abusive relationships. It is also possible that the malevolent behavior of hybrids toward abductees is necessary. Perhaps they need to generate fear, intimidation, guilt, shame and humiliation to fulfill the objectives of their agenda. An alien seemed to reinforce the hypothesis that sexually violent behavior was part of their program after a particularly violent assault upon Beverly onboard a UFO. When it was all over, she asked the alien why he allowed the hybrids to do that to her. He replied, 'The expression is necessary.' This could mean either that it was a necessary part of the program for all hybrids or that some hybrids must express their sexually aggressive tendencies in this way because they are unable to express them in the controlled society in which they live."

Jacobs also allows for the possibility that the human genes in the hybrid might be responsible.

"Because the late-stage hybrids are mainly human," he suggests, "they have strong sexual drives but little conscience. It is as if they have human attributes but lack human controls. Even if they do have a conscience, they know that the human victim will immediately forget what has happened to her. The hybrid might assume that there is no lasting effect upon the human and he therefore can do and say anything he pleases with impunity."

A FURTHER COMPLICATING FACTOR

Cases where the abductor is human-looking and seems to be a member of the military have been kicked around for decades. In an online posting called "Evidence for Military Kidnappings of Alleged UFO Abductees," abduction researcher Dr. Helmut Lammer, credited with coining the acronym "MILAB" for "military abductions," explains why he feels there is such a strong case to be made for this particular type of "alien" encounter.

"Some UFO abductees told that they were also kidnapped by military/intelligence personnel," Lammer writes, "and taken to hospitals and/or military ground/underground bases. Not many of the popular books on the subject of UFO abductions mentioned these experiences. Especially disconcerting is the fact that abductees recalled seeing military and/or civilian personnel together with alien beings working side by side in these secret facilities.

"The presence of human military and/or civilian personnel inhabiting the same physical reality as the aliens," Lammer continues, "exceeds the mindsets of the skeptics and the open-minded researchers by several orders of magnitude. Researchers in the field of mind control suggest that these cases are evidence that the whole UFO abduction phenomenon is staged by the intelligence community as a cover for their illegal experiments."

Lammer contrasts the two kinds of experiences – totally alien or human/

military – thusly:

"MILABS involve the following elements: The activity of dark, unmarked helicopters, the appearance of strange vans or buses outside the houses of abductees, exposure to disorienting electromagnetic fields, drugging, transport with a helicopter, bus or truck to an unknown building or an underground military facility. Usually there are physical aftereffects, like grogginess and sometimes nausea after the kidnappings. There is also a difference when the abductors appear. In most UFO abductions, the beings appear through a closed window or wall. Or the abductee feels a strange presence in the room.

"Most abductees report that they are paralyzed from the mental power of the alien beings. At MILABs, the abductee reports that the kidnappers give him or her a shot with a syringe. MILAB-abductees report that they are examined by human doctors in rectangular rooms and not in round sterile rooms, as in the descriptions of UFO-abductees. The rooms, halls and furniture are similar to terrestrial hospital rooms, laboratories or research facilities and have nothing to do with UFO furniture."

Lammer says that the human "doctors" seen during a MILAB also implant small devices into their abductees in a way similar to what we have come to expect from the grays. He gives a brief overview of human implant technology, including the work of Dr. Daniel Man, who holds a patent on a biochip that operates as a transponder after being surgically inserted under a person's skin.

DAVID JACOBS ON MILITARY ABDUCTIONS

Returning to David Jacobs and his book, "The Threat," we get a very different understanding of MILABs.

"Late-stage hybrids may also dress in military-like clothes," Jacobs writes, "such as one-piece jumpsuits that resemble flight suits. Because they look so human, it is easy to mistake them for American military personnel, and many abductees have linked military personnel to their abductions. Over the years, abductees have reported that soldiers are involved with the abductions or that uniformed males, sometimes in military-type surroundings, are present during abduction events.

"Hybrids will sometimes abduct people," he continues, "and bring them to abandoned military bases or even to unused areas of active military bases. Abductees will occasionally see actual armed forces personnel in the process of being abducted, still wearing their uniforms. All this, in conjunction with the longstanding and widespread suspicion of a "cover-up" by the American government, has led many abductees and researchers to conclude that the government is secretly conspiring with the aliens. Some abductees have even petitioned the Secretary of Health and Human Services to investigate the military's abduction activities."

UFO HOSTILITIES AND THE EVIL ALIEN AGENDA

But in Jacobs' opinion there is no evidence that the American government or any foreign government is involved with abducting people. Abductees are most likely remembering fragments of experiences with hybrid aliens who took them to military-like settings.

In either case, whether we believe Jacobs or Lammer, the fact remains that abductions are a frightening experience for which the abductee has not given his or her consent. It is the fact that the experience is involuntary and not sought after by the "victim" that makes it comparable to a rape scenario – or maybe even an act of war.

WEBSITES TO VISIT:

FOR DAVID JACOBS:

International Center for Abduction Research
www.ufoabduction.com/

FOR WHITLEY STRIEBER:

Unknown Country: The Edge of the World
www.unknowncountry.com

FOR BUDD HOPKINS:

http://www.simonandschuster.com/authors/Budd-Hopkins/702623

Right:
Whitley Strieber (seated)
and
Sean Casteel

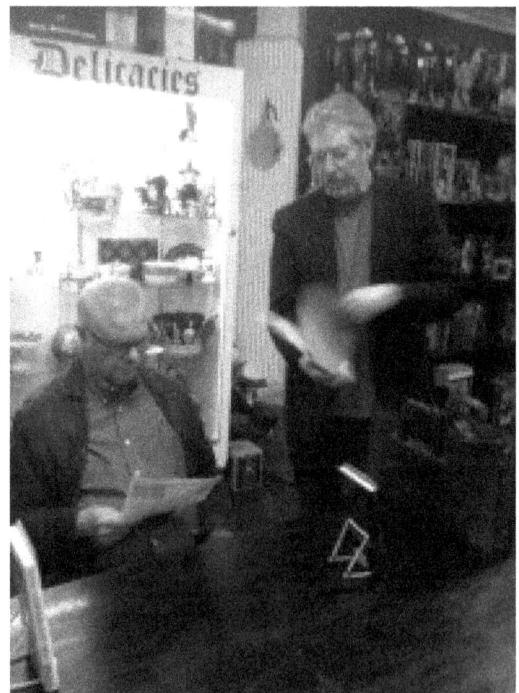

Chapter 30

CELESTIAL SECRETS OF THE GODS:
PEACEFUL EMISSARIES FROM OTHER REALMS?
By Hercules Invictus

Some thoughts on the Planetary Spirits and the Nature of their Actions

About two thousand years before modern contactees began interacting with alien visitors and recording their cosmic messages, Emissaries from the Realms Above were interacting with our ancestors who dwelt near the Mediterranean Sea. They too sought to harmonize human activity with the will of Heaven and revealed many of the same doctrines their counterparts are sharing with us now.

In antiquity, the Vault of Heaven, also called the Firmament, was conceived as a dome that encircled and contained our world. Both of the Luminaries (our Sun and Moon), and the five Wandering Stars (the Classical Planets) dwelt beneath this barrier and each exerted their influence upon the Earth and all its inhabitants. Our atmosphere, and all it contains, served as the Aether that bridged our terrestrial realm with the abodes of the Gods.

The ancients knew the Aether to be populated by Aerika, the faery intelligences known to Western Occultists as Sylphs. These denizens of Elemental Air are invisible to our eyes unless they choose to be seen. When in view, Aerika are usually perceived as comely and majestic humanoids, animal/human hybrids or fabulous beasts.

The Celestials, a very high order of Sylph, are beings of light who can assume various forms to facilitate interacting with us. They are the Olympians of the Greco-Roman world, the Pagan Gods in the pantheons of old, the Virtues in angelic lore as well as the Nordic aliens of the modern contactees. They claim to have created us, they guide our evolution and by all accounts they are intimately involved in the everyday affairs of their favorite mortals. They intermix with and sometimes incarnate through us.

Since the days of Babylon, the Seven Heavenly Bodies were equated with these entities and each was honored and approached on their corresponding day of the week. Their ordering of the celestial spheres demonstrates understandings which differ from our own. These Tables of Correspondence were subject to change as the population's knowledge of astronomy, astrology and other arcane sciences increased.

Sunday is the day of the Sun and often associated with Hercules, Apollo, Helios and Hyperion.

Monday is the day of the Moon and equated with Athena, Diana, Artemis and Seline.

Tuesday (Anglo-Saxon Tiu's Day) is dedicated to the Gods of War, ergo Ares and the Planet Mars.

Wednesday (Woden's Day) is sacred to the Powers of Magic, hence it is dedicated to Hermes and the planet Mercury.

Thursday (Thor's Day) honors Zeus and his planet Jupiter as well as the Open/Visionary Mind/Intellect.

Friday (Freya's Day) celebrates Love, the Goddess Aphrodite and the Planet Venus.

Saturday is the day of Saturn (aka Kronos), who imposes Limits and Structure.

Cacodaemon
(Bad Demon)

These deific correspondences served, and still serve, the Theosophical world view that emerged during the Hellenistic Era (the days between the conquests of Alexander the Great and the fall of the Deified Caesars). This is not surprising, as Theosophy is actually a body of lore given to us by the Celestials themselves. Theosophy's syncretic and universalist approach focuses on similarities rather than differences between seemingly irreconcilable spiritualities and postulates an Ancient Wisdom that unites them all.

The ancient Gods are served by a variety of Daemons, or In-Between Spirits. They too are dwellers of the Aethers, fellow Aerika. Like their Masters, these entities are neither good nor evil. But their actions

Halphas: Great Count

affect us in ways that are sometimes to our benefit and at other times to our detriment. Aerika sometimes also serve as Guardian Angels or Spirit Guides. The Higher Selves of certain individuals (like Socrates) have also been described as Daemonic. The Nephelae, like other Aerika, can readily assume many forms. They are masters of disguise, can become quite solid, and can impregnate/be impregnated by mortals. The Nephilim, or Giants/Heroes of old, were said to be the offspring of such unions.

Daemon is a neutral term, like Human or Angel. A Daemon who helps us is called an Agathodaemon (Good/Pure Daemon). One that does us ill is labeled a Cacodaeman (Bad Daemon).

The Gods themselves (including the God of the Old Testament) are often credited as the authors of both the good and evil in our lives. In the planetary schema, Zeus/Jupiter and Aphrodite/Venus are considered Benefics as they are likely to be beneficial and generous in most of their interactions with humanity. Kronos/Saturn and Ares/Mars, on the other hand, are thought of as Malefic, as their influence usually manifests in what we consider negative ways.

In fact, according to some interpretations of this astral information, Kronos/Saturn (the Great Malefic), to whom the Sabbath Day (Saturday) is sacred, is one and the same with God the Father in the Abrahamic traditions. He is also equated with the Angel Samael and sometimes even Satan himself. The Gnostics favored this viewpoint and it has caused them much woe as Yahweh/Jehovah still has many dedicated followers who are quick to smite in his name.

According to some Greeks, Ares/Mars (the Lesser Malefic) actually slew the resurrected Christ and assumed his Messianic mantle. Christianity took root in Greece when Paul addressed the Athenians from Aeropagus (Ares' rock) at the Akropolis and the new religion of Agape (Love) became dominant, aggressive, oppressive and war-like in just a few centuries.

According to the *Arbatel of Magick* and other arcane texts, Ares (here known as Phaleg, Phalec or Pharos) rules 35 provinces of Aerika. Mythic lore informs us that he is served by several formidable martial Martian Daemons, including:

Lyssa, who is Mindless Fury personified. Her name is the Greek word for Rabies and she spreads madness and irrational violence. She can manifest as one entity or several beings at once and seems to prefer wearing short dresses when in human form. Lyssa is extremely dangerous and keeps company with Mania and other spirits of irrationality and insanity.

Eris, Ares' daughter (or sister) and the Goddess of Strife/Discord. She is often portrayed as being winged and one of the Ker (Ladies of Pestlilence). Eris causes disharmony by appealing to our desire to be better than others. She inspired contests of one-upmanship. Eris has a modern following, the Erisians (aka Discordians). Her present-day scriptures include the Principia Discordia.

UFO HOSTILITIES AND THE EVIL ALIEN AGENDA

Enyo, another of Ares' daughters (though sometimes she is referred to as the War God's sister or wife). She delights in battle and is honored as a War Goddess. Some consider her to be identical to Eris, an aspect of Eris or Eris' twin sister. Enyo also has modern-day worshipers, warrior-acolytes with dedicated shrines.

Phobos, who is a son of Ares. His name means Fear and, with his twin brother Deimos, he accompanies his father into battle.

Deimos, Ares' son and Phobos' twin brother. His name means Terror and very little is actually known about him. The brothers were honored in Sparta and the moons of Mars are named after them.

Other Attendants: Trembling, Panic and Dread.

Quite a formidable crew! And on the surface quite Malefic! Yet, in truth, Mars and his War-Band are humanity's Mentors at this stage of our collective evolution. In some of the surviving Etruscan lore Martus (Ares/Mars) is the beloved son of Hercules (Hercle) and Athena (Mnerfa). He is the Crown Prince of the Heavens and is not seen as a negative figure at all. He is an exuberant youth who seeks to awaken the Heroic Spirit in Humanity. He provides us with situational challenges to test our mettle and helps us achieve greater life-mastery through conflict resolution.

KRONOS

Can we learn to quiet our minds when circumstances suddenly spin out of control? Can we avoid playing the petty and hurtful games that keep us stuck in unpleasant situations? Can we recognize when we are being manipulated by those who smilingly egg us on? Can we face our greatest fears without trembling? Can we remain centered when all seems lost? Can we confidently confront challenges head on rather than seeking to avoid them? These are a few of the questions that can be answered (and lessons that can be learned) by being mindful of what the Lesser Malefic and his crew can teach.

What of the Greater Malefic? According to Renaissance era texts (where he is called Aratron/Arathron) he commands seventeen million six hundred and forty thousand Daemons and rules forty nine astral provinces. He is the Lord of the Titans and remains one of humanity's best known and most powerful deities.

Despite an epic rebellion against authority during my youth, I currently have no quarrel with Grandfather Kronos. I have learned to value the lessons and discipline he imposes. Having spent a lifetime exploring the mysteries of Mount Olympus, I know that the Powers, who are very real, cannot be contained by our limited and simplistic categorizations. They exist within us and outside of us simultaneously. They are us and we are them, but not exactly. Our lives are variations of their timeless tales and through us they grow and develop.

Kronos is the Master of Time, our most precious commodity. How will we spend the moments we are granted? They are most certainly finite and we have no idea when they will run out. Each moment is fleeting and can neither be grasped nor replaced. Saturn imposes limits, but limits can be transcended.

Although the tale does not survive, the Theban Hercules' encounter with Geras, known to us as Old Age, is preserved on ancient pottery.

Geras was said to be a child of Erebos and Nyx, or sometimes the issue of Primordial Night alone. He is a powerful Daemon in the service of Saturn.

As can be expected, in some depictions our hero is engaged in staving off, or actively fighting, Old Age. The shriveled and diminutive Geras doesn't seem to stand a chance.

In other depictions the encounter between Hercules and Geras seems remarkably free of conflict, almost friendly.

How do we reconcile these contradictory images? Perhaps they are different points in a greater tale. Perhaps they depict different incidents altogether.

Alas, we may never know.

Now that I am approaching 60, with at least half of my life behind me, I would like to venture a guess:

At my age, the impulse to mitigate the effects of aging, which can no longer be ignored, is quite strong. Exercise, supplementation and diet are recruited. And a variety of cosmetic, emotional, mental and spiritual techniques are gradually added to our arsenal of weapons until Geras is kept at bay. Some fight harder, with drugs, surgery and fringe (or cutting edge) science.

Through this struggle, acceptance eventually emerges, and with it peace. Embracing Old Age and facing the finality of all mortal life allows you to appreciate the preciousness of each and every passing moment.

When I found myself in Kronos' domain confronting Geras, Hades became my new Mentor and Cerberus joined my circle of living Totems. This sounds dark and grim, and at first it certainly seemed so, but I now find it apt and oddly comforting. I resolved to learn as much as I could from them.

I also made a promise: Until I die, I resolve to fully live.

Onwards!

ABOUT HERCULES INVICTUS

Hercules Invictus is a Lemnian Greek, a proud descendant of Argonauts and Amazons. He is openly Olympian in his spirituality and worldview, dedicated to living the Mythic Life and has been exploring the fringes of our reality throughout his entire earthly sojourn. For over four decades he has been sharing his Olympian Odyssey with others.

Having relocated the heart of his Temenos to Northeastern New Jersey and the Greater New York Metropolitan Area, he is now establishing his unique niche locally and contributing to his community's overall quality of life. Hercules is also recruiting Argonauts to help him usher in a new Age of Heroes.

Hercules currently hosts *The Elysium Project* and *Voice of Olympus* e-radio shows on the *Spiritual Unity Radio Network*. He currently writes for *The Magic Happens* and *Paranoia Magazine*, has published two e-books on Kindle, *Olympian Ice* and *The Antediluvial Scrolls,* and has been contributing to Timothy Beckley's awesome anthologies.

Hercules founded or co-founded *Mount Olympus LLP*, *Olympian Heroic Path*, *Olympian Shamanic Path*, *Cosmic Olympianism*, the *Regional Folklore Society of Northeastern PA* and the *Center for the Study of Living Myth* here in NJ. He also spearheaded many of the real-world Age of Heroes initiatives and the fictive Mythic Adventure tales.

For more information please Friend him on Facebook or visit his website: http://www.herculesinvictus.net

Spanger-Bellona against the Turks

Chapter 31

THEY HAVE LANDED - AGAIN AND AGAIN AND AGAIN – THERE ARE MYTHS AND MYTHS
A VARIETY OF ABDUCTION EXPERIENCES – SIN VS. SALVATION
By Allen Greenfield

The always insightful Allen Greenfield leads us on a rapidly-paced tour of the fantastic and the phantasmagorical, showing us how the Ultra-Terrestrials have often lead us astray in our endeavors to get to the bottom of the UFO mystery, leading some to think in terms of an intractable foe whose self-serving interests are sometimes unhappily revealed.

****

"The abduction stories form a continuum with old legends and beliefs ...They do contain a message ...given to us by the hidden parts of our being."
– John Rimmer – "The Evidence for Alien Abductions" – 1984

And gradually she lost her fear, and heOffered his breast for her virgin caresses,His horns for her to wind with chains of flowersUntil the princess dared to mount his backHer pet bull's back, unwitting whom she rode.Then—slowly, slowly down the broad, dry beach—First in the shallow waves the great god setHis spurious hooves, then sauntered further out'til in the open sea he bore his prizeFear filled her heart as, gazing back, she sawThe fast receding sands. Her right hand graspedA horn, the other lent upon his backHer fluttering tunic floated in the breeze. - Ovid Metamorphoses

"In a chapter of 'The Varieties of Religious Experience' called 'The Reality of the Unseen,' William James attested to the existence of a 'sense of reality,' distinct from the other senses, in which 'the person affected will feel a "presence" in the room, definitely localized, facing in one particular way, real in the most emphatic sense of the word, often coming suddenly, and as suddenly gone; and yet neither seen, heard, touched, nor cognized in any of the usual "sensible" ways.' As evidence, James produces several firsthand accounts from people who were visited by 'presences' late at night. These have a familiar ring: They sound just like stories from alien abductees, minus the aliens. Objects of belief, James says, may be 'quasi-sensible realities directly apprehended.'" - Karen Olsson

"Modern tales of alien abduction share similarities with ancient stories of 'fairy abductions.' 'The Encyclopedia of Fairies' (Briggs, 1976) explains, 'Those who were taken by the fairies were almost always given a special drink described as a thick liquid previous to any sexual encounter. The victims (most commonly women) were then paralyzed before they were carried (levitated or flown) away into "fairyland," which is always located nearby, although it cannot be perceived under "normal" conditions. The paralysis plays a central role in fairy lore as without it the abducted humans cannot enter fairyland. The word "stroke," which we associate with conditions of paralysis, originated from the ancient terms "fairy-stroke" and "elf-stroke."'" - Vidya Anand

* * * * *

TO CATCH A PARANORMAL PREDATOR:
REPRODUCTIVE PARASITISM AND WHAT FAERIE KIDNAPPINGS
CAN TELL US ABOUT ALIEN ABDUCTION

[Extract from the EsotereX blog and posted by EsoterX in Anthropology, Folklore, History, Mythology]

"I was amazed as people must be who are seized and kidnapped, and who realize that in the strange world of their captors they have a value absolutely unconnected with anything they know about themselves." – Alice Munro

Mostly in the interest of avoiding such a fate, I've been reading up on alien abductions and faerie kidnappings, and have come to the inescapable conclusion that aliens and faeries need an online dating service. Jacques Vallee, in his seminal work "Passport to Magonia," helpfully pointed out the detailed similarities between historical accounts of contact with faeries and modern extraterrestrial encounters, ranging from the broader instances of missing time, hypnotic control, and accompanying strange lights to the minutia of ritual offerings of unfamiliar food.

But by far the most disturbing overlap is the fact that both faeries and aliens seem to have an unhealthy obsession with kidnapping people and a fetishistic fascination with human reproduction (admittedly nothing weirder than you see on Craigslist e.g. "Short, balding Chinese gentleman seeks tall African-American woman with passion for leather and Brahms."). Now, the hypothesis that both faeries and aliens are reproductive parasites, having trouble perpetuating their respective races due to some freak genetic accident, disaster, or horrible twist of fate, and are conducting organized efforts to ensure their own survival through, in the case of aliens, wacky and seemingly unnecessary medical experimentation and involuntary gamete donations, and in the case of faeries, generally demand-

344

ing child care services for various sorts of hybrid faerie/human babies or some such domestic activity related to child rearing (I guess faeries just aren't nurturers), has a certain odd resonance.

Here's the problem. Examine the adult classifieds in any periodical, or visit any dating website, and you'll rapidly discern that the services required by aliens and faeries would be easy to obtain simply by asking. A lot of people would probably pay them to participate. In addition, people on average are mostly decent when it comes to lending a hand (or other appendage), so not caring whether aliens and faeries are part and parcel of the same phenomena, we can be reasonably sure that if a paranormal population appeared and explained that they were having a little trouble getting it on, charitable contributions of everything from human reproductive cells to Viagra and pornography would come pouring in.

The fact that alien and faerie abductions are conducted so surreptitiously, suggests (1) they are associated

The Monarchy of FAIRIES once was great
As good old Wives, and Nurses do relate:
Then was the golden Age, from whence did spri
A Race of Fairies, dancing round a Ring,
Who in the Night-time did inform Mankind,
Of what the following Tales will bring to mi

with a degree of shame, and (2) there isn't some vaguely logical, concerted motivation like combating genetic impairment behind the effort. By examining the commonalities in faerie and alien abductions, one comes to believe that we are not dealing with cultures (supernatural or otherwise) that we simply don't understand, rather with the sociological outliers of those cultures, that is, the sexual predators of the universe with an erotic fixation on humans. We're not being visited by the sophisticated ambassadors of galactic civilization and the luminaries of alternate realities. We're being visited by the degenerates. The rest of them probably think we're a little dirty.

Of course, monsters have always exhibited an unnatural attraction to humans (sometimes for food, sometimes for companionship to fill the lonely hours of monsterhood between midnights, and sometimes just as a chew toy), but their fondness for us was more instrumental, logical, and overt. We tasted good. We're good conversationalists and collect cool stuff. We're an ideal rag doll when all our bones are broken. Monsters and gods abducting humans have a proud history, rooted in ideas about transformation (of monster or man). And there is an underlying order to the anomalies of the universe based on an organized otherworld, which might not have rules we recognize as legitimate for daily terrestrial life, but certainly recognize as representative of a social structure. Kidnapping an unsuspecting human and dragging them off gives us glimpses of that otherworld. Traditionally, the assumption has been that human abduction was simply the monster oeuvre, and that kidnapping by faeries represented a standard operating procedure that served some benefit for both abductor and abductee, i.e. mortal man had some utility for supernatural folk and symbolized a hero's journey for us that invariably led to some sort of positive character development.

Motifs of kidnapping or abduction by supernatural creatures are numerous. Whether the abductors are gods, spirits, faeries, goblins, witches, or aliens from outer space, there are striking similarities. The seduced or captured abductee is always taken to another place, used or manipulated in some way, and, if released, returned in an altered form or with continuing aftereffects...Perhaps the earliest forms of abduction are those associated with shamanism or priestly initiation rites, which may be compared to magical transformative journeys. Shamans, who gain power from their intercourse with the spirit world where, in trance or semi-trance, they are tested and tried, returning to the ordinary world with new powers and a new way of life...As traditional belief dwindled or was rationalized, kidnapping faeries from the otherworld became aliens from outer space who had much the same function and nature. The fairy-fascination subsided (though it did not die) in the 1920s, but before it was completely over the new kidnappers had appeared on the scene (Garry & El-Shamy, 2005, p380).

Scholars of faeriedom have frequently pointed out the differences between the Seelie (the reasonably friendly faeries, who may prank us, but are not essentially malign) and Unseelie (the critters that hate us with unbridled, psychotic rage). Faerie courts went so far as to label the Seelie faeries as "the Good People," who rewarded kindness and charity, and generally regarded us as pleasant and amusing. Seventeenth Century folklorist and the grandfather of faerie anthropology, Reverend Robert Kirk (1644-1692), minister of Alberfoyle and author of what can only be described as the first rigorous ethnography of faeries in the Scottish Highlands, called "The Secret Commonwealth," diligently collected firsthand accounts of human-faerie interactions, noting that many of the stories involved abduction of both adults and children.

Kirk believed in the ability of the Good People to perform kidnappings and abductions, and this idea was so widespread that it has come down to us through a variety of channels. We can therefore examine in detail four aspects of fairy lore that directly relate to our study: (1) the conditions and purpose of the abductions; (2) the cases of release from Elfland and the forms taken by the elves' gratitude when the abducted human being had performed some valuable service during his stay in Elfland; (3) the belief in the kidnapping activities of the fairy people; and (4) what I shall call the relativistic aspects of the trip to Elfland (Vallee, 1993, p100).

Of particular note is the fact the Reverend Kirk emphasizes that faerie society is not a homogenous quintessence of goodness, and in fact that they differ in opinion on notions of fairness, justice, and relations to human beings, which suggest that just as there are weird and obsessive people, there are similarly weird and obsessive faeries (the entire species of which he refers to as "Subterraneans"). This gives us telling hints that most faeries looked askance at human abduction.

"These Subterraneans have Controversies, Doubts, Disputes, Feuds, and Siding of Parties; there being some Ignorance in all Creatures, and the vastest created Intelligences not compassing all Things. As to Vice and Sin, whatever their own Laws be, sure, according to ours, and Equity, natural, civil, and reveal'd, they transgress and commit Acts of Injustice, and Sin, by what is above said, as to their stealing of Nurses to their Children, and that other sort of Plaginism in catching our Children away (may seem to heir some Estate in those invisible Dominions) which never return. For the Inconvenience of their Succubi, who tryst with Men, it is abominable; but for Swearing and Intemperance, they are not observed so subject to those Irregularities, as to Envy, Spite, Hypocrisy, Lying, and Dissimulation." (Kirk, 1893 trans. Lang, p27).

The prominence of kidnapping as a motif in faerie-human relations would not seem to result so much from this being a normal aspect of fairy society, as it is clearly denoted as a transgressive act that other faeries regarded as a sign of poor adjustment; rather it makes for a more exciting folk narrative. If you meet one of the fey, and he turns out to be a pretty cool and levelheaded dude, odds are there isn't much of a story to tell. If he throws a bag over your head and hauls you off to the netherworld, you might one day be able to sell the film rights.

A simplified version of Anthropologist and Folklorist Vladimir Propp's morphology of folktales is that invariably a fairy tale consists of a set of functions and moves: (1) An initial situation is laid out (midnight in rural West Virginia; rainy day on the Scottish moor), (2) An interdiction emerges (don't drunkenly wander in the forest; stay away from the ancient standing stones), (3) the interdiction is violated (rules were meant to be broken?), (4) the violation leads to a consequence (alien abduction; faerie kidnapping) and typically some sort of test/competition (aliens need something from you; faeries need you to perform some service – curiously both often seem to be reproductively related), followed by (5) "victory" (mission accomplished, release from captivity), (6) a further test of the hero in the form of a sustained ordeal, and finally (7) a return. The dramatis personae of such accounts are equally important, and our abductees fall firmly into the intriguing category of the "victimized hero." There's not much to be learned from a folktale that says you were nice to a faerie, and the faerie was nice to you. Well, okay, it does point out a certain value in being a decent person, but it doesn't have the rollicking, action packed excitement of crisis and dénouement. Propp, as a committed structuralist, elucidated the "victimized hero" theme in Slavic fairy tales, which is clearly applicable to Celtic fairy abduction.

UFO HOSTILITIES AND THE EVIL ALIEN AGENDA

This function brings the hero into the tale. Under the closest analysis, this function may be subdivided into components, but for our purposes this is not essential. The hero of the tale may be one of two types: (1) if a young girl is kidnapped, and disappears from the horizon of her father (and that of the listener), and if Iván goes off in search of her, then the hero of the tale is Iván and not the kidnapped girl. Heroes of this type may be termed seekers. (2) If a young girl or boy is seized or driven out, and the thread of the narrative is linked to his or her fate and not to those who remain behind, then the hero of the tale is the seized or banished boy or girl. There are no seekers in such tales. Heroes of this variety may be called victimized heroes...There is no instance in our material in which a tale follows both seeker and victimized heroes (Propp, 1928, p21).

If, as Propp proposed, folklore is meant to guide us through the declaration of a violable rule, the violation of said rule, and the consequences of the violation (doom or redemption), it is difficult to fit the theme of faerie or alien abduction into the category of moral lessons, beyond being able to point to the doll and say, "the bad alien touched me here." This suggests that the faerie and alien abductions don't easily fit into widely observed patterns of the mythological. Don't get kidnapped for the purposes of pleasuring paranormal perverts or providing free childcare is not a useful message. Most of us are fairly avoidant of this anyway. Which leads us to the reproductive aspects associated with both faerie and alien abduction.

Throughout the history of abductions the beings have shown a keen interest in reproduction, an interest expressed in the old-fashioned way for Villas Boas and with greater technological sophistication for the Indianapolis woman whose fetus was removed, but present in some form or other in many accounts. Male abductees may report sperm samples taken and females a long needle inserted into the navel or abdomen for what the beings identify as a pregnancy test, while both sexes experience some inspection of the genital area. More than evenhanded scientific curiosity appears to motivate this attention, since the examiners devote a disproportionate amount of time and effort to the reproductive system at the relative expense of other equally significant bodily systems.

In one case the abductors say outright that their mission is to build a better being by combining their qualities with those of humans. In other cases abductees have seen nurseries and hybrid children. Less direct references, where one man was rejected because of a vasectomy and another as too old and infirm for the beings' purposes, indicate a depth of consistency in this theme.

The reproductive theme so prominent in abduction reports recurs as a staple throughout the lore of supernatural contact and kidnap. Greek mythology portrays Zeus as a notorious lecher prone to carry off mortals, and tells of Pluto taking Persephone to Hades while her mother Demeter deprives the earth of its fertility.

UFO HOSTILITIES AND THE EVIL ALIEN AGENDA

Genesis 6:4 says that "the sons of God came in unto the daughters of men, and they bare children unto them. ..." Demons seduce humans in their sleep as incubi and succubi, La Llorona lures men with sexual attractiveness, and the devil as a handsome man may seduce and carry off a young woman. Even the dead, or at least the undead, feed on the lifeblood or sexual energies of the young and virile.

Of course sex is always a topic of interest in or out of narratives. What makes faeries especially significant in this connection is the fact that they are somehow not reproductively self-sufficient and depend to a degree on humans. Faeries mate with humans, carry off women to fairyland, exchange their elderly for human babies, and seek a mortal midwife to assist at a fairy birth. This reproductive parasitism bears a close resemblance to the relationship between humans and aliens depicted in abduction accounts (Bullard, 1989, p156).

Scholars have tried to equate beliefs about faerie abductions with memories of ancient conflicts between pre-Celtic peoples and Celts or secret Druidic recruiting drives, in which kidnapping of children was probably more prevalent, as would have been the use of caves as hideouts by a dwindling minority. But this seems like the typical intellectual stretch engaged in by those who wish to be entirely dismissive of the possibility of alternate realities and do so without blatantly referring to those who hold such beliefs as morons.

Some writers have argued that the changeling belief merely reflects a time when the aboriginal pre-Celtic peoples were held in subjection by the Celts and forced to live in mountain caverns and in secret retreats underground, occasionally kidnapping the children of their conquerors. Such kidnapped children sometimes escaped and told their Celtic kinsmen highly romantic tales about having been in an underground fairy-world with faeries. Frequently this argument has taken a slightly different form: that instead of unfriendly pre-Celtic peoples it was magic-working Druids who — either through their own choice or else, having been driven to bay by the spread of Christianity, through force of circumstances — dwelt in secret in chambered mounds or souterrains, or in dense forests, and then stole young people for recruits, sometimes permitting them, years afterwards, when too old to be of further use, to return home under an inviolable vow of secrecy (Evans, 1911, p245-246).

Ultimately, faerie and alien abductions seem to be a product of a paranormal fetishism, where creatures that are themselves anomalies among anomalies are obsessively-compulsively satisfying their own marginal, prurient interests. Ironically for our supernatural abductors, there is a certain segment of society that is probably turned on by the notion of being a sexual aid for aliens or faeries, but as Fran Lebowitz observed, "If your sexual fantasies were truly of interest to others, they would no longer be fantasies." Perhaps these faerie and alien abductors aren't all that odd after all, given the broad range of things that get humans all hot and bothered. They just need to form a club. The ones mutilating cattle and stealing their reproductive organs? They're just plain weird.

References:

Bullard, Thomas E. "UFO Abduction Reports: The Supernatural Kidnap Narrative Returns in Technological Guise". The Journal of American Folklore, Vol. 102, No. 404, (Apr. – Jun., 1989), pp. 147-170Evans-Wentz, W. Y. 1878-1965. The Fairy-faith In Celtic Countries. London: H. Frowde, 1911.Garry, Jane & El-Shamy, Hasan. Archetypes and Motifs in Folklore and Literature. Armonk, NY: M.E. Sharpe, 2005.Kirk, Robert, 1641?-1692. The Secret Commonwealth of Elves, Fauns & Faeries: a Study In Folk-lore & Psychical Research. London: D. Nutt, 1893.Propp, Vladimir 1895-1970. Morphology of the Folktale. 2d ed., Austin: University of Texas Press, 1968 (reprint of 1928 translation).Vallee, Jacques. Passport to Magonia: On UFOs, Folklore and Parallel Worlds. Chicago, IL: Contemporary Books, 1993.

* * * * *

REPTILIAN ALIENS: THE MASTER SHAPESHIFTERS

[The following is extracted from an essay by Christine Aprile on gaia.comblog]

No other alien species strikes as much fear in the human psyche as the Reptilian. These beings, snake-like in appearance and malevolent by nature, are the stuff of nightmares. Is it possible that Reptilian humanoids are the source of the devils and demonic entities who have tormented humanity since early history? Many alien researchers and contactees postulate that these lizard creatures may have been the mythological characters spoken of in numerous ancient religious texts and folk beliefs.

MYTHOLOGY

CHINESE

In Chinese mythology there exists a special reverence for Reptilian creatures. Dragon Kings symbolize the power of the four elemental corners, shapeshifting into humans at will, pulled by celestial dragons in their heavenly chariots.

ISLAMIC

Within Islamic mythology, the Jinn are creatures of smokeless fire who sometimes appear as snake-like beings; the Jinn were created by God and exist under the same rules as mankind.

SUMERIAN

Some researchers believe that extraterrestrial entities have influenced humans since the beginning of human history, creating cultural practices around their likeness. Zecharia Stitchen believed that the Anunnaki of Sumerian mythology was an ancient ET race, controlling humans and using them as slaves to do their bidding.

BIBLICAL

There are also claims that the snake from Genesis was in fact a Reptilian being, who convinced Eve to break her oath to God by tasting the forbidden fruit of knowledge.

Could these myths be interpretations of Reptilian humanoids, suited for the time, place and circumstance around ancient moments of contact?

ORIGIN

These creatures are thought to originate from the star system Draco.

CHARACTERISTICS

PHYSICAL CHARACTERISTICS

While some of these mythological creatures play benevolent roles within their given society, the Reptilians encountered in modern abduction scenarios are generally cruel and negative beings. Standing anywhere from 6 to 8 feet tall, their most recognizable feature is their snake-like head, skin and eyes.

Abductees report a variety of skin colors ranging from brown to green, red, and sometimes white. These colors and the presence of wings are said to signify rank amongst the Reptilians, with the white skinned beings viewed as the elite class.

Their webbed hands have three fingers, tipped with long, sharp talons, and they are often seen wearing armor or cloaks. Some abductees report scratches and bruising after an encounter. Reptilians are fourth and fifth density beings.

COMMUNICATION METHODS

Many agree that a hallmark of the Reptilian alien is an almost sadistic tendency towards eliciting human drama and fear. These beings utilize psychic communication, and abductees report that Reptilians seem to intentionally manipulate human emotions. This is achieved by using the emotional field created by trauma as an energetic source that the Reptilian supposedly feeds from. Some can implant screen memories upon their subjects, creating false scenarios to hide an abduction occurrence. It is also reputed that Reptilians can access the human dreamscape, attacking people on the astral plane. Reptilians are known to be master shapeshifters, able to assume human form.

MODE OF TRANSPORTATION

Reptilians generally travel in disc-shaped craft.

INTENTIONS ON EARTH

David Icke, a popular Reptilian researcher, has accused presidents, kings and queens of being shapeshifting aliens, intent upon controlling the resources on planet Earth for their own benefit. Although the Reptilian is often approached as a physical creature, some claim that these beings exist outside of our dimension. This would make their shapeshifting an immaterial manipulation of human consciousness. Reptilian creatures seem to be warlike beings, bent on conquest and control.

COMMON ABDUCTION SCENARIO

Positive Reptilians are not the norm, and many abductions involve forcible acts upon the abductee. Some of these intrusions are sexual in nature, leading researchers to the conclusion that a hybridization between human and Reptilian may be in progress. Abductees have reported encountering strange amphibian-like beings while abroad Reptilian ships, but perhaps the truth of their intervention dates back even further than recent encounters. Researchers hypothesize that human beings may have been genetically altered by these entities for thousands of years, torn from a peaceful evolutionary path by otherworldly forces and subsequently enslaved.

The darkest rumor goes beyond genetic manipulation. Some believe that Reptilians are farming humans as cattle to satiate their apparent taste for Earthling flesh.

NOTABLE RESEARCHER: DAVID ICKE

Abduction scenarios and contactees of the Reptilian species seem to share similar experiences of psychic violence and manipulation. The most notable researcher of these creatures is David Icke, a former BBC sports correspondent turned alien researcher, who states that the Reptilians are the source of many of the problems that currently plague societies around the world. Their goal is the domination of resources and all aspects of human life, from the physical to the psychosocial.

Icke believes that this elite ruling class, also known as the Illuminati, carry the vestiges of Reptilian DNA from an ancient hybridization program.

These powerful rulers seek to remain in power through any means necessary. Icke's views on the Reptilian Agenda are quite controversial, and have been criticized by some as concealing racist undertones.

ARE REPTILIANS INHERENTLY EVIL?

There are some who claim to have been in contact with peaceful Reptilians, which leads to the question of whether an entire species can be classified as inherently evil. After all, humanity has its share of cruel and manipulative individuals, yet many people on Earth are good natured and value love above hate and destruction.

UFO HOSTILITIES AND THE EVIL ALIEN AGENDA

ABOUT ALLEN GREENFIELD

A past member of the Society for Psychical Research, he has twice been the recipient of the "UFOlogist of the Year Award" of the National UFO Conference. His book Secret Cipher of the UFOnauts discusses UFOs in terms derived from Carl Jung. His The Story of the Hermetic Brotherhood of Light includes discussion of the Hermetic of Light vs Helena Blavatsky. His book The Complete Rite of Memphis is a comprehensive history of an Egyptian Rite of Freemasonry. Greenfield has devoted the last eight years to the worldwide Free Illuminist or Congregational Illuminist Movement which currently has at least three thousand members.

SUGGESTED READING

Silver Bridge by Gray Barker, Introduction by Allen Greenfield.

Angel Spells

Saucers and Saucerers. PANP Press, 1975.

The Complete Secret Cipher of the UFOnauts. Paranoia Publications, 2016.

The Book of Lies: The Disinformation Guide to Magick and the Occult. Edited by Richard Metzger.

Dr. Karla Turner and husband, Elton Turner

Chapter 32

ARE ALIENS NEGATIVE FROM OUR POINT OF VIEW?
By Karla Turner, Ph.D.

Dr. Karla Turner (1948-1996) was an English teacher and ufologist. She wrote books and gave speeches about UFOs and the alien abduction phenomenon. She investigated claims where various contactees said that the aliens were friendly. Turner called into question how "friendly" these aliens were and became a respected voice of anti-alien protest in the UFO community. When Turner died of cancer in 1996, her death was suspected of being foul play, possibly perpetrated by the Men-In-Black. What follows are Turner's findings, portions of interviews she gave to sympathetic journalists and the testimony of her husband Elton on the "lawlessness" of the alien agenda.

****

If "abduction" reports can be believed – and there is no reason to doubt the honesty of the reporters – the abduction phenomenon includes the following details:

1) Aliens can alter our perception of our surroundings.

2) Aliens can control what we think we see. They can appear to us in any number of guises, and shapes.

3) Aliens can take us – our consciousness – out of our physical bodies, disable our control of our bodies, install one of their own entities, and use our bodies as vehicles for their own activities before returning our consciousness to our bod-

ies.

4) Aliens can be present with us in an invisible state and can make themselves only partially visible.

5) Abductees receive marks on their bodies other than the well-known scoops and straight-line scars. These other marks include single punctures, multiple punctures, large bruises, three- and four-fingered claw marks, and triangles of every possible sort.

6) Female abductees often suffer serious gynecological problems after their alien encounters, and sometimes these problems lead to cysts, tumors, cancer of the breasts and uterus, and to hysterectomies.

7) Aliens take body fluids from our necks, spines, blood veins, joints such as knees and wrists, and other places. They also inject unknown fluids into various parts of our bodies.

8) A surprising number of abductees suffer from serious illnesses they didn't have before their encounters. These have led to surgery, debilitation, and even death from causes the doctors can't identify.

9) Some abductees experience a degeneration of their mental, social, and spiritual well-being. Excessive behavior frequently erupts, such as drug abuse, alcoholism, overeating, and promiscuity. Strange obsessions develop and cause the disruption of normal life and the destruction of personal relationships.

10) Aliens show a great interest in adult sexuality, child sexuality, and in inflicting physical pain on abductees.

11) Abductees recall being instructed and trained by aliens. This training may be in the form of verbal or telepathic lessons, slide shows, or actual hands-on instruction in the operation of alien technology.

12) Abductees report being taken to facilities in which they encounter not only aliens but also normal-looking humans, sometimes in military uniforms, working with the alien captors.

13) Abductees often encounter more than one sort of alien during an experience, not just the grays. Every possible combination of gray, reptoid, insectoid, blond, and widow's peak have been seen during single abductions, aboard the same craft or in the same facility.

14) Abductees—"virgin" cases—report being taken to underground facilities where they see grotesque hybrid creatures, nurseries of hybrid humanoid fetuses, and vats of colored liquid filled with parts of human bodies.

15) Abductees report seeing other humans in these facilities being drained of blood, being mutilated, flayed, and dismembered, and stacked, lifeless like cords of wood. Some abductees have been threatened that they, too, will end up in this condition if they don't co-operate with their alien captors.

16) Aliens come into homes and temporarily remove young children, leaving their distraught parents paralyzed and helpless. In cases where a parent has been able to protest, the aliens insist that "The children belong to us."

17) Aliens have forced their human abductees to have sexual intercourse with aliens and even with other abductees while groups of aliens observe these performances. In such encounters, the aliens have sometimes disguised themselves in order to gain the cooperation of the abductee, appearing in such forms as Jesus, the Pope, certain celebrities, and even the dead spouses of the abductees.

18) Aliens perform extremely painful experiments or procedures on abductees, saying that these acts are necessary but give no explanation why.....Painful genital and anal probes are performed, on children as well as adults.

19) Aliens make predictions of an imminent period of global chaos and destruction. They say that a certain number of humans...will be "rescued" from the planet in order to continue the species, either on another planet or back on earth after the destruction is over. Many abductees report they don't believe their alien captors and foresee instead a much more sinister use of the "rescued" humans.

"In every instance from this list, there are multiple reports from unrelated cases, confirming that such bizarre details are not the product of a single deranged mind. These details are convincing evidence that, contrary to the claims of many UFO researchers, the abduction experience isn't limited to a uniform pattern of events. This phenomenon simply can't be explained in terms of cross-breeding experiments or scientific research into the human physiology....... Before we allow ourselves to believe in the benevolence of the alien interaction, we should ask, do enlightened beings need to use the cover of night to perform good deeds? Do they need to paralyze us and render us helpless to resist? Do angels need to steal our fetuses? Do they need to manipulate our children's genitals and probe our rectums? Are fear, pain, and deception consistent with high spiritual motives?"

ALIEN BEHAVIOR: CONCEPT OR PRECEPT?

By Elton Turner

"186,000 M/Second - it's not just a good idea, it's the law," the bumper sticker read. Something about that statement irritated me. Here we were in the midst of a UFO conference and someone was selling an old reality!

The mixture of our notions of physical reality and our concept of law are keeping us in the dark ages of human thought. Modern science has brought us many new ideas about the nature of the universe, but those ideas are constantly being challenged and changed as our powers of observation sharpen and equipment improves. I thank our scientists for their contributions; I love air conditioning, airplanes, and the television waves that travel our air. What bothers me, is that we have not stopped to consider, "The Law."

What laws do the invaders (and I use that term intentionally) of our world abide by? I posit that we have no idea of the rules of the navigable universe by which these otherworldly entities operate. We continue to develop ideas of their intentions based on our own social rules and written laws. I asked a prominent author and researcher of the UFO phenomenon the other day why he thought the aliens could be trusted - why we should believe what they tell us. His reply was sincere, I think. He said they have demonstrated their truthfulness by predicting some future events, and, lo and behold, what they said came true. He said they have told us our planet is in ecological crisis and we know that is true. And, although they seem to have been here for thousands of years, they have not invaded us. What wonderful creatures they must be!

In the few years that I have been studying my own personal invasion by these creatures, I have come to understand that the invaders do not tell the truth unless it serves THEIR purpose. They play on our fears, using pollution, war, nuclear holocaust and greed as backdrops for their warnings. But every day since I was a small child, I have been aware that those things are part of our world. We all know these things. It does not take a zillion-plus-IQ creature from the planet Orlon to make me aware that we have problems in our world that we must face. We have very human problems to deal with - problems that we can deal with.

The problems we cannot yet overcome is that of outside interference in our affairs. Some people may call it "benevolent intervention" and point to positive results. I respect the scientists and laymen of all disciplines who have been studying the alien phenomenon and artifacts for the past 50 or 60 years. It appears that they have made some progress, as witnessed by the rapid developments in the aerospace industry, medicine, communications, etc., a great deal of which seems to have come from such research. What is missing is a thorough and public study of the mission and rules of engagement in the war for our world.

I believe that our very thoughts and consequently, our behavior as a race of sentient beings are being UNDERMINED through the power of insinuation and the implantation of controlling devices in our bodies by non-human (most of the time) entities. This is truly the most effective way to invade and conquer. I do not trust such creatures no matter what I have been told about their altruistic motives.

We, as a race, have never been free to discover our own true identity. Every social advance we attempt is thwarted by some maniac who springs up with almost divine grace to lead us into madness. Saint Paul, for instance, seems to have taken the real message of Jesus and his earliest followers and distorted it into something that we kill, lie and cheat for. And, in spite of all that, we still aspire for redemption of our souls. The followers of that doctrine - Christians, they call themselves - are not the only ones who behave in such a manner. Every major religion has managed to find an excuse in its teachings to destroy non-believing fellow human beings. A part of me shudders every time I hear of yet another killing based on 2000-year-old hatreds.

What law allows us to continue with such atrocities? What influence keeps such hatred and fears alive? Why are we abductees so afraid to ask for real help from our own society?

We have been INVADED - but I do not yet believe that the battle is over. Invasion with sticks, knives or guns is a human reality, not necessarily a universal one. There are very sophisticated mechanisms being used in the invasion of our world. Why should our invaders use pointed sticks against us when they can get us to sharpen sticks and use them against each other? We provide them with everything they want from us, and they take none of the blame for our misery. They just zip around in their wonderful flying machines, dazzling us with their magical abilities and filling us with awe at their insight.

Can there be a more successful military campaign than one in which no shot is (apparently) fired and in which the conquered populace gladly and openly welcomes their enslavers? We are being programmed mentally and socially to accept our invaders as saviors, not a conquering force. I truly believe we are being deceived by smoke, mirrors and sleight-of-three-fingered-hand movements. Are we going to sell our birthright to some sneaky beings who appear on our shores in marvelous ships and offer us a few glitzy baubles?

The researcher asked if I personally knew of harm that has come to anyone at the hands of, or because of, the aliens. Yes, harm has come. My early youth was damaged severely by the unconscious knowledge that I was being used by some non-human agency. It took me 40 years to recognize that the fears which guided me daily were not of my own making and that the rebellion I constantly felt was engendered by my contempt for the powerful invisible agents that forced me to do things that I knew were wrong, even as I was doing them.

For example, I did not want to marry the person who became my first wife, yet I had no control over the decision. Before we were married, we were jointly abducted and subjected to severe programming. The results brought no happiness to either of us. We both starved for love and companionship, even though we tried with all our might to find them. My son (now 25) was also one of their subjects, and is miserable and lost. He is an artistic person with so many unknowable fears that he is paralyzed. I know of abductees MURDERED by mutilations (reports of which are suppressed immediately and completely), by cancers that no physician has ever seen before, and by madness that has led to suicide. In my opinion, these acts were not caused by "brothers" of any sort.

I do not believe that all is lost, however. I have felt a guiding hand that helped me to discover happiness and inner peace amid all this chaos and misery. What I have come to understand is that that hand is only there when I take responsibility for my own happiness and do something about whatever is bothering me.

Reality left in the hands of the invaders is neither what we need nor what we want. It is time that we think hard about ourselves and what we have on this gem of the universe, our home - our planet. There are laws governing the actions of our invaders, rules guiding their actions and patterns of behavior we can discover if we will make a concerted effort to discern them. We humans have something valuable that is desirable to, and usable by the alien forces acting on us. I feel it is time we take back that which is ours, that we use all our resources to discover the laws that govern reality and become the beings that we intrinsically know we are.

This article was originally published in *Contact Forum*, September/October, 1994.

AN INCREDIBLE INTERVIEW WITH AN INCREDIBLE KARLA TURNER PH.D.
From Contact Forum, May/June 1995

Some editorial notes added by the mysterious Branton.

(**Note:** Karla Turner was a dedicated and compassionate woman who helped many abductees come to terms with their abusive alien experiences, and was not afraid to tell the truth, no matter how bizarre it might have sounded. Karla has appeared on the Montel Williams show and other programs to discuss her research. She passed away as a result of a particularly deadly form of terminal cancer. She will be missed. Some suspect that the cancer was induced via radiation received from earlier abductions, but this cannot be confirmed. This is a rather extensive interview, so we will deal with only some of the highlights. Once you have finished reading the following you, like myself, may also become a firm advocate of "draconicide"! - Branton):

INTRODUCTION:

CF: You are widely regarded as one of the leading experts in the field of UFO and "alien-abduction" research. How did you get started in your study of these things?

KT: Our family knew nothing about the phenomenon when we started having UFO sightings and abduction encounters. Being a researcher, I turned to the UFO literature for an explanation. When I absorbed what was available, I found no answers that I felt were trustworthy. I decided that this was a crucial situation for my family (if not globally), and the only way I could get answers was to do the research myself. The only way to do the research, in this case, was to go out into the field and deal with abduction cases.

UFO HOSTILITIES AND THE EVIL ALIEN AGENDA

PERSONAL HISTORY — "INTO THE FRINGE":

CF: Was "Into the Fringe" the first result of that? [Karla's first book]

KT: Actually, "Into the Fringe" was not a result of research to gain answers. It is more of an account of my family's awakening to, and coping with, these experiences during the first year and a half, when they were very intense. It was not until after that that I started to branch out and work with other people. I worked with Barbara Bartholic on our case, and began working with her on other cases. Many times she would come to Texas (where I lived) and we would set up a four-or-five-day work session, during which people in that area who wanted to work with her would come to my home. She would interview them and place them under regressive hypnosis there. I began to learn by acting as her assistant. (If Ph.D.'s were available in this field, Barbara should certainly have one. Working with her proved to be much more educational than my academic career.) Then Barbara's caseload got so heavy that she was no longer able to handle it. It was no longer enough for me to assist, and I had to being doing preliminary investigative work myself. And that was how my involvement developed.

GENERATIONAL ABDUCTIONS:

CF: We have been finding, in a lot of cases, that experiencers' parents, sometimes their great-grandparents have had the same types of encounters that they have. Is that what you found in your family?

KT: Yes, it is definitely "transgenerational" in Elton's family. [Elton, Dr. Turner's husband, was given the pseudonym "Casey" in the books 'Into the Fringe' and 'Taken'. They no longer feel it is necessary to protect his identity.] Before Elton's grandmother died, in 1990 or 1991, the family knew she was near the end of her time here, so they asked her to tell some of the old stories, and videotaped her response for posterity... My mother refuses to say anything, because it is just too frightening to her. She has not yet even finished reading "Into the Fringe". Each time she reads a page or two, she becomes so upset that she can't go any further — which tells me that there is probably a reason for her feelings. I remember that, in 1965, when I was a senior in high school, a big flap was making national news. It was one of the few times that I had ever paid attention to the UFO thing. One day, Mother and I were listening to the TV while doing something in the kitchen. Walter Cronkite was talking about the UFO flap, and I told Mother that if a UFO landed in the backyard, I probably would go get on it. My mother, who is extremely gentle, and who never raised her voice or hit me, stopped what she was doing, grabbed me by both shoulders and shook me until I felt as if my teeth would fall out. All the while, she was saying, forcefully, "You swear to me, you will not ever, ever, ever get near one! Don't you dare even say that!" It was the only outburst I have ever known my mother to have in my entire life. I now know — from research — that extreme responses like that to this phenomenon are often indicators that a person has had experiences.

HYPNOTIC REGRESSION:

CF: You mentioned the use of hypnosis, which has been the subject of a lot of controversy. Some of the other researchers have said that people under hypnosis can come up with scenarios that did not happen, in order to please the hypnotist. Some have said that the multiple levels of experience — where one can break through screen memories and ferret out buried memories that are different — are artifacts of the process of hypnosis. What are your opinions about these issues?

KT: I think those positions are completely untenable, they grow out of what I call armchair research. I don't conceive you will find them being espoused by anyone who has actually had the experiences. If they have been through them and want to come back and talk about what happens when they undergo hypnosis, to look at what they consciously remember, then we can have a dialogue. Right now, they are speaking without knowledge. They are speaking hypothetically, and their opinions are based on erroneous understandings of the phenomenon, of the experiences, and of the control exerted upon abductees during these experiences. It is easy to philosophize any number of explanations, but that does not mean that those explanations have any relationship to what is really going on. Also, there are bad hypnotists and good hypnotists. (Note: A bad hypnosis is one who in essence takes advantage of the individuals psychological vulnerability by engaging in what amounts to 'psychological rape'. A good hypnotist will NOT violate a persons' will, or try to alter their perception of reality, or inject unsolicited post-hypnotic suggestions in the individuals mind, or attempt to make them do or believe things that they would NOT consciously accept while in a waking state. THIS is the danger of hypnosis. While in this state, a hypnotic subject is completely at the mercy of the hypnotist. This is why a hypnotist dealing with suppressed memories must be someone WHO CAN BE ABSOLUTELY TRUSTED, and one should NEVER be "put under" unless there is at least one other trustworthy individual in the room monitoring the hypnotist. Hypnotism is serious business and should NOT be attempted by anyone who is not qualified. - Branton). A bad hypnotist probably can foul up a number of things. I know that people who have gone to hypnotists for smoking or dietary problems have sometimes suffered more after hypnosis. Obviously, some things can be mishandled. But my experience with hypnosis and the veracity of what is recalled has, in several cases, been proven to me to be accurate. I have been able to investigate these cases. At times erroneous material does surface, or is created because of the situation, but that is not typical. I conclude that hypnosis is, by and large, one of the most excellent tools we have. Used properly, it may be the only tool we have to get certain pieces of information (or levels of information) back up to the conscious state. I have been able to test a number of hypnotically recalled memories against externally verifiable evidence, and they have proven to be correct.

UFO HOSTILITIES AND THE EVIL ALIEN AGENDA

SCREEN 'MEMORIES', CHAMELEONS AND 'NAZIS':

CF: You have found, have you not, that sometimes there are multiple levels, like the layers of an onion? An experiencer undergoes hypnosis and comes up with a scenario, then, when he is regressed to a deeper level, he breaks through the first level (you find out that it was a screen memory), and a different scenario emerges.

KT: Yes, and it seems to me that, in some cases, a bottom level can be reached.

CF: How many layers are there; how deep can you go; and what's at the bottom?

KT: We have not done enough research to answer any of those questions without being an armchair philosopher. Typically (not always) the first recall deals mostly with conscious information. When the subject is taken to the next deeper level of the trance state and asked to focus, often what will be reported is that what was seen was not the same as the conscious recall. Then a groping process begins. The subject thinks, "This was inaccurate; I feel that something was wrong; and when I focus, I see that it was not what I thought it was." That is a transitional level. There may be only a couple of levels — as opposed to, say twenty levels — but there certainly is a cover level, underlain by a more solid foundation. If the subjects are helped to program their mental computers to penetrate illusion and to speak only truthful, accurate statements, to, as Barbara has often said, "clarify vision," then they will recall radically different scenarios — not expanded versions of the firsts scenarios, but something quite different from what their conscious memories had left them with. There are at least two levels, and possibly three.

CF: People have told us that they can break through screen memory after screen memory until they get to a scenario involving reptilians, and that is as far as they can go. Have you found that to be the case?

KT: In the few cases that I am very familiar with, when the "base line" was reached, reptilians were involved.

CF: Are the greys always involved in the top level?

KT: Sometimes the first level involves greys, sometimes humans, sometimes Pleiadians, sometimes strange animals.

CF: Abductees tell stories of seeing beings — angelic Nordics, for example — and then, when they concentrate and try to focus on their memories of those beings, they disappear, and behind them are these "lizard people."

KT: I am not familiar with a number of cases. I have heard other researchers talk about the same thing. In one case that I recount in "Into the Fringe," James had mostly conscious recollections and almost no hypnosis. He remembered being drawn into the proximity of a beautiful "Pleiadian" woman, who was very alluring and tender, and almost seductive. She wanted him to come into her embrace. When he got into the embrace, and thought she was going to kiss him, she disappeared entirely, and what was left in her place was a purplish-black, bumpy, almost slimy-looking character with fairly asymmetrical features. I have encountered this same type of creature in a couple of other cases. The entity was very strong. Instead of embracing James, the creature threw him down on the ground and shoved a two-foot-long tube down his throat, into his stomach, and pulled up stomach juices. The next day, he still had some of the bile taste, the interior of his throat was sore, and he discovered claw marks around both sides of his neck, where he had been held down. Whatever the entity was, there was something claw-like about it (which, of course, matches reptilians). Maybe, as close as he was to it, he could not perceive the whole figure. But he could see a bumpy covering, which could equate to the rough, scaly exterior sometimes reported to be reptilian. It is described as bumpy, ridged, bony, strong, clawed.

CF: Apparently these beings have the ability to project different images.

KT: Some people say that they transform — that they mutate or change their own real forms. I don't accept that as accurate. I don't believe they really look like a blond, and they do something to trick you and then they suddenly look like a reptilian. I think that what they alter is human perception. They certainly can project false images — just as Ted's [Ted Rice's] grandmother was shown her dead husband, so that she would consent to have a sexual encounter. Ted's grandfather had been dead for six years. And in the middle of having the encounter with what she thought was her restored husband, the image disappeared — I suppose because the aliens wanted to get the "emotional juice" from her — and she saw a 'reptoid' on top of her. We also have heard stories about military people being present during abduction, and when people focus on them, they change. Budd Hopkins tells a story about a person who saw a military policeman. He wondered why on Earth the MP was there, and tried to focus very carefully on him. When he did so, the MP changed, before his eyes, into an officer of high rank, and then into a NAZI officer. (Note: Under regressive hypnosis Barney Hill — who along with his wife Betty claimed to have been abducted by reptilian 'grays' from Zeta Reticuli — also reported seeing a "German Nazi" military officer on the craft, working with the aliens. This is suggestive that the CIA more-or-less inherited the alien collaboration agenda that was initiated by the 'Baverians' in or around 1933. - Branton) The aliens cannot allow us to be involved with them in our normal state of mind because we would be under our own control, and that is not what they desire."

[Further statements from Karla Turner by subject]:

THREATS THROUGH THE USE OF CLONES:

"...She was told, 'If you don't cooperate, we'll replace you with this and nobody will know the difference.' When Ted was a teenage, he and a number of other teenagers were abducted together and shown copies of their bodies. In this instance, too, the clones were used as threats."

ALIEN USE OF CLONES:

(Note: The cloning of lambs and monkeys is a documented FACT in our society, so it is not beyond the limits of possibility to suggest that a hi-tech alien culture might, unfortunately, succeed at cloning human beings. - Branton)

"Ted recalled a process whereby his original body was killed. They first gave him a glowing, green, fiery substance to drink. It made him extremely nauseated. He vomited it immediately, and then they cut off his head [and his vital fluid was drained from his body into a container]. When his soul energy — or whatever you want to call it — came up out of his body, it remained attached to the body at this lumpy, glowing, green liquid area. It appeared to be unable to get free of that. They sucked it into a little black box, which was set on a counter while the aliens readied his new cloned body. Then they put probes into the shoulders, neck and feet of the new body to activate it. Once it began to breathe, the soul energy could be put into it. His soul energy, which had been stored in the little black box ever since they killed his first body, was introduced into the new body, and because the body was breathing, it was trapped there."

(Note: During this abduction Ted encountered other 'clones', a young boy and a girl. This took place inside a MASSIVE space station, dark green in color with 'spikes' reaching out of it like one might see on a old World War II floating 'mine'. Smaller ships would enter and exit via these 'tube' extensions. He only encountered a few humanoids on board the massive ship, a red-haired women with little or no emotional expression, a dark haired hybrid type 'man' with a malevolent disposition, and a 'kind' man with short blond hair who attempted to nullify his apprehensions throughout it all. Although there were some 'Bigfoot' type people, some dwarfish hairy humanoids, large 'Mantis-like' insectoids with remarkably 'human-like' facial features, and numerous bizarre or grotesque beings which seemed to be genetic hybrids composed of both human-like and animal-like features within this massive station, the greater majority of the aliens were of the common reptilian-insectoid 'Grey' variety. - Branton)

DRACOS 'NEED' EMOTIONAL ENERGY AS FOOD?

"Evidently, a body that has had soul energy in it — one that absorbs certain kinds of emotional energy — is more nutritious than a 'flat' body that has not been through life processes. Let's say that, at eight or nine years old, you have lived some already, and you have some emotional storage going on already. If they kill your nutritionally 'yummy' body for their use and put your soul material into a cloned body that has not had life experience, then they can let you go through several more years in that body and build up emotional energy in it. Then it will be ready for them to eat, and they can continue the process again, many times."

(Note: The reason the reptilians feed off of human vital fluids in a vamperial-parasitic manner, it seems, is because they are in symbiotic relationship with astral parasites which use the reptilian bodies as 'hosts'. The astral parasites have absolutely no connection to the Creator or the Divine, who is the source of the essence of all life force/manna energy, and so they exist in an anti-life field/state and must steal life force second hand, much like a black hole or a sink hole or wallstreet or a cancer tumor vampirizes everything around them to feed their insatiable hunger... in so doing they grow ever larger and deformed, moreso than God or nature intended, and therefore 'require' ever more energy, which is usually stolen from the weak and defenseless. In a sense, these kinds of symbiotic aliens operate as individual cancer cells within a collective alien 'tumor', and they should be dealt with accordingly!!! - Branton)

PSYCHIC VAMPIRES AND INDUCED EMOTIONAL TRAUMA:

"Abductees are put through other kinds of programming and compulsions. The only commonalty that I can see in all of them, so far, is that they all cause great upheaval and produce great amounts of emotional energy. And maybe tastier bodies. But certainly immediate energy production."

HOW THE ALIENS MIGHT CONTROL THE LEADERS:

"Personally, I believe one could control political leaders more easily with implants. Now, if they wanted to use one of their own souls (although some would debate whether these reptilians have souls at all, other than the astral entities possessing them - Branton), perhaps to inhabit the body of a politicians and work full-time through it, that could be done. Perhaps they could simply take the soul out and stick another soul in. They have the ability to retrieve what we call the soul, to store it in a container, and to put it back into another body. They can put it in any body they wish."

ANY POSITIVE (ETHICAL) ETs?

"I do accept there are intelligent forces that can contact and inform us — perhaps to help us help ourselves."

(Note: Contactee Alex Collier says that there are. He was told by his alleged friends' from Zenatae Andromda - who like most federated human-occupied worlds are descended from the ancient Lyran-Terran alliance in the times of Muria and Atlantis, to which most if not all human-like ETs can trace their ancestrage - that three federated carrier ships, Andromedan, Pleiadean and Tau Cetian, had intercepted a huge Dow-Gray alien carrier ship that had emerged from the ocean floor and was trying to skip-planet with a load of hybrids, human children in cryogenic freeze, and literally thousands of human 'souls' that had been 'harvested' and which were discovered trapped-contained WITHIN SMALL ELECTROMAGNETIC BOXES, apparently to be used as ectoplasmic energy sources by the aliens.

UFO HOSTILITIES AND THE EVIL ALIEN AGENDA

So we may very well have friends out there, which is probably why some reptilian ETs [for instance the 'greens' and the 'greys' within the Dulce - Los Alamos - Kirtland AFB megabase that are working with the 'OMEGA AGENCY'] insist that there are no 'nordic' or 'human federation' ETs out there. Of course the LAST thing these dracs desire is for us Terrans to join up with these federated human starfarers, especially in an alliance AGAINST their reptilian collective. - Branton)

UNDERGROUND ALIEN BASES:

"Abductees report seeing other humans in these facilities being drained of blood, being mutilated, flayed, and dismembered, and being stacked, lifeless, like cords of wood. Some abductees have been threatened that they, too, will end up in this condition if they don't cooperate with their alien captors."

(Note: We've had "Operation DESERT Storm"... anyone for "Operation DUNGEON Storm"!? - Branton)

HUMAN CHILD ABDUCTIONS:

"Aliens come into homes and temporarily remove young children, leaving their distraught parents paralyzed and helpless. In cases where a parent has been able to protest, the aliens insist that 'The children belong to us.'"

RAPE AND FORCED SEXUAL ABUSE:

"Aliens have forced their human abductees to have sexual intercourse with aliens and even with other abductees while groups of aliens observe these performances. In such encounters, the aliens have sometimes disguised themselves in order to gain the cooperation of the abductee, appearing in such forms as Jesus, the Pope, certain celebrities, and even the dead spouses of the abductees."

CHILD SEXUAL ABUSE:

"Children abductees sometimes show a new and obsessive interest in their own genitalia after alien encounters, saying that their abductors who come at night have been touching these parts of their bodies."

PAINFUL AND TERRIFYING MEDICAL PROCEDURES:

"Aliens perform extremely painful experiments or procedures on abductees, saying that these acts are necessary but giving no explanation why. Abductees' eyes are painfully removed from the sockets, allowing the aliens to scrape the area or implant devices into the area before the eyeballs are replaced, for instance. Some abductees are subjected to painful constrictions, often around the head, chest and extremities. Painful genitalia and anal probes are performed, on children as well as adults."

SO-CALLED 'ALIEN PROPHECIES':

"Aliens make predictions of an imminent period of global chaos and destruction. They say that a certain number of humans — and the number varies dramatically from case to case — will be 'rescued' from the planet in order to continue the species, either on another planet or back on earth after the destruction is over. Many abductees report that they don't believe their alien captors and foresee instead a much more sinister use of the 'rescued' humans.

(Note: This reminds me of the vision/dream one man related of a mass alien landing and 'harvest' amidst global chaos, with aliens encouraging humans to board their craft so as to escape to a better place, however in the vision/dream this man had the distinct impression that those who boarded these craft would be entering a living hell... Satan's rapture? - Branton)

COLLECTIVE HALLUCINATION OR COLLECTIVE EXPERIENCE?:

"In every instance from this list, there are multiple reports from unrelated cases, confirming that such bizarre details are not the product of a single deranged mind. These details are convincing evidence that, contrary to the claims of many UFO researchers, the abduction experience isn't limited to a uniform pattern of events. This phenomenon simply can't be explained [exclusively] in terms of crossbreeding experiments or scientific research into the human physiology.

SPIRITUALLY ENLIGHTENED?:

"And it becomes clear from these details that the beings who are doing such things can't be seen as spiritually enlightened, with the best interest of the human race in mind. Something else is going on, something far more painful and frightening, in many, many abduction encounters. There is a theory current in ufological research that says abductees who perceive their experiences in a negative way only do so because they themselves aren't spiritually or psychically advanced. Persons with higher cosmic development have positive alien encounters, so the theory goes, and those who have painful or frightening experiences are merely spiritual Neanderthals. This is a pet theory of researchers who claim that aliens, whether objectively real or not, serve as 'mirrors' of our spiritual nature, on an individual or a species-wide basis. Streiber has voiced this theory, for instance, in Majestic, where he says, 'In the eyes of the others [the aliens], we who met them saw ourselves. And there were demons there'.

(Note: On one segment of the STRANGE UNIVERSE TV show, Whitley Streiber commented on a reputed video-tape of a military interrogation of a reptilian 'grey' alien at Area-51. Streiber showed extreme rage and stated in essence that he was astonished that one of these "wonderful people" could be in such a situation, and that the interrogation was "one of the most evil" acts in human history. Streiber did relate time and time again in his writings about the absolute stark terror and evil he encountered in relation to the aliens. He also stated in essence that the terror and pain consumed him to the degree that he felt that the 'old' Whitley disappeared and something else took its place. Could it be that 'Whitley Streiber' is no longer home, in a certain sense, and that the 'person' who reacted to the video was NOT exactly the same 'Whitley' who resided there before his major abduction experiences began? The use of pain, torture and terror is a common and well known means of mind-control and mental fragmentation among 'insiders'. In the throes of pain and emotional intensity or even sexual stimulation it is easy to induce hypnotic programming or even an alternate or "alter" personality within a mind-controlled subject. - Branton).

Having worked with so many decent, honest, positively oriented abductees, however, I believe this theory is wrong. It is worse than wrong — it is despicable, as despicable as blaming a rape victim for the violence committed against her. This attitude [that they have negative experiences because they are not spiritually developed enough] leaves many abductees feeling doubly violated, first by the aliens who took them and then by the UFO researchers to whom they turn for explanations and help. But it is easy to understand why such a theory would be so popular. Humans have a deep need to believe in the power of good. We need for the aliens to be a good force, since we feel so helpless in their presence. And we need for some superior force to offer us a hope of salvation, both personally and globally, when we consider the sorry state of the world. I think the aliens know this about us — they know that we want and hope for them to be benevolent creatures — and they USE our desire for goodness to manipulate us. What better way to gain our cooperation than to tell us that the things they are doing are for our own good? But looking at the actions, the results of alien interference such as the long list above. There is a great discrepancy between what we desire from them and what they are doing to us."

BENEVOLENT KIDNAPPING?

"Not all abduction reports are filled with frightening or painful events, of course. Many people say that their alien encounters felt benevolent, that their ABDUCTORS treated them kindly or at least with a scientific detachment. Some abductees recall being told that they were 'special,' that they were 'chosen,' and that they have an important task to perform for the benefit of humanity. Given such a positive message, the abductees may ignore the fear and the pain of their encounters and insist to themselves and to others that a higher motive underlies the abduction experience. And, in some cases, all that an abductee REMEMBERS is a benevolent encounter and so has no reason to assume any negative action has occurred."

MORE ON SCREEN MEMORIES:

"Intensive research shows that at the core of the human-alien interaction there is a clear pattern of DECEPTION. We know, for instance, that "screen memories" are often used to mask an alien abduction. Such accounts abound, in which a person sees a familiar yet out-of-place animal, like a deer or owl, a monkey or a rabbit, and then experiences a period of missing time. The person often awakens later to find a new, unexplained scar on his body. Uneasiness about the encounter will persist, however, and far different memories may start to surface in dreams or flashbacks, and then the person seeks help to explain the uneasiness. Quite often, hypnotic regression is used to uncover the events behind the "screen memory," and that is when a typical alien abduction surfaces. The most recent research in which I've been involved has turned up yet a second sort of screening process. If it turns out to be accurate, then thousands of abduction cases are in urgent need of re-examination. The typical scenario of undergoing the regressive hypnosis usually results in penetration of the initial blocked memories. The abductee then recalls an encounter, hitherto unremembered, such as undergoing a physical examination of some sort, perhaps having body tissues removed or having a gynecological exam. Other typical reports include the taking of sperm and ova, of being told of an important task to be carried out, or of receiving a warning of upcoming disaster. And in most cases, both the abductee and the investigator come away from the hypnosis session feeling that they have discovered the truth about the experience. Rationalization leads them to believe that the aliens' purposes must be scientifically objective or benevolent. The less threatening and more benevolent the hypnotically recalled event seems, the more satisfied are the investigator and the abductee. "That wasn't so bad, now, was it? These beings are our friends, or at least they are not our enemies." And everyone goes away with a sense of relief. I have yet to hear of a researcher who actually questions the uncovered scenario. But from several recent cases, it is apparent that these recovered memories may well also be yet another screen, masking events that are much more reprehensible. I will explain one such case, to make the point clear."

A STRANGE REPORT.

"A man in his late 40's came to us to explore several alien-related events in his life, and in the interview he told of a strange, although not apparently alien-oriented, episode that had haunted him since childhood. When he was ten years old, his grandmother came to visit in his home, and since the house was small, she shared his bed on the first night of her visit. During the night, the boy was awakened by a loud male voice. He couldn't understand what the voice was saying, but it sounded angry and was addressing the grandmother lying beside him. The next morning, he asked his grandmother, "What was that voice in the bedroom last night?" His grandmother, with tears in her eyes, pulled him tightly to her and said, "That was the devil." She said nothing more about the episode, but she did insist that her son take her back to her own home immediately. It was an unreasonable request, and her son tried to talk her out of it. But the grandmother was adamant, and finally her son agreed to take her home the following day. The entire family made the trip of over a hundred miles back to the grandmother's farm, and within an hour of their arrival, the grandmother suffered a massive stroke and died. Ever since that event, the man had felt a heavy burden of guilt associated with his grandmother's death. Yet there was no conscious reason for him to have felt that way. The entire event was poignant and mystifying, but in all the alien encounters he had subsequently undergone, he had felt that the aliens were his friends and were helping him by expanding his psychic abilities."

[Interview continued. In response to a question about the 'psychic vampire' nature of the aliens, Karla states]:

KT: ...Yes. Now you are getting to why may be the crux of the 'harvest.' That may be that they not only need emotional energy, but also at least one faction [and I would be tempted, if I were to guess, that they would be reptoids] actually uses the physical bodies. They are trying hard to get us detached from our bodies by telling us that they are "only containers." Why? Because they eat our bodies. If a cow knew you were going to eat it, you would want to tell that cow [if they could understand], "Your body is not important. It doesn't matter."

[In response to a question about the spiritual nature of the Greys and the Reptiloids]:

KT: I think it is something to think about. I don't believe the grays have souls, but are more like 'Frankensteins' or 'zombies' or whatever term you want to use for the "living dead." When I have been with them, I have had an overwhelming feeling that they are not alive — that they are dead.

Dr. Gregory L. Little has added the following in regards to Karla Turner's experiences:

Karla told of walking into her home at night when a being grabbed hold of her arm. The creature told her it was her mother, but Karla stated that it looked like a giant grasshopper. Another time, Karla was coming home through her back yard when she felt like she "...had hit an electric fence. I wasn't feeling right...wasn't moving right ...there was a glow everywhere...I stopped...and saw four gray beings standing side by side in my backyard. I assumed I was having a hallucination (but) I'm awake — why? I felt I could see through them and they talked to me telepathically.

"'Greetings, we are your ancestors,' they said. 'We are a part of you, but we are real.'

"I couldn't move as I normally do," Karla continued. "Then two females behind me came up close — they started buzzing."

SUGGESTED READING, BOOKS BY KARLA TURNER
TAKEN: INSIDE THE ALIEN-HUMAN ABDUCTION AGENDA
MASQUERADE OF ANGELS
INTO THE FRINGE: A TRUE STORY OF ALIEN ABDUCTION

Bizarre Mental, Psychological And Physiological Effects

UFO HOSTILITIES AND THE EVIL ALIEN AGENDA

Chapter 33

ULTRA-TERRESTRIALS ARE HERE SCREWING WITH OUR MINDS
By Timothy Green Beckley

DO ALIEN VISITORS from outer worlds walk among us?

According to researcher Milton L. Scott, at times the popular 1960s television show, "The Invaders," starring Roy Thinnes, seems to be just "a bit more than a concoction of a science fiction writer's vivid imagination; the kidnappings, murders and sabotage being done by David Vincent's adversaries with the opposable pinkies are the same things being done by their true-to-life counterparts." Or so says Mr. Scott, a Philadelphia-based theorist who was known to release "position papers" to his peers.

Furthermore, Scott points out that the difference between the television invaders and the real aliens is that there are no funny looking fingers to make them easy to spot. Thus, in order to carry on their various acts of mayhem while on Earth, it is necessary for them to "resemble us as closely as possible." In order to do this, they need not "involve themselves" with using some fantastic hypno-screen that "clouds" our minds to "their true gruesome appearance," because they are no more gruesome looking than Chinese, Japanese, Vietnamese or Ainu.

Because of their Earthling-like appearance, they are able to carry out various acts of "espionage" without being detected.

How does a nation wage an undeclared war with yet-to-be-invented weapons against an enemy who is not supposed to exist? Milton Scott speculates: "You might begin by alerting the nation's power complexes to the ever-present danger of sudden blackouts and advising those stations to close each and every switch within reach when they detect a gigantic surge of power racing down the lines from an unknown source."

Or: "You might make noble-sounding pronouncements of peace and friendship to the world—hoping that whoever is listening outside our civilization believes it. You could sign treaties among nations to ban wars in outer space; clutching at the powdery straw contactees have left to us in hopes that flying saucers invading our skies really are big brother-type angels who shed tears over our savage nature.

"You could even build giant radio telescopes to send messages beaming across space: 'We are really nice fellows. We can't hurt you. We only hurt each other. Don't hurt us.'"

Scott asks: "Then what do you do when the blackouts keep occurring and the deaths and the kidnappings continue to mount?

"You could explode nuclear bombs high in the atmosphere in the hope that the radiation will disrupt the machinery of the saucers or even kill their occupants. You could even test your theory by having the bomb explode in the vicinity of a satellite – like the Transit 4-B satellite – and when the satellite suddenly stops sending signals, you could congratulate yourselves on a theory well proved. Then, what do you do when the satellite suddenly comes back to life FIVE YEARS LATER – and starts broadcasting again?"

Obviously, Scott is convinced that the UFOs are here on a non-peaceful mission. We asked him how he reached this important conclusion. His answer was as follows:

"For years, the public has been fed false information from both the government and newspapers who scoffed at anyone who dared to report a flying saucer. It's gotten to the point that the Air Force and the CIA can expect more information from a Martian than they can from John Doe. Ol' John just won't tell anybody anything.

PSYCHOLOGICAL WARFARE

"That's where my tale begins: the first stages of the war that the flying saucer occupants have been waging against us have been psychological. They have spread confusion, fear and doubt from one corner of the globe to the other in order to keep their movements and their purpose a secret until they are ready to make their move. It's a fantastic tale of ghosts, ESP, thought control, liars, dupes and murder.

"The most important battle in the war of the worlds was waged and won in the minds of the people. If there can be one glaring fault on our part that led to our defeat, it was the view that we were the supreme result of billions of years of a thing called evolution – that we were the only intelligent beings in the universe.

"Our scientists, philosophers and clergy boosted our egos by telling us, endlessly, what great works of God we were – so complex (and so stupid) that there couldn't possibly be anyone else as grand and as wonderful as we were. It was a perfect setup, and the characters from the flying saucers exploited it to the fullest extent: they made a few flights over villages, countrysides, swamps and cities to shake up the public and drive a wedge between belief in what the government says and what our own eyes say."

In order to keep their mission a secret and make reporting UFOs look like the work of idiots, Scott claims that they used ships of varying shapes, sizes, colors and methods of propulsion to spread confusion among the few investigators. They even used an effect that produces a number of images from a few actual saucers so that the viewers would think there were whole fleets of saucers tooling about the skies. Thus the subject of flying saucers soon became a tiling for disbelief and tired old jokes.

Scott contends that the hundreds of little men, gods, beautiful spacemen, winged monsters and surplus Atlantean biplanes and talk of the UFOs being from the bowels of the Earth, a fifth dimension, an antimatter universe, etc., was nothing more than lies and false leads, implanted in gullible earthly minds by the aliens.

In reality we were lulled to sleep while more land was taken. We nodded and dreamed while more men and materials from other worlds were flown in. We giggled as things rushed rapidly toward a point of no return. The government refused to believe the abundance of evidence before its eyes until it was too late.

Adamski and all the other contactees, Scott says, did have real enough experiences, but they were selected for their gullibility, and they wrote books that were just as naïve and as gullible as they were. The books got the large heehaw from the public that the aliens had expected, and the case for flying saucers was laughed into obscurity for ten more years while the aliens went about their plans uninterrupted.

If, as Milton Scott says, the aliens' purpose for being here is other than peaceful, and they look almost exactly like us, how then can they be identified? The answer may lie in the scientific analysis of a "suspected" alien once he has been captured.

"Perhaps the answer is in the chemical balances of the body or in the theory that the little DNA molecules only have a limited number of types to choose from among earthlings."

Interesting theory? So much so that upon hearing of Mr. Scott's opinions many months later, Dr. Edward U. Condon, of the ill-famed University of Colorado UFO study, requested that we send reproductions to him of several newspaper columns which had carried these ideas.

UFO DIVERSIONS – SEEK TO FIND!

In harmony with many of the opinions expressed by Mr. Milton Scott is another famed UFO investigator who we need to bring into the conversation, the late John A. Keel. Besides being one of America's foremost authorities on flying saucers, Mr. Keel had a long history of objective scientific study of ether phenomena. He was a reporter for decades and authored several bestselling books dealing with both offbeat and more conservative topics.

Commenting on the various diversions inherent in the UFO enigma, Keel told us before his passing that: "From 1897 on it has been a common practice for the UFOs to leave behind ordinary debris such as newspapers, pieces of metal, articles of ordinary clothing, mundane chemicals, etc. Investigators who had discovered such items have often been led to believe that the whole incident was a human hoax or prank of some kind. It is also quite common to find ordinary tire tracks in inaccessible fields where the landings have been reported."

Keel warned us that we should not permit ourselves to be misled by these "negative factors." Keel points out that even in these cases a thorough and investigation should be made.

"We have discovered that a multiple grouping of these negative factors often leads to positive proof that a UFO event DID OCCUR."

Other odd factors inherent in UFO contacts is that "ancient Greek is often employed by the UFO occupants. Greek names and phrases are frequently used for their nonexistent planets. Many of the entities adopt Greek nouns as their personal names. The witnesses very rarely realize this or understand it. Prepare yourself by obtaining and studying a book on Greek mythology."

Keel also suggested that we should also study "our own techniques for psychological warfare, since they are often by employed by the UFOs."

Diversionary landings or seemingly important incidents frequently are staged a few miles from an area where a truly significant UFO activity is taking place. The diversion wins all of our attention and publicity and the important activity goes unnoticed.

Like Scott, Mr. Keel informed us that we should discard all preconceptions.

"You must learn to accept only the correlative evidence and ignore the assorted speculations which have dominated UFOlogy. We are interested only in hard facts. All of these facts indicate that we are dealing with an environmental phenomenon, but that we have been misled into believing the extraterrestrial thesis."

Thus, unlike Milton Scott, John Keel was convinced that the flying saucers, although very real, ARE NOT from other planets.

"So long as we accepted the ET concept, the phenomenon and its source was safe and free from interference. Deliberate hoaxes were executed to sustain skepticism and convince government agencies that the phenomenon was 'nonreal.' The UFO buffer was convinced of the ET thesis, which was unacceptable to both the general public and the scientific community. And by loudly advocating it, they succeeded in heaping ridicule upon the subject. Thus the UFO source was able to operate unhindered for decades."

HALLUCINATORY EFFECTS

UFO believers usually rebel at any suggestion that the UFO phenomenon may be hallucinatory or psychological. However, Keel pointed out: "In the past three years many psychological factors have been discovered, and various groups of psychologists and psychiatrists are now actively engaged in UFO research. Unfortunately, very few UFOlogists are trained or equipped to understand or even to investigate the underlying psychological factors. You should read at least on good book on psychiatry and/or psychology."

As far as the contact stories are concerned, Keel told us that "at least some of these cases in the past three years have proven to be hallucinations because it seems that the effects were produced in the witnesses' minds by an exterior influence. These effects are similar to hypnosis. While the witnesses' bodies undergo one sequence or experience, false memories of another sequence of experiences are planted in their minds.

"Frequently, the true (but forgotten) experience surfaces from the witness's subconscious later on in the form of a dream or nightmare. We cannot outline the whole process here, but it must be considered as a very important factor in many cases."

IMPORTANT FACTORS IN UFO SIGHTINGS

Some of the important factors to look for in UFO sightings, according to Mr. Keel, include:

"**EMOTIONAL REACTIONS** – In low-level sightings, auto pursuits, etc., the emotional and psychological responses of the witnesses are extremely important. Get them to explain in detail how they felt immediately before, during and after the sighting. Did they suffer fear, nausea, dizziness? Did they have unusual dreams afterward? In some cases, these reactions are more important than the sighting itself.

"**SOUNDS** – The sounds accompanying the objects can be of great importance. Many of these sounds have proven to be mental in nature. That is, they were not audible movements of air but were electrical responses in the brains of the observers. Beeping sounds frequently indicate that the witness was subjected to an unconscious experience. Such witnesses may find that they are unable to explain lapses of time or geographical transfers during such sightings. Such witnesses should be examined by a qualified psychiatrist whenever possible.

"**EYE BURN** – Witnesses who suffer from burned or inflamed eyes after viewing a UFO should be examined immediately by a professional doctor and a full medical report should be obtained. In those cases involving 'eye burn' weeks or months previous to the investigation, the investigator should get the witness to draw up a full statement explaining in full the reactions suffered. Medical documentation is most important.

"**DREAMS** – Many witnesses suffer unusual nightmares weeks before their UFO sighting. Others have strange nightmares for weeks afterward. These dreams are important, and you should obtain full descriptions of them. Some witnesses begin to have prophetic dreams after their UFO experience.

UFO HOSTILITIES AND THE EVIL ALIEN AGENDA

"In landing cases, when definite marks are found on the ground, they should be photographed and measurements carefully made. For the past 20 years, hundreds of landings have been neglected even though the markings are always similar in size and formation. If we had collected and documented photos of all these landings, we would now have an impressive body of correlative evidence.

"In further investigating important sightings, landings and contact experiences, under no circumstances should any witness be hypnotized by anyone other than a qualified psychiatrist. Amateur hypnotists have ruined several important cases in recent years."

MEN IN BLACK

During this same period which John Keel speaks of, there has been an increasing number of cases which involve the MIB, or Men In Black. These strange individuals have been known to warn UFO witnesses not to reveal what they have seen long before the case is ever made public.

Keel commented on the activities of these MIB, pointing out that many different investigators in "flap" areas have now had confirmatory experiences with the MIB and only a small percentage of these cases have been published.

There are several different types of MIB. One group appears to be more psychic or hallucinatory than real. They appear and disappear suddenly in bedrooms; the witnesses often experience paralysis or a sudden rise in temperature while in the presence of the MIB. We now have dozens of such cases in our files.

Another type common throughout the U.S. is represented by men who travel in pairs. The same description is always given. One man is tall, blond (usually has a crewcut), has a fair complexion and seems to be a Scandinavian. His companion is shorter, with angular features and a dark olive complexion. The blond usually does the talking while the other remains in the background. There seem to be several identical pairs of these individuals operating simultaneously in several states.

Other types of MIB include men with Asian features, dark complexions, slight stature and heavy undefinable accent. These men sometimes pose as salesmen or some other form of business occupation that normally requires a suit and tie.

MIND MANIPULATION: THE NEW UFO TERROR TACTIC

There are mounting indications that UFOs have a long-term plan of operation in store for Earth and its inhabitants. Data, meticulously collected in a worldwide research effort, would seem to support that stunning theory.

UFO literature is filled with hundreds of cases in which unsuspecting observers have been subjected to continuous harassxcments following an encounter with a flying saucer. Many times the witness finds his home plagued by a host of inexplicable phenomena. In other cases, eerie, mechanical-sounding voices, purporting to be "messages" from an alien source, begin emanating from their radios, TV sets or telephones. In addition, mysterious strangers dressed in dark clothing, commonly referred to as the Men In Black, or MIB, visit the often confused eyewitness and warn them not to speak about their sighting to anyone.

Many observers, however, endure far more harrowing experiences than these. As terrifying as these incidents may seem, they cannot be compared to the instances which appear to be actual cases of "UFO possession."

Often, while interviewing a UFO witness or contactee, I find myself face-to-face with an individual who is convinced he is slowly – but surely – losing touch with reality. Having come that close to the unknown, the individual feels his very existence is being threatened by an alien force bent on gaining total control of his body and soul.

Some witnesses persist in believing that they are being "haunted" day and night by an invisible specter whose main objective is to capture the witness's free will and make them the "property" of someone – or something – else.

Cases of UFO possession are often quite common! Yet very little is known about it because of the scarcity of research into the subject. Investigators have remained extremely cautious about digging too deeply into this particular area. Their hesitancy, however, may be justified.

An exhaustive study of my own shows that accounts of UFO possession are almost always identical. Frightfully so! The following patterns have emerged, again and again:

*** After a close encounter with a UFO, the eyewitness goes through a period of anxiety during which he is unable to consciously remember certain aspects of the incident.

*** Within months – sometimes weeks or even days – the personality of the observer actually changes. Eventually the observer's personality may alter to the point where he finds it impossible to get along with coworkers, close friends or even family. Personal tragedy seems to strike many of those who have had ground level encounters with UFOs. Much could be written about individuals whose entire personal world crumbles around them following such an experience.

*** In some cases, the eyewitness discovers he has developed certain "gifts" or abilities. Though they may appear to be beneficial at first, too frequently this is not the case. Among these unusual abilities are extraordinary powers of ESP, precognition or psychokinesis. In addition, a heightened intelligence level or an unusual increase in physical strength may be noticed. Such peculiarities will often manifest themselves shortly before a person is about to be possessed. Shortly after this, he may begin slipping into a "trance," during which it appears as if an alien intelligence has "taken over" his body and is using his brain.

It was during my in-depth investigation of an extensive UFO wave in the U.S. Southwest that I met Simon Swagger, a tall, slimly-built man in his mid-20s. (Because of the seriousness and the possible repercussions this may cause, we have decided to change the names of those individuals involved.)

Simon's story is one of the most believable accounts of alien possession that I have ever heard. And I'm convinced it is not a hoax.

During the course of three rather lengthy conversations with Simon, I felt I learned much about him. Like so many other American boys, he spent his teenage years playing baseball, listening to music and chasing girls. Simon never paid close attention to his schoolwork, with the result that his grades were "just average." Nevertheless, he was well liked by his classmates and also managed to get along amicably with his elders.

Now, at 25, this same "average young man" feels destitute, as if he doesn't have a friend in the world. Nearly a total recluse, he shuns any activity which might expose him to public scrutiny. He is divorced from his wife of four years and has quit or been fired from numerous jobs.

Of course, these radical changes in his life did not occur overnight but rather were a painfully slow period of moral and physical deterioration.

Simon was more than willing to tell me the details of his ordeal. As we talked, it became apparent that he was anxious to get the matter off his chest. The problem had obviously been weighing him down for too many years.

"For the longest time I thought I was going insane," Simon said. "Often my best friends would accuse me of behaving irrationally, and I wouldn't have a clue about what they were referring to. My mind, on those occasions, was an absolute blank. I found myself going to doctors and psychiatrists, but even they couldn't offer me an explanation that could account for these amnesia attacks."

FIRST THE THUNDER AND THEN AN OMINOUS ENCOUNTER

Whatever the cause of his trouble, it was obvious that it was rooted in an eerie confrontation with a visitor from outer space! The event took place on a Friday night in August 1967. At the time, Simon lived with his parents on a rather secluded ranch near Waco, Texas, which was surrounded by trees, dense thickets and bramble. Here is his personal account of what happened on that fateful summer evening:

"The weather had been unbearably hot all day, with temperatures soaring into the 90s. I got permission from my folks to spend the night outdoors, camped in back of our ranch house with a couple of friends.

"We set up a makeshift shelter, turned on a portable radio and proceeded to shoot the breeze. The sky was as clear as I'd ever seen it, with stars twinkling against a background of absolute blackness.

"Around 10 P.M., the air began to gradually get cooler. In the distance we could hear an occasional rumble of thunder, and once in while the sky would light up with a flash of lightning. It was a great sight."

UFO HOSTILITIES AND THE EVIL ALIEN AGENDA

Unfortunately, the beauty of the night was short-lived. It was shattered less than two hours later.

"Shortly after midnight, we lowered the flame of our kerosene lantern and retired," Simon Swagger continued, a slight trace of tension building in his voice. "Immediately I turned over and closed my eyes. Before long, however, a peculiar high-pitched whine woke me up. The nearest I can come to describing this would be to say it sounded like a million bees buzzing."

Sitting up, Simon peered into the darkness and saw nothing. With a few moments, however, he managed to pinpoint the source of the noise. It was coming from the woods near the ranch. With his curiosity now aroused, Simon decided to investigate.

"I didn't want to wake up my friends, so I tiptoed over to the area, hoping to catch a glimpse of whatever was causing the noise. I recall wandering aimlessly farther and farther away from our backyard camp, as if I was being pulled by an invisible rope. All around me, the whine continued to grow in intensity until finally it encircled me on all sides."

At this point, Simon sighted his first UFO.

"Up ahead of me, between the trees and bushes, was a glowing light the size of a basketball. As I approached to within 25 feet of it, I could see the light was actually a pulsating sphere."

In an attempt to block out the loud, irritating noise that continued to grow in intensity, Simon put his hands over his ears. This had little effect, however.

"My head began to swim, and my eyes started to water. Next thing I knew, I was on my hands and knees – somehow I must have fallen without realizing it – crawling on the ground, trying to get back to the safety of my friends."

He was unsuccessful.

Upon "coming to," Simon found himself in his parents' living room. His head was pounding from "the worst headache I've ever experienced." Standing around him were his mother and father and his two friends.

Simon says he found it difficult to understand what they were trying to tell him.

"It was as if they were talking to a complete stranger," he said. "I had, for all intents and purposes, lost my identity. I had no idea who I was or where I was."

While he tried to calm his nerves and gather his thoughts, Simon's friends filled him in on what had happened.

"They said they had suddenly been awakened by a brilliant flash off in the woods. They noticed that I wasn't in my sleeping bag, nor did they see me nearby, and they became worried. Considering all the possibilities, they felt I might have wandered off in my sleep and fallen into one of the many ravines in the area,"

Simon said.

Using a flashlight to guide them through the underbrush, Simon's friends began calling out his name. Their worry grew into fear because he did not respond to their cries.

Five minutes later, their search ended when they found Simon stretched out on the ground face down.

"Lifting me to my feet, they explained how I seemed to be in another world, dazed and looking right through them. My eyes, they claimed, were rolled back in their sockets and my skin had turned as white as a sheet. In addition, they said my flesh felt ice cold, like that of a corpse."

On the way back to the house, they noticed something else. Simon's head had swollen like a balloon. His forehead appeared enlarged and extended several inches beyond normal.

"It was 'puffed up,' as if I'd been stung by a mass of bees."

This condition rapidly disappeared and Simon's head returned to normal by the time the three boys reached the safety of his parents' quiet ranch house.

For weeks afterward, Simon felt worn out, "as if I'd been drained of all my energy." He found it extremely difficult to concentrate long enough to do even the most mundane chores. All he could do was mope around the house and he spent a good portion of the time sleeping.

As the months went by, Simon regained his strength. However, even as he returned to normal physically, he couldn't help but wonder about what had really happened on that late summer night.

"My friends came up with a rather logical explanation. They concluded that I'd been walking in my sleep – I'd never done that before, to my knowledge – and that a thunderstorm had come up in the middle of the night and I had barely missed being struck by lightning. They figured a bolt had struck near where I stood, and, after traveling over the surface of the ground, had reached me. Along the way, the lightning must have lost a great deal of force. Otherwise, they theorized, I would surely have been instantly killed."

Though their explanation seemed reasonable, Simon couldn't shake the persistent feeling that a lot more was involved.

"I recalled a bright light, dizzy spells, headaches, and fainting spells. And a strange buzzing that literally ran through my skull."

He knew there had to be another answer – even if it was an unpleasant one.

After several years had passed, Simon became engaged to and later married his high school sweetheart, Irene.

"After we got married, I took a job as a ranch hand near Calvert, Texas. And though the week was tiring, it paid pretty well. Each week, I was able to put some money in the bank, figuring someday I would have enough saved to buy a small place of our own and perhaps even start a small cattle business."

Since that night in back of his parents' house, Simon suffered both mentally and physically. The dizzy spells, headaches and fainting became common.

"I'd be seated at the kitchen table and all of a sudden my wife would be applying cold compresses to my forehead. I'd pass right out reading the newspaper or eating."

Gradually his condition deteriorated. During this difficult period which followed his UFO experience, Simon became keenly fascinated with science and began reading books on physics and engineering, subjects in which he had never before shown even a mild interest.

"It was as if I were furthering my education," he said. "I didn't know why I found these topics so fascinating. My mind seemed to be developing – expanding – at a rapid clip."

Coinciding with this heightened curiosity and intelligence, Simon found himself growing extremely moody. As the months passed, and after discussing his suspicions with his wife, that ufonauts were trying to control him, it became more difficult for Simon to be around other people – including his wife, parents and coworkers.

"I had a hard time keeping my thoughts together."

He started showing up late for work and then not showing up at all. Finally, he quit, not wanting to wait until he was fired. A string of lesser paying jobs followed, but they all ended the same way. Then things went from bad to worse.

"My mind was incapable of thinking straight. It was always a million miles away, toying with some advanced mathematical formula or scientific equation. The funny part of all this was that I still didn't know the reason why I was so hung up on these things. After all, I wasn't a scientist or an engineer, just a simple country boy."

During this period, personal tragedy struck the Swaggers. Their year-old son suddenly died. Doctors concluded that the infant had succumbed to a cerebral hemorrhage. In early 1973, Simon's wife left him and filed for divorce. One of the reasons she gave was that she felt the boy's death was somehow related to Simon's condition. She felt that the UFO issue had broken up their previously happy marriage. This same pattern is often repeated as UFO witnesses have found their formerly normal lives turned into nightmares.

Simon didn't even bother to contest the divorce action.

"Even though I loved Irene dearly, she didn't matter that much to me anymore. It was as though I had a special mission on Earth. It was 'beyond me' to lead an ordinary existence."

UFO HOSTILITIES AND THE EVIL ALIEN AGENDA

Irene's decision to leave came after Simon's second encounter with mysterious unidentified objects. Not knowing what to expect – or what her husband was capable of doing – she decided to leave.

"Again, I must tell you what happened as seen through another person's eyes. My mind is almost a total blank when it comes to the events of that night."

Simon and his wife were driving home from Belmont, Texas, where they had spent the evening with relatives.

"It was around 1:30 A.M., and I was speeding along the darkened back roads to avoid traffic, when suddenly a large, yellowish ball of fire appeared on the road ahead. Immediately I slammed on the brakes because otherwise I would have collided head-on with the object."

The UFO slowly lifted from the road a few feet above the pavement and began drifting toward the side of the road. About 30 seconds passed before it stopped and hovered next to a grove of trees. It was then that a frightened Simon Swagger insists he was directed, as if by magic, to leave his car and walk toward the UFO, which now remained stationary. Just as in his first UFO experience, the nighttime air was now filled with an eerie, loud whine, similar to a shrill scream.

"My wife pleaded with me not to leave the car, but I was no longer in control of my movements. It was as if my body was being made to react, pulling me in the direction 'they' wanted."

Walking toward the light, Simon says he heard a voice inside his head. This "inner voice" demanded that he walk straight ahead and not look back.

Meanwhile, inside the parked car, Irene Swagger was almost hysterical.

"She considered going for help but was afraid the police might think she was daffy," Simon said. "So, in desperation, and because there wasn't anything else to do, she decided to 'sit tight,' hoping I would return soon."

When he did, havoc followed.

"Somehow, I wound up back in my car. Opening the car door, my wife says I looked like a 'monster' – that's exactly the way she put it. My face was more alien in appearance than human. My features had changed grotesquely, eyes bulging out of their sockets."

She compared his face to the creation of a master makeup artist on the set of a science fiction movie. Shaking nervously, Mrs. Swagger tried to get Simon to climb into the backseat where he could lie down and remain calm while she drove to the nearest hospital. Irene thought a wild beast or perhaps a poisonous snake had attacked her husband. Instead of complying with her wishes, Simon pushed her aside with a "violent shove" that sent her sprawling against the opposite door. Simon slid into the driver's seat and grabbed the steering wheel in a rage.

"Supposedly I was talking incoherently, as if in a trance."

UFO HOSTILITIES AND THE EVIL ALIEN AGENDA

When he gripped the wheel, it bent out of shape like it was made of putty. Within a minute of this remarkable feat – one that would require extraordinary strength – Simon slumped against the dashboard with his eyes shut and his forehead dripping perspiration.

To Simon, it was all a dream. "I don't remember a damn thing after leaving the car and hearing the hypnotic sounding voice and seeing the lighted object. If it wasn't for the steering wheel being twisted, I'd say my wife probably made up the whole crazy story."

Since his second meeting with a UFO, Simon feels more strongly than ever he is being influenced by alien beings. Since his last encounter with ufonauts, Simon Swagger's life has stabilized somewhat. He has learned to cope with the "force" trying to control him. At this time, this "average student" is on the threshold of obtaining a degree in electrical engineering!

THE CIA VERSUS "THE SPACEMEN"

While Simon's narrative is intriguing in itself, he is by no means the only person to have been selected to receive such "special" treatment at the hands of ufonauts. Even high-ranking government officials have received "communications" and been "manipulated."

Somewhere in a locked file cabinet, hidden in some obscure office in the Pentagon, is a two-inch thick file that contains perhaps the best-documented UFO possession case of the decade. The episode actually involves an Air Force officer, the Office of Naval Intelligence and the CIA.

For a long time, this manila folder was closely guarded – stamped "Top Secret." Its contents were finally leaked to an enterprising scriptwriter, Robert Emenegger, on assignment from Sandler Institutional Films, producers of a syndicated documentary on UFOs. The source of this "leak" was, surprisingly enough, Lt. Col. Robert Friend, USAF, former head of Project Blue Book and well-known "staff debunker" for the government.

Now long retired, Friend seems to have done an "about-face" on the question of UFOs. Not only does he think they exist, but he also seems to give serious consideration to the even more puzzling UFO contactee cases.

A most revealing interview with Friend appears in the book "UFOs, Past, Present and Future." In this interview, the former Blue Book spokesman describes a case which contains all the typical elements of a "UFO Possession."

While head of the Air Force's UFO project, Friend says he was informed, as a "matter of courtesy," that a well-respected Rear Admiral was especially interested in a woman living in Maine who claimed to be receiving highly advanced and technologically correct information from extraterrestrial beings. These entities were said to contact her while she sat in a trancelike state. The admiral, with the approval of the Air Force, sent two of his most responsible and trusted men to investigate.

Relaxing in a chair before them, the woman expressed her willingness to answer any questions they might have. At this point, she no longer seemed to have control of her physical self. Her body was ostensibly "taken over" by members of an intergalactic organization referred to as the "Universal Association of Planets."

A few minutes into this unprecedented "conversation," one of the officers present, a Navy commander, was told that further questions would be directed through him. The officer was instructed to hold a pen lying on a nearby table. The "spaceman" then took control of his hand and proceeded to respond to questions through a process known in parapsychological circles as "automatic writing."

Colonel Friend notes that news of this highly provocative experiment reached Washington almost before the men returned. Top officials at the Central Intelligence Agency also heard about the episode and demanded to know more. It was Friend's duty to find out what he could.

"CAN WE SEE A SPACESHIP?"

"It was in 1959," he told researcher/scriptwriter Emenegger, "when I was invited to attend a meeting in the security section of a government building in Washington. I was briefed on an experiment that had been conducted with this same Navy commander before a group of CIA members and military personnel. It was described how, after going into a trance, the commander contacted a supposed extraterrestrial being. Several questions were put to him, and answers came back, such as 'Do you favor any government, group or race?' The answer was 'No.' 'Can we see a spaceship?' The commander, still in a trance, told the group to go to the window and they'd have proof. The group went to the window, where they supposedly observed a UFO. I was told that when a call was made for a radar confirmation, the tower reported that that particular quadrant of the sky was blanked out on radar at that time."

Friend says that after being briefed on all the details he asked if the officer could attempt a contact for him personally. While he watched, the commander went into a deep trance.

"Questions were put to him, and he printed the answers in rather large letters, using rapid but jerky motions very unlike his natural handwriting. During the course of the questioning, we were told the names of some of the so-called extraterrestrials. One was 'Crill,' another 'Alomar,' and another 'Affa,' purportedly from the planet Uranus."

The former head of Blue Book admits that he was puzzled.

"All those involved were found to be highly credible and responsible professional government men," Friend affirmed.

After turning in his report, Friend was told by a superior to forget the entire affair. He was informed that the CIA was making their own study, and therefore the Air Force was being instructed to "lay off."

What was his reaction to this command? As might be expected, it was a military one.

"Well, when a general tells a colonel to forget it – you forget it!"

Friend later discovered that every witness present in that government office on the day the Navy officer went into a trance was relocated or transferred to other duty.

"To this day," concludes the ex-Air Force officer, "it's an unresolved incident to me. I just don't know what to make of it. It seems totally unique in all my experience with investigations of UFOs."

Had he cooperated to any degree with civilians, Colonel Friend probably wouldn't have been so awed with this case. For many years, private organizations have patiently gathered and investigated similar cases. Indeed, whole sects have been founded based on similar "trance" messages.

Chapter 34

THE SUPPRESSIVE FORCE – THE LONG-TERM CONTROL OF HUMANITY
By Susan Reed

INTRODUCTION

SUSAN Reed (real name Jeannie Gospell), is the author of the book *"The Body Snatchers,"* which was published in 2006. Susan originally sent this article to Timothy Green Beckley in 2004 in which she detailed her experiences with an extraterrestrial Reptilian who was part of a larger, invading force that is engaged in a long-term agenda that is threatening the continued existence of the human race.

In this original article, Susan refers to her reptilian/human lover as "Steve." She had requested that his name be changed from "Brian" (as he is called later in *"The Body Snatchers"*) because she was worried that she was putting her life in danger from revealing too much sensitive information. Her fears may have had some validity as she allegedly drowned in October 2009 while on holiday in Nassau, Bahamas.

Even though Susan refers to the Reptilians as being an "invading force," she is told that the Reptilians have been on the Earth for at least 400,000 years. This implies that rather than being invaders, the Reptilians are actually "controllers" (farmers?) of the human race. It seems as if their agenda is not to invade and conquer; that has already happened. They live among us, and for reasons known only to them, they exert complete control over our lives.

***

HERE are transcripts of conversations I had in November 2003 with a man who was a member of what he termed "the Suppressive force," also called the "Illuminati" or the "global elite." I am making this information public for my protection, as they don't harm those who go public, as it will add credibility to the information. I believe the information to be true. I will first say that Steve is a junior in their organization and that is the reason for some of the errors that he has made where I am concerned.

I discovered who Steve was by accident in May 2002. I was his girlfriend, we had fallen out and he fired weapon number 2 (see weapons) at me from a distance. He does this routinely in his personal life to get back at any slight done to him. This time he was found out. I am more sensitive than most and instantly knew that he was responsible. I confronted him and he panicked and used weapon number 1 on me. Steve was 20 miles away at the time! I ended up in a hospital and almost died.

Thinking that I was going to die, he told me he worked for a suppressive force and that looked at from a higher perspective bad is OK. Three neighbors had taken me to the hospital and I had told them that Steve was responsible. When he found this out – and to this day I do not know how he did it – he removed the effects of the weapon. I have always admired him for "saving me," although his motive was keeping himself out of jail. We continued to see each other; I am not great at picking men. Unpleasant mistreatment is the norm for me.

I did not want to go public with this information but I have to stop Steve. He has been trying to harm me for months since we split up so that I would lose my credibility or shut me up for good. His ego blinds him and he has to win. I have even had to make a Will last December with instructions to publicize all my information in the event of my death or disappearance. The information on the Greys and NASA has resulted in me going on their death list. I know their ways and I recorded the conversations to protect myself in the event that we finished our relationship. Soon afterwards we did.

When we split up he pursued me out of Spain using a colleague in Leon, then out of a good job in Rochfield, England, again using a colleague there. I had not told a soul about what I knew for 18 months nor had I any desire to do so.

I am not interested in conspiracies and would much rather be advising on nutrition. I knew that I could stop him by going public but I didn't want to do it. Instead I just kept avoiding him by moving, at considerable cost to myself. I had pleaded with him on the phone, but he wouldn't stop. I sought refuge in America and he even got a colleague near Middletown, California, to find me and then harm me. I had nowhere else to go and was backed in a corner I told him that I would go to the Spanish Press with who he really was unless he stopped, He did. Just the threat of this exposure was enough to have Steve transferred out of Spain this January where he had settled for two years to, I think, Germany.

The suppressive force is a secret organization that is taking control of the planet without us knowing about it. They place themselves in positions of power and they are also found in all walks of life and there is a ground force covering geographical areas. This was Steve's role. His area used to be the Costa del Sol, Andalusia, Spain, that he called his "turf." He lived on the road to Mijas. Steve termed his organization the Suppressive Force because they are using techniques to suppress us all so that we don't realize what is going on and so we are too caught up in our problems to even care. Suppression would mean suppressing our intellect as described under nutrition, suppressing our abundance as described under debt, suppressing our health as described under viruses and nutrition and the weapons that they use, suppressing our consciousness as described under anti-higher consciousness program, art and architecture.

The transcripts take the form of conversations we had and I have also included evidence provided by my own experiences. I have not included all of my questions, which were probing, to reduce the size of the document. The man does swear a great deal socially so excuse the language.

About the man in question: There is no conspiracy or even political books in his house. He lives a façade. His persona is a jovial cockney (a term for a Londoner) a 42-year-old tall black man from Walthmonstow, London. He has a hidden strength and wisdom that didn't fit with his persona. He has extraordinary mental abilities; he was able to memorize long telephone numbers with ease.

He was an expert on the computer although he did not use this for his "day job." His thoughts are stronger and clearer than other people's and I was able to pickup his thoughts although I am not telepathic. He even used advanced mental processes to seduce women – I know, because he did them on me. His abilities extended to such things as remote viewing, astral projection and mind scanning. All these abilities he disguised. Even the way he spoke would change remarkably on the phone to work colleagues. Basically he was trying to disguise his real identity.

He is extremely knowledgeable on all subjects. His eyes have a steeliness about them that was unexpected. He avoids talking about himself and when he does it was always positive. He was paranoid about other people being psychic; I believe he was frightened that they would discover who he was.

He is a Jekyll and Hyde Character, both extremely harmful and helpful, with a desire to harm and help. He was like two opposite people rolled into one. I was harmed and also given help. I found this very confusing.

He is extremely cruel and I have evidence of his cruelty towards me. We were going out and he kept me subjected to some of the weapons when he could have so easily stopped them. I accepted this behavior, a flaw in myself.

He openly displayed animal cruelty by cutting a ducks neck with a blunt knife and watching it slowly bleed to death for all to see. I am an animal lover and did what I could to help the poor animal.

He has amazing success with women. Remarkable, considering he is not an attractive man, nor even the smooth charmer. He admires female beauty very much and his expectations in this area go far beyond his own attractiveness and yet all his girlfriends are very attractive. I was astonished.

I believe the reason for this is the seduction techniques he uses. I experienced these seduction techniques and they involved advanced mental processes. I suddenly became very attracted to him and yet I was always tense around him. His womanizing was rampant, above anything or anyone else I knew had ever experienced before, and his interests were only there until the women stayed the night. Once this was achieved, he completely lost interest.

One of his duties, as odd as it may sound, is the collection of DNA samples; mine were taken along with many of the women I knew. Once his initial seduction ended, I found him to be negative about humanity, and he hated the Spanish. He would constantly pull me down with phrases such as: "You're hopeless." I was never any good at anything. He became a bully, unpleasant and critical, and yet my attraction persisted.

Despite all the remarkable abilities that he had, there are hidden ego problems that have resulted in this document being written. He has told me, "I've got an ego and it gets in my way. I push too far and don't let off when others would have seen reason. My feelings are...I want to get that bitch – I want to squash her down – I get blinded and can't see reason."

Q - This is my question. **S** - This is his answer.

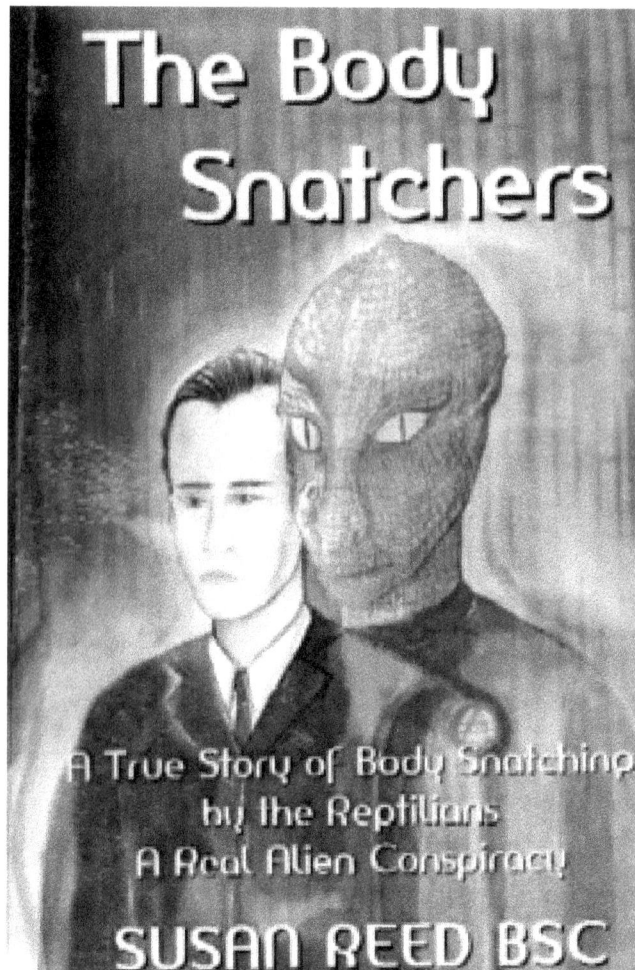

The Body Snatchers

A True Story of Body Snatching by the Reptilians
A Real Alien Conspiracy

SUSAN REED BSC

UFO HOSTILITIES AND THE EVIL ALIEN AGENDA

THEIR OVERALL PLAN

Q - Why are you here?

S - To take over the planet, girl. We are in control right now. We'll have you all under wraps

Q - How?

S - With an implant

Q - Is it a microchip?

S - Maybe, maybe not.

Q - What about that lens of the eye identity thing?

S - We want to do that - so we can identify you all - no fake passports, we'll know who you are.

Q - What is the implant?

S - It's a little device, a bit like a part to a computer - a few millimeters long.

Q - What's it made of?

S - Some kind of plastic and some metal and this and that.

Q - Are you using it already?

S – Yeah, we've got a few people tagged up.

Q - Who?

S – Prisoners. Yeah, we've tried it on animals.

Q - How will people agree to this chip?

S - They'll want it, because they'll think it's a good thing

Q - Why?

S - Because we are rigging it right now - dinero (money) girl - it ain't secure - theft, fraud, money laundering, wallets getting nabbed, credit card fraud, lost pins. A microchip will seem a better option - you'll be totally secure, no theft whatsoever.

Q – Is money the only part of it all?

S – That's our second option in it. Our numero uno is the identity card, plan A. The identity card will replace everything - your passport, NHS number, driver's license. And that's another way we can put the chip in, social security number, because they'll get lost and it will makes things awkward for you - no holiday abroad for the next year because there'll be delays in replacing it - if that's lost everything's gone hasn't it - they'll have no way of doing nothing until it's replaced and we'll probably say we'll replace it at a drop of a hat but then things will change and it ain't going to be replaced so easy - extremely convenient after the deed. We may try both and see what works out.

Q - What are you doing with plan B.

S - We're rigging things right now.

UFO HOSTILITIES AND THE EVIL ALIEN AGENDA

Q - How?

S - Money ain't going to be money quite soon - the euro is part of it - mass facile (easier).

Q - What will happen to the euro?

S - It will lose its value to force you all in to it - then when everyone has got the euro there ain't going to be no euro - it will be all on your cards - no dinero at all.

Q - What will happen to the euro?

S - I ain't telling, yeah - the euro will keep dropping to about 1.1 euro per pound - so when the euro's strong you'll want it.

S - There will be incentives (for no money) - maybe we'll just withdraw it all - then it is all on cards and then the card ain't no good, they ain't secure and you are going to want something different - because you'll get ripped off - credit card fraud – it's all being set up right now.

Q - How?

S – No, we don't need no fucking pin numbers for credit cards - if you have to give your pin then it's secure, isn't it? So we'll stop that. There will be no pin usage and we will make it easy to steal from you.

S - Yes I do credit card fraud - I'm encouraged to - on the computer - I purchase things on the computer using fake credit cards – well, they are other people's.

S - We can't do much with switch cards (it will be credit and switch together) other than make it difficult for you - its already happening - they get lost and they ain't replaced so quickly - they ain't secure, till receipts. Credit cards are the best for you, put you all in debt - then you are all going to be in debt and then we'll up the stakes, interest rates - they'll shoot up and then we'll come up with some better deal and you'll want it.

S - Credit cards are easy to come by and that's us and we're the credit cards; we make the rules and the rules are geared so you spend more than you earn, you're all in debt. We're in control of most of the credit cards and they are fucking easy to come by.

S - We're trying to stop pins and anyone that opposes us gets it - we killed a geezer because he didn't support us - yeah he was in parliament and he snuffed it - heart attack - it was us - it was this year (2003) - our whirly thing (see weapons) - during one of our meetings - early in the year- March - we'd been working on him before, his blood, making him more susceptible - it did not look natural, we take our chances. Our heart attacks are a bit different - because the bloke don't have no clot, no fucking ischaemia - we don't fucking allow an autopsy.

Q - Why?

UFO HOSTILITIES AND THE EVIL ALIEN AGENDA

S - Because they would find nothing wrong with him. It's always at the same fucking time of day, between two and three a.m. (By four a few geezers are getting up) normally at 2:40 am is the exact time and we all work on people together.

S - If you had pins our plan would not work and we'd have to rethink.

S - Anyone with any sense would want pins at the till -and some countries use it and we're going to stop it.

S - Tony Blair wants a fucking pin - the fucking shit rag.

S - Tony Blair ain't going to last long if he promotes pins.

Q - Credit card fraud- right now I get all the money back.

S - That will change, we'll stop all that, when you have no choice - when there's no dinero (money), we'll change the rules, maybe we'll pay a bit but not the full whack, the new rules are that you ain't going to get your money back if it's stolen - we just can't afford it - we'll reduce the time period down - four hours to report it or maybe we'll lower the amount.

Q - What about protection?

S - We'll try and make insurance protection hard to come by, lots of exclusions, things like that.

Q - How to avoid the microchip?

S - Get yourself out of society, don't pay your dues, no bank account - that's basically the main way of doing it.

Q - What do you want?

S - We want you, the planet. No two ways about it, that's all we want.

S - You are all going to snuff it one day and then we'll take over and this is our planet, not yours.

Q - How are you going to kill us?

S - We'll have you all implanted and then we'll set this timer ticking and you'll all go off. The implant will wipe you all out.

S - We'll rig it so it kills you - it will electrify you -we're not sure yet exactly how we are going to do it - no we can't use viruses. It ain't effective - some sort of electrocution.

S - Right now we haven't got the technology- we are working on it.

Q - Are you destroying the Peruvian rainforest?

S - Yeah we are. We ain't in there so we've got to do a bit to harm them, haven't we? We ain't got Peru have we? And if we were in there no rainforest would be touched but right now we've got to do it. We've got to get those natives out of Peru so we can get them civilized - we don't want no-one left who ain't chipped up. Natives who ain't civilized can't be part of our plan so we do what we can to destroy their culture and their habitat so they have to tow the line - we want them all to be under our thumb - they don't have no banks and that's the fucking problem isn't it? No banks means no microchip.

WHAT CAN WE DO TO STOP YOU?

S - If you all refuse or a lot of you refuse to take part in our control mechanisms then our plan won't work. You could all refuse to use credit cards and bank cards and then we ain't got a leg to stand on.

Q - What is the most important thing anyone could do to stop you and keep the planet to themselves?

S - Riot like crazy and that will do it. Refuse all credit cards. Campaign for a pin if you use credit cards. Refuse to send your children to school. Expose the top dogs. Refuse to vote for any president whatsoever, because it will always mean we're in control. Take all your money out of the bank and put it in a safe deposit box. Expose us basically and then we won't hang around. If you don't use a credit card then we ain't going to be able to do away with money and if money is still there we can't chip you and without a chip we can't destroy you. Right now there's still money around but when we need to withdraw money then you need to prevent us doing that by not wanting it and the way to do that is not use credit cards.

Q - How are you going to get rid of money?

S - It will seem more sensible to have everything on card

Q - Why?

S - It's going that way already. It's so easy to use a card. Why bother with money? There'll still be cheques, cheques or a card. The protests/riots need to be when we propose getting rid of money, because once we've done that we're on a roll - more and more stuff that is on cards will simplify things.

Q - What will children use?

S - Children will have cards too – through their parents, with a spending limit on it. At the till there won't be no cheques - it will all be on card. The shops won't accept anything but cards soon.

Q - Why?

S - Because it's easier. We have to remove money in gradual stages, so it's harder to use, cash machines ain't safe. People will start getting mugged at cash machines; the machine will swallow their card more easily; they've got no card for another 20 days; the banks are open at shit hours; you've got to take time off work to get there - huge queues at lunchtime. We'll make it a lot harder to get your cash. They'll invent something else and it will seem like the ideal way of purchasing, like an identity card - it will give you everything you need in one card but then of course it will be converted to a chip because they'll get lost

UFO HOSTILITIES AND THE EVIL ALIEN AGENDA

WARS

S - It's not about the money. It's about getting into power and a bit more besides - keep the masses down and poor.

S - Iraq. We didn't have Iraq. We worked on that one and got the U.S. and Great Britain to get Iraq for us. Other countries that we don't have, we'll start having wars so we can take control. The Middle East is falling under our control with the use of ISIS and other terrorist groups. We create wars so we can take control. We create wars because we have to, but we'd rather not have them because they wreck the environment and they cost an arm and a leg. Still, wars help us get in power.

OTHER MEMBERS OF THE SUPPRESSIVE FORCE - NAMING NAMES

Q - Name some members of the suppressive force for me? What do you look like?

S - We want to look normal, like everyday people, so no one suspects us of being anything other than who we portray. We never look out of place or dress differently. It's the last thing we want to do.

S - Franklin Roosevelt. But we prefer not to be in the spotlight. We are grooming some good candidates for the future. Think crooked businessmen who make money off of hotels and gambling. They will work to undermine all the good things past Presidents did to help people. Whenever a President does anything to help the regular people and not the rich, we will bring in a President that will change all that. Why do you think that leaders always favor the rich and big corporations? It is all because of us. Seventy percent of the U.S. government is us. We also control the media. We need them everywhere, because certain films shouldn't be made. We promote specific types of movies, war and violence. We try to stop New Age films. We don't want alien films made unless they are pure fantasy, *Close Encounters of the Third Kind*...that film we never wanted made. We don't want people thinking of aliens and UFOs except in a negative way.

WEAPONS

Weapon Number 1 is used to kill and it appears as though the victim has had a heart attack, always at night, time of death 2:40 am or soon afterwards. This weapon has been used on me twice. Once by Steve (see intro) and the second time by Steve's colleague and both times at 2:40 am. Steve has told me that three or more of them fire this in unison to kill. In my case, only one person used it on me. He has also told me that they use this weapon at night, as there is less interference.

This weapon constricted my chest to the point that I almost died. Evidence that Steve uses this weapon is that he would get up at this time several days a week and work on his computer. (He does not need to be near his victim to do this.) He has told me that the best protection is to tell everyone publicly that there is no problem with the heart and have it checked out by a doctor and keep home addresses unknown.

S – It is like a missile aimed at a certain point of your body - I tell the machine where you are and what part of your body

Q - How?

S - Via the computer (Internet?) I can't communicate with that machine telepathically; it's got to be done via the computer. We don't always need to all get together. I type in your location in co-ordinates, and once I've typed them in the machine sends you that missile, its harmful energy, particles of energy that are vibrating at the speed you don't want. They suck you in, they cave you in, your energy collapses and harms your organs.

S - The whirl (a sensation I felt when I received this weapon as it was being fired) is me and the machine and I have a piece of equipment for it and it ain't fucking invisible. And you could find it if you wanted, it's a joint effort, the missile gets fired and I encircle it around your energy - I direct it as it heads through the ozone layer. I have to help it on its way with my old weapon, which is physical. The ozone layer gets in our fucking way. We ain't doing nothing to help it, that's for sure, and we'd like it a bit thinner. I attend a meeting twice a year in the Royal Albert Hall, England, and kill, using weapon Number 1, those standing in their way.

[He described how he killed a scientist in England this October. These deaths occur simultaneously and the usual reason given is a heart attack. Surely, this should be investigated.]

S - One of the ways we do it is we get together in one of our big groups and then we zap people and you could be on that list, people in our vicinity. We meet in Royal Albert Hall. We wouldn't mind meeting in a dome (the millennium dome) in London. We take it over for a night, all night, no one else is there and that's including the people that work there unless they are one of us and some of them are. We arrive about 1:00 am and leave at 4:00 am. We do this a few times a year, a society meeting. I am a Freemason when I go to these meetings – we are Freemasons, but we ain't. Fucking Freemasons – we just call ourselves that for the purpose of coming up with some excuse for why we want to turn up like we do. We are the Freemasons, they aren't all us but a lot of them are. It's a certain part of the Freemasons that meet up; the other Freemasons ain't got a clue. I've just had one in the UK in October, it's twice a year, we kill a few people that are well known, politicians and the like. We kill a few scientists now and again. There was one scientist we killed in October, quite a well-known scientist. I tell you that now we've been known to kill a few journalists if they snoop around. We killed this scientist in October, he's been in the papers, he was a young geezer - it wasn't no heart, sometimes we use CVA's to kill people, (I think it was a mysterious death) he was into research, he was a bit into physics - he was working on something we didn't want him to work on – yes, I helped kill him.

WEAPON NUMBER 2

This weapon is an everyday weapon the Suppressive Force uses. I experienced it as though out of the blue I was swamped by what felt like a thick cloud. I was tense, disorientated. Evidence that energy was used is that water (a lengthy power shower) removed it.

S - The bullets are me (toxic energy fired at a person). I have a little piece of equipment in my pocket so to speak which I carry around with me and use whenever I fucking want to - that ain't just at 3:00 am - I can use it during the day too - as you found out. Then I visualize the person and then send it to their location and I just fire it off - we make it - the UK makes some of them and other places make them, too.

Q - What size?

S – It's little - hand held - a bit bigger than a pat of butter - not a lot - its circular - its solid but its slim - I keep it in my wallet sometimes - I do it when I'm driving. I just press a button and there it goes - when I'm queuing in traffic whenever I want. Yeah, its flat alright - thicker than a credit card - if anyone found it they wouldn't have a clue what it is – it's hard as nails and there is no way you could have snapped it in two

WEAPON NUMBER 3

This weapon I received for eight months. Steve put in an implant and I received what he terms as toxic energy from midday until 10:00 pm. They use this on antigovernment writers to reduce their work output. The effects are tension, inability to focus, and heaviness. Water removes this weapon. A long bath (at least 20 minutes, preferably an hour) in a Jacuzzi/hot tub (the jets speed up the removal) is preferable. This again is evidence that toxic energy is involved.

S - It makes people feel heavy, depressed, not clear thinking, can't function as well, the afternoon blues, that's what it does. So we can control you better, to suppress you. So, if you are under the weather, you ain't geared up for anything and we get the better of you.

S – It's toxic energy - our machine manufactures it and shoots it at the person - as long as they are within our zone - way out to sea they'll be OK.

S - Hours – 10:30 am-ish until 9:00 pm right now (winter) and 11:30 pm-ish until 10:00 am in the summer.

UFO HOSTILITIES AND THE EVIL ALIEN AGENDA

Q - Who receives this weapon?

S - Fucking common, we use this big time so you can bet you bottom teeth that anyone we don't want has got one.

S - Prince Charles, Dalai Llama, Tony Blair - most politicians receive it unless they are us. The Queen.

Q - Why?

S - Because we want her to snuff it. This weapon brings you bad health, you snuff it earlier. If we want someone to die a bit earlier than they should then we put it on, don't we? Princess Margaret got it. We want them all dead so we can take over. President Bush gets it so we can control him better. Any powerful bloke gets it; all the politicians have it.

S - It's our machines that give you these weapons. The timer, it's all done by a machine - our machine is fucking incredible.

DNA SAMPLES

My own DNA sample was taken unknown to me. My intellect was assessed and my cancer risk found to be moderately high.

S - I've collected at least 1000 samples.

S - I take a brain sample.

Q - What with?

S - A piece of equipment.

Q - When?

S - When you are fucking snoring and I see to it that you are snoring.

Q - How?

S - The women have to fucking stay the night - the sex bit is up to me.

S - I knock them out, about 3:00 am is the usual time.

Q - Where do you access the brain?

S – In between your soft tissue and your skull. I don't need to go through the skull; there will be a slight hole. Sometimes I choose the nostril or the hairline so you don't notice. I get a miniature sample and I give it to one of my mates, we meet and he takes it - in the UK - in the London area.

EDUCATION

S - Education one of the main things we are into behind the scenes.

S - We make the rules, we dish it out, we're on the board of directors in top universities, and we write the national curriculum. We bog you down with it as much as we can get away with.

Q - Why?

UFO HOSTILITIES AND THE EVIL ALIEN AGENDA

S - To control your minds. We don't want you thinking about other things. We don't fucking like any English - we'd rather you didn't do it - so you can't read and write - can't put pen to paper. We can't have you thinking for yourself and complaining. If people thought for themselves, the Republicans in the U.S. would never get elected. We control the neoconservatives as well.

Q - How should education really be?

S - They used to teach kids in medieval times to read and write and that's all you need, the rest you can learn other ways. Numbers - a bit of math, maybe a language or two, the rest they can learn if they are interested e.g. geography - you learn what you want to learn and not what's forced on you. Send your kids to an alternative school, or keep them at home and teach them yourself. If one shows interest in academic things, then look at sending him or her somewhere where they can get normal qualifications so they can get the job they want.

Q – What about teaching religion?

S – This is hilarious for us. Especially in the U.S. We have worked hard to make science look bad in favor of religion. Schools will be teaching creationism over science. This way, kids will have no early science training that could be a threat to us when they are adults. They will believe that God is telling them that science is in league with the devil and they will eat it up and be like lambs to the slaughter.

CROP CIRCLES, ALIENS AND UFOS

Q - Crop circles, they're caused by spaceships aren't they - the Greys?

S - We use it a bit. We fabricate them so none of you believe - we put fakes in and then expose it to put you off the scent, but some of them are real - the old ones are real. But the new ones ain't. Well, some are, but not many.

S - The small aliens (the Greys), we're not at war with them or anything, and it's none of their fucking business. We slaughter them if we can. They are pretty quick on the draw, nippy bastards, and they can defend themselves - they want us to leave.

Q - How are the Greys involved with the force field? (He had previously mentioned how the Greys are trying to prevent them erecting a force field.)

S - The Greys are trying to prevent us doing it, aren't they?

Q - How?

S - They're negotiating with us. They want us to hold off and they'll leave us be. They are there for you and wish you fucking realized it, slagging them off, Christ almighty. They're getting slapped in the face for doing their bit for you.

Q - So what's going to happen?

S – So, I suppose we're going to restart it because right now we're doing

403

nothing because we've agreed kind of thing, but then we're going to restart when they fucking sod off because that's what they said they'd do.

Q - How long on hold for?

S - It's been a while - we would have completed it by now if it weren't for them.

Q - Why 20 years? (He had mentioned a force field being erected within 20 years).

Q - So you're developing these force field machines?

S – Yes, that's it - in America. It's top secret, one of those places - so the Greys haven't got a clue.

Q - Is it NASA?

S – Yes, it is in NASA - the space places, we're developing them right under their noses.

Q - What do you say to the people working on them?

S - We tell our workers that they are space equipment.

Q - What do you tell them about the force field?

S - They think it is some way of holding other beings in place.

Q - What beings?

S – Well, we tell them it's for the reptilians but obviously it ain't. They know there are reptilians, the NASA people.

Q - How?

S - Because they were attacked by them - some of their spacecraft have run into reptilian places and been zapped because we don't want them nosing around.

Q - What NASA bases?

S - We've got it going in a few places in America - sometimes it's NASA, sometimes it's not. There is one in New Mexico; I think it's called Dulce.

Q - What about the aliens in the abduction books?

S - Their planet has been wrecked like ours - they have already got a cross of them and you - they have been experimenting and they have got a crossbreed that is hideous. (Insectoids). We haven't got a crossbreed yet - we're working on it.

Q - They use humans to experiment on?

S - A great deal - they are not great, they highjack people like we do, using them for their own experiments. I have been told that they usually do it to humans who were one of them. An agreement has been reached. They are not here to harm you - we don't mind them - if they want to do it they can as long as they don't interfere with us.

Q - Why?

UFO HOSTILITIES AND THE EVIL ALIEN AGENDA

S - Because they've got weapons, so we've got no choice. People get abducted by them and it is for experimentation. They are good beings and would help if they were with you and they reincarnate as some humans and help mankind if they are able to. They know that someday humans will not be around and that's when they will do it. They're peaceful enough; they just need a place to live

Q - What do they need to learn?

S - Emotional stuff. And a lot of them have a hard time getting to grips with it. They have got their crossbreeds stored until the time comes. We haven't got a crossbreed yet; we're working on it. They're good when they are here, but they aren't often here, we see to that. They're into the environment and will help out that way once they've sorted themselves out.

Q – Get what sorted?

S - Their emotions - one life can do it and they can reincarnate again and help you lot out.

Q - How do they do that?

S - They help keep the planet green and more besides, higher consciousness/enlightenment, and that kind of thing.

Strange clay figurines with heads of serpents

THE DINOSAURS HAD A 64 MILLION YEAR HEAD START
By William Kern

Stenonychosaurus has been credited with being the most intelligent dinosaur. Compared with most others, it had a relatively large brain, although the excess brain volume was probably not concerned with reasoning and other activities that could be called "intelligence." Stenonychosaurus had large eyes, slender flexible fingers, and a light body. The brain was probably concerned mainly with its highly developed senses, fine control of its limbs, and fast reflexes, which were used in hunting small and elusive prey.

In 1982 Dale Russell and R. Seguin (Ottawa) published an article on Stenonychosaurus. A new partial skeleton had been discovered in 1967 which provided the basis of the first skeletal and flesh restoration of Stenonychosaurus. The detailed work of building the model was illustrated in their paper.

In addition to the restoration, they indulged in an imaginative experiment, posing a question: What might these intelligent dinosaurs have evolved into had they not become extinct near the end of the Cretaceous period about 64 million years ago?

Stenonychosaurus proved to be an interesting choice for the experiment because it was one of the largest-brained and therefore presumably one of the most intelligent of all dinosaurs. The result of the experiment was a creature named "dinosauroid."

One interpretation of the habits of Stenonychosaurus is that they were lightly built active hunters of small prey—perhaps small lizards and mammals. The long grasping hands, and the large eyes which pointed partly forward and therefore gave reasonably stereoscopic vision, may indicate that these were nimble predators which were active at dusk or even at night when many small nocturnal mammals would have been active.

Dinosauroid was constructed by extrapolating from these attributes. It was visualized as a highly intelligent and manipulative dinosaur. What it might have lacked in speed, it would have made up for by its superior intellect. This would have allowed it to avoid potential predators by outwitting them rather than by running away.

As a predator it may have been able to catch prey both by endurance running and perhaps by making simple weapons, much as primitive homo sapiens would do 64 million years later.

But let's take the evolution of the Saurian a step further. Let's say, for the sake of argument, that Stenonychosaurus did not become extinct at the end of the Cretaceous period and actually had a chance to evolve into something close to the Russell-Seguin model.

It is remarkable to note that the Saurian creature bears a striking resemblance to descriptions given by witnesses during a number of UFO encounters! Long, clawlike fingers, large, elongated eyes, reptilian nostrils, three-toed, clawed feet, lizard-like skin, small stature and absence of ears are all features people have reported as belonging to UFO occupants.

Scientists do not know most dinosaurs became extinct; they assume it because few relatives exist today in forms recognizable as dinosaurs. But what if one or two examples actually survived and managed to evolve into highly intelligent creatures capable of building not only simple weapons, but sophisticated craft to explore the cosmos?

The Sauroids would have a 64 million year head start on homo sapiens. They could have built their empires and space craft and disappeared among the stars millions of years before humans ever evolved to walk upon this planet!

Or if the Saurians are not from planet Earth, why not from another where evolution might have followed a similar pattern with similar, if not identical, creatures, including dinosaurs, at about the same cosmic time: 50 to 70 million years ago?

While it is fairly certain life as we know it does not exist on any other planet of our solar system, we cannot rule out the existence of life on any of the billions of other planets that have revolved about billions of other suns in billions of other galaxies for billions of years before humankind ever existed.

Let's face it: When we finally arrive on some distant planet inhabited by sentient beings, whether more or less intelligent than ourselves, scientists and intelligence agencies are going to insist that specimens be returned to Earth, dead or alive, for study. Knowing this, should we be surprised or outraged if creatures from other worlds arrive here and begin taking specimens of earthlings for their own scientific studies?

The requirement for the complete and successful examination of any living organism is to reduce it to its smallest parts and look at each cell or atom under a powerful electron microscope or vaporize small samples in a spectrometer to determine the elements of which the creature was comprised when it was alive. Sample parts might suffice for some studies but whole creatures, alive and dead, will be required for others.

These sample creatures will be acquired for study by abduction and murder. Period. Those in government agencies whose business it is to plan and coordinate these missions have known it all along. It is possible that they are practicing and honing their skills by abducting and dissecting their fellow humans from time to time. At the same time, they may be building their own secret parts bank for the generations of space travelers who will need spare kidneys, livers, eyes, hearts and lungs on Mars about three decades hence.

It's a thought, isn't it?

Another thought is that the creatures who crew UFOs might not be from another planet, but might have been genetically built and incubated right here on earth in one of those secret underground laboratories, and not by aliens, but by human tinkerers. Suppose the future astronaut is not a warmblooded mammal (human), but a cold-blooded intelligent reptile (saurian) who can tolerate cosmic radiation better than humans and who have shown to be able to survive mass extinctions with little change or effect in their subsequent behavior and evolution.

Suppose the saurian is not a creature who lived before us, but is the creature, by genetic manipulation, some of us are soon to become. Some reptiles, remember, have an uncanny ability to regenerate lost parts, often two or three parts. This would prove a real benefit for explorers on a planet several million miles from home base where spare arms and legs are not readily available. Some reptiles can survive days, months or even years between meals while warmblooded mammals can hardly exist more than a few hours!

Some reptiles appear to be unaffected by cosmic radiation that is killing human beings by the thousands. Some reptiles can hibernate for months and years at a time in Arctic conditions without suffering adverse effects.

Cruel experiments on humans in Germany during the second war helped put American and Russian Cosmonauts into space only two decades later. Those experiments haven't stopped simply because we have no announced plans to return to the lunar surface soon.

This country is planning a journey to colonize Mars in less than 30 years! Imagine what it will take to get us there and keep us there!

Dinosauroid.

Have witnesses seen this creature in one of its evolved forms? Is this what contemporary Stenonychosaurus might have looked like had it continued to evolve to the present day? Russell and Seguin assumed for it a large brain, and the short neck and upright posture were arrived at as a way of balancing the head more efficiently. In turn, the vertical posture removed the need for a tail. The legs were modified by lowering the ankle to the ground and the foot was lengthened. It would have stood upright at about five feet tall. Given the proper conditions and time, this evolution would be quite possible.

THINKING OBJECTIVELY
Striving for truth and the scientific method.

Sir Arthur Conan Doyle, creator of Sherlock Holmes, postulated that *"...when you have eliminated the impossible, whatever remains, however improbable, must be the truth."*

This, of course, is a dangerous dictum in the hands of the wrong people for it allows all sorts of crackpots to attribute impossible interpretations to perfectly logical, if not totally understood, phenomena.

Ideally, we should be open-minded when dealing with such phenomena as UFOs and TLOs. Most zealots demand of others a broadminded view, yet they are

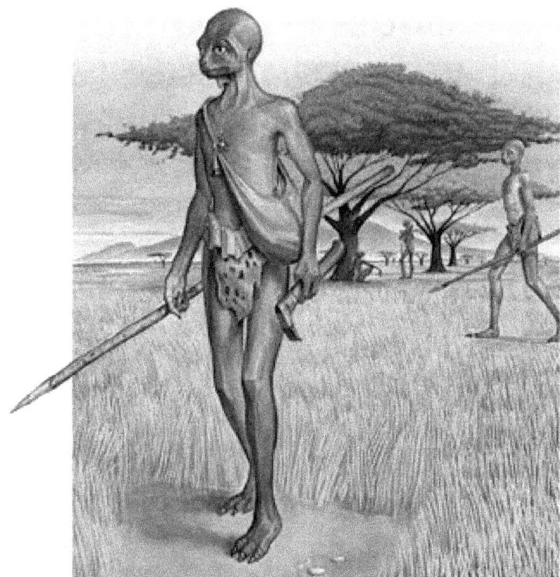

often as closeminded as the scientists they disparage, rejecting as "conspiracy" any theory or expression that does not support their own.

There is always a danger of going too far the other way, of accepting as fact things that are really only popular beliefs, or of parroting imaginative and theatrical explanations for phenomena that are really appallingly mundane.

Yet a point can be made for Holmes' dictum. There does come a time when evidence in favor of a conclusion becomes overwhelming and it would be perverse not to admit it, even if it does not support one's own theory. It would be as foolish to reject out of hand the existence of UFOs as to embrace their existence without question for it is that very debate that will ultimately uncover the truth of the matter. After having eliminated the impossible relative to UFOs and TLOs— viz. that all reports of all such craft and crew result from improper recognition of aircraft, manic delusion or hallucination—we are left with the initially improbable but ultimately inevitable conclusion that UFOs, TLOs and "Flying Saucers" do, indeed, exist.

It would be irrational to disagree with the validity of the argument that something is there, even if it ultimately proves to be not exactly what many of us now believe (or hope) it is.

While I cannot with certainty state that "Flying Saucers" from other worlds are traversing the skies of planet Earth on a regular basis, I would be surprised if they were not. Similarly, while I cannot state with certainty that hundreds or thousands of people appear to have been abducted and carried off in craft of some sort, the corpus of evidence and testimony indicates that they have been and, again, I would be surprised if they have not. Certainly I would have no confidence in expressing the opinion that neither of these events have ever occurred or that they are not now occurring regularly on a worldwide basis.

SUGGESTED READING:

UFO—POV: The Unexplained UFO Story From Roswell To The Montauk Time Travel Experiments—by William Kern

http://therealmidori.com/billbooks.pdf

Chapter 35

NOTHING MORE THAN INSECTS IN THE COSMIC CUPBOARD
By Tim R. Swartz

THE idea of mind control by the government or clandestine groups is in itself a disturbing concept. Yet the possibility that humans are being mentally and spiritually manipulated by an evil intelligence not of this world is a scenario far more horrifying to contemplate.

A number of investigators have suspected that UFOs may be responsible for somehow controlling the minds of some witnesses and abductees. UFO literature is filled with hundreds of cases in which observers have been subjected to continuous harassments following an encounter with a UFO. Some witnesses report strange, ghost-like activity in their homes. In other cases, weird, mechanical-sounding voices, purported to be "messages" from extraterrestrials, begin emanating from their phones, radios and televisions.

Some witnesses persist in believing that they are being continuously harassed and controlled by evil UFO entities. Cases of UFO mind manipulation are actually quite common. Yet very little is known about it because of the scant research being conducted.

Investigators have attempted to distance themselves from cases of alien mind control and UFO entities behaving in an evil and menacing fashion. Most feel that the witnesses who complain of such attacks are probably mentally ill. However, what research that has been done on the subject shows that accounts of UFO/human manipulation are eerily similar in nature.

The pattern that emerges usually follows a close encounter with a UFO. The eyewitness goes through a period of anxiety, during which he is unable to consciously remember certain aspects of the incident. Within months, the personality of the observer actually changes. Eventually, it may change to the point where he finds it impossible to get along with co-workers, friends or even family. Personal tragedy seems to strike many of those who have had UFO experiences.

In some cases, the eyewitness discovers he has developed certain "gifts" or abilities. Though they may appear to be beneficial at first, too often this turns out not to be the case. Among these unusual abilities are powers of ESP, precognition, or psychokinesis. In addition, a heightened intelligence level or an unusual increase in physical strength may be noticed. Such peculiarities will often manifest themselves shortly before a person is about to be controlled. Shortly after this, he may begin slipping into a "trance," during which time it appears as if an alien intelligence has "taken over" his body and is using his brain. There are hundreds of so-called "mental contactees" who claim to receive information and data of a highly advanced scientific and philosophical nature.

During the 1950s and 60s, this method of communicating with UFO occupants (better known as channeling) became so popular that entities calling themselves "Ashtar," "Agar," and "Monka" were heard from daily somewhere in the world.

There is no doubt that this phenomenon is widespread and it is by no means limited to the United States. Cases of mind-altering UFOs seem to be occurring at an alarming rate. There have been reports of entire communities being placed under a strange "spell," with the simultaneous appearance of UFOs in the area.

MENTAL INVADERS

A large-scale attempt to invade and seize the minds of human beings occurred on April 29, 1967, when a coastal village on the outskirts of Rio de Janeiro became the target of a mysterious malady that may have been perpetrated by a strange craft sighted overhead.

In an hour's time, the citizens of Barra de Tijuca, Brazil were literally forced into establishing contact with an unearthly intelligence, which quickly subdued many people in the town. The series of disturbing events began at noon, when an emergency telephone call reached Dr. Jeronemo Rodrigues Morales, chief physician at Barr de Tijuca's general hospital. An excited voice explained how a man in his late 60s had fallen unconscious on the beach near town.

Dr. Morales immediately drove to the scene. Upon arriving he found the man brushing sand from his clothes and talking to a crowd of people who had gathered to offer help. "I was merely walking about the sand dunes," the man explained. "I had been watching the gulls high above the water, when suddenly I blacked out."

An examination ruled out the possibility of a heart attack and Dr. Morales decided that the man had suffered a mild case of sunstroke. Within minutes, another call came in with the news that a fisherman had been discovered in shallow water beneath a nearby bridge, and was said to be trembling from shock.

Dr. Morales quickly drove to the area and arrived just in time to see the "stricken" fisherman casually drying himself off and asking what all the excitement was about.

When the doctor explained that he had blacked out, the man seemed insulted. "I'm not sick," he argued. "I feel perfectly well." He assured Dr. Morales that he had been tossing his nets into these waters every day for twenty years without any difficulty, and would do so for twenty more.

Within a short while, Dr. Morales received word of six other "stricken" individuals. All followed the identical pattern: People keeling over and then reviving themselves without aid, and, after a flurry of excitement, insisting that nothing was wrong.

While Dr. Morales was treating a mother and her young son, who had both collapsed together on the beach, he noticed something high overhead. Glistening in the sun, the doctor observed an enormous disc-shaped UFO over the town. The craft was darting about in the sky at tremendous speeds. Several other physicians and nurses on the hospital staff reported that they had seen the UFO suspended over the town since noon. Shortly after that, the object disappeared along with the strange illness. Still, the town's people had not heard the last from their strange visitor.

Three days later, another UFO, similar to the first, appeared over the city. Once more, a number of people dropped unconscious to the ground. During these two days, many other individuals were treated at the hospital for headaches and dizziness. Some even reported hearing strange voices talking to them in an unknown language.

VOICES FROM THE SKY

In the weeks after the strange incident at Barra de Tijuca, people who had experienced the mysterious malady began to speak openly about what happened to them. Most reported a strange voice in their head that spoke in a guttural language no one understood. Others said the voice was clearly understandable and kept repeating the phrase "do not be afraid" over and over. One man said the voice told him not to tell anyone what had happened to him, and promised that it would return soon. What has not been reported are the continuing strange incidents that have plagued many of the town's people of Barra de Tijuca in the years after their initial event.

One Brazilian UFO investigator wrote that: "The people of Barra de Tijuca continue to be haunted by the insistent voices in their heads. Most will no longer talk to outsiders about their problems. Those that do tell frighteningly similar stories of voices that control every aspect of their daily lives. The voices, the towns people say, originate from alien beings hovering high overhead in their UFOs."

While some people say that they have learned to "tune out" the constant chatter in their heads, others have not been so fortunate. The suicide rate in town is staggering. Some try to drown out the voices with drugs or alcohol. Others try and leave Barra de Tijuca for good. Nothing really seems to work against the continuing torment.

UFO HOSTILITIES AND THE EVIL ALIEN AGENDA

Strangely enough, when asked what the voices talk about, most town people say they can't remember, that the voices didn't want them to remember. Could there be other towns across the globe experiencing the same harassment? Are the inhabitants of these towns being prepared through mind control for some kind of unknown situation or mission in the future? Might we be faced someday with an army of hypnotically controlled humans, ready through years of mental manipulation, to do the bidding of their otherworldly controllers?

In his book **Passport to Magonia**, Jacques Vallee writes of a chilling account of possible alien mind control in the former Soviet Union. "In 1971, an eminent scientist in the field of plasma research died under suspicious circumstances. He was murdered by a mentally disturbed woman who pushed him into the path of a train at the Moscow subway station. The accused women claimed that a 'voice' from space had instructed her to kill this particular man, and she felt unable to resist the order."

Vallee has also stated that he has heard from "trustworthy sources" that Russian police are disturbed about the recent increase in cases of this nature. "Quite often," Vallee maintains, "mentally unstable people are known to run wildly across a street, protesting they are being pursued by Martians, but the present wave of mental troubles is an aspect of the UFO problem that deserves special attention."

Ukrainian UFO researcher Anton A. Anfalow reports that after the fall of the former Soviet Union, dozens of UFO research groups sprung up in an attempt to finally investigate the thousands of UFO reports that had been suppressed by the government.

Because of their efforts, many prominent researchers soon found themselves being harassed and physically attacked by unknown assailants. These assailants would often act like muggers, but would then forgo easier prey to target UFO investigators.

One such attack led to the murder of well-known Russian scientist and UFOlogist, Dr. A. Zolotov. Dr. Zolotov was attacked by a knife-wielding stranger in the town of Tver. Russian authorities say that the attacks are being carried out by individuals suffering from a "type of mental illness where the person claims that voices from alien beings are ordering them to kill certain people." Cases such as this have led some to speculate that the wave of alleged abductions of humans is part of an agenda by extraterrestrials to control mankind with the help of electronic implants.

Physical implants may be used for long-term efforts by unknown intelligences, but there are many reported UFO incidents where people were influenced mentally without any apparent physical connection.

In his book, **UFOs: The Psychic Solution**, Jacques Vallee related an amazing case that happened on the night of November 17, 1971. Two men, Paulo Gaetano and Elvio B. were driving near the town of Bananeiras, a municipality in the state of Paraíba in the Northeast Region of Brazil.

Gaetano noticed that the car was becoming difficult to steer and mentioned this to Elvio. His companion reacted by saying that he was tired and wanted to sleep. Next, the car suddenly died and Paulo had to pull off onto the shoulder. He then saw a strange, egg-shaped object hovering over the road.

The UFO projected a red beam of light at the car and, at the same time, several small beings materialized and took Gaetano out of the stalled car. The man was taken into the craft and placed onto a small table. After tying down his arms, the entities lowered down a device that looked like an x-ray machine. With this device, the beings collected blood from a cut near his elbow.

Next, Gaetano was shown two pictures; one was a map of the town of Itaperuna, the other was a photograph of an atomic explosion. At this point, Gaetano doesn't remember what happened or how he got back to the car. He did tell investigators later that he remembered being helped by Elvio, but did not recall how they got home.

Elvio's story on what happened that night is very different. He said that near Bananeiras, Gaetano had begun to act nervous, saying there was a flying saucer following them. Elvio didn't see any UFO behind them, there was only a bus. Elvio added that the car had slowed down and stopped, and that Gaetano had gotten out and collapsed behind the car, with the door on the driver's side remaining open.

Elvio managed to get Gaetano on his feet and boarded the bus that had been behind them. The pair went to the town of Itaperuna, where Gaetano was examined by the first-aid station. The police went to the site and found the car still on the side of the road. Elvio could not explain what had happened to Gaetano and why the car door was open. He did not remember when Gaetano had got out, and could not explain why they had left the car behind and taken the bus. The police found no trace on the car that could explain the wound on Paulo's arm.

Vallee comments that some experiments with microwaves suggest it is becoming technically feasible for sensory impressions to be projected into people's minds at a distance. He asks: "Is this part of the technology that is involved in the UFO phenomenon? Are we dealing with a technology that systematically confuses the witnesses?"

Another possibility is that instead of being influenced by some kind of advanced mind-control devices acting on the physical brain, the mind could be influenced on the astral level without the use of any physical technology. Many UFO abductions do seem to have a physical component. There is certainly a great deal of evidence that UFOs can manifest physically and leave physical traces. In some cases people may have been physically taken on board these vehicles, and there are a few abduction cases in which the abductee was apparently dropped off miles from the pickup point.

If humans are occasionally taken on board materialized craft, then a physical medical examination is not inconceivable, though it may only be a simulated one, conducted by paranormal entities rather than by extraterrestrial scientists. However, many aspects of abduction experiences sound like visions or dreams.

Abduction cases with definite physical elements seem to be rare compared with the numerous cases where there is no hard evidence of anything extraordinary. Many aspects of abduction experiences sound like visions or dreams. In these cases the entire experience could be taking place on the mental plane and reflect a variety of influences. Some of these cases could be generated during the hypnosis session itself, while others may originate in an actual unusual experience.

UFO encounters may actually take place on several different levels...a physical level, a mental level and an astral or spiritual level. Whatever the source is for the intelligence behind the UFO phenomena, it apparently can operate in ways that are completely outside of the realm of known science. This is why many religious leaders over the years have warned about avoiding any contact with UFO intelligences. The fact that these unknown entities can influence people on a spiritual level is frightening and is reminiscent of the ancient mythologies of demons and other malevolent spirits.

ALIEN IMPLANTS

In recent years, hundreds of people claiming to have had contact with aliens also believe they have been implanted with strange electronic devices. The exact purpose of these microchip-like implants, reportedly found embedded in the skin of abductees, remains unknown. Until recently their existence has only been supported by anecdotal evidence. However, as the abduction phenomenon gathers momentum, more physical evidence is being gathered and studied by doctors and scientists.

According to UFO folklore, implants are usually located in the nasal cavity. In some famous cases, such as the alleged abduction of author Whitley Strieber, brain scans have shown disturbances in an area of the brain close to that part of the body.

Some abductees have reported experiencing nose bleeds, believing that implants were forced into their nostrils so that their brains could be monitored and controlled.

In recent years, however, implants have begun appearing in different parts of the body, sometimes in the back of the neck, behind an ear or in the hands and feet. Hard evidence of purported alien technology has been very hard to come

UFO HOSTILITIES AND THE EVIL ALIEN AGENDA

by. On August 19, 1995, Ventura, California, surgeon Dr. Roger Leir and his surgical team, along with Houston alien contact investigator and Certified Hypnotherapist Derrell Sims, removed three "implants" from two people, a man and a woman who had experienced what they believed to be UFO-related incidents in their life.

Two of the implants were removed from the woman's toes. The third was in the back of the man's hand. All three were attached to nerves where no nerves are known to exist. So far, two additional surgeries have been performed. Three out of four patients turned out to have nearly identical, highly anomalous iron alloy objects involved.

In all cases, ultra-hard metallic, highly magnetic "cores" were surrounded by an ultra-dense dark gray membrane which couldn't be cut with a brand new scalpel. The membrane somehow prevented any sign of inflammation or rejection. Dr. Leir noted that: "If the implants can teach us how to prevent tissue rejection, we could revolutionize surgery."

Interestingly, the membranes on these objects turned out to be made of a tough matrix of proteins from skin and blood. This could explain why the body accepted the objects so readily. It might also explain the very common "scoop marks" that abductees often find on their bodies.

The removed tissue could be wrapped around an implant to "fool" the body into believing the object is part of the system. Also not so easily explained is how the implants got into these people's bodies. Even with a powerful magnifying glass, Dr. Leir could find no sign of a scar or other evidence of a point of entry for objects which had come to be placed deep in the victims tissues.

If implants are actually electronic devices of some kind...what is their purpose? The most prevalent explanation for implants is that they are used to tag an individual to make sure they can be found again. Others believe that the implants are bugging devices, used to monitor conversations and actions. Another theory is that the implants are a means of mentally controlling human subjects.

Often, victims of UFO abduction complain of the feeling that their minds are being influenced by aliens. Abductees report a number of experiences that could be induced by the implants: Buzzing, beeping and strange voices, missing time, inexplicable emotions in inappropriate circumstances, loss of self control and telepathic communication. Many report the disturbance of electrical objects in their presence, perhaps a side effect of such implant technologies.

THE PURPOSE OF IMPLANTS

One certain group, who refer to themselves as "The Light," stated in an e-mail received by the author that people worldwide have been implanted with devices to allow certain kinds of control by extraterrestrial entities.

"Extraterrestrials are currently living among us as humans to monitor human development and assist mankind. Through metabolism cloning, they have the capability to transform their body into a human form taking several minutes. By choosing a desired path, they discreetly & consciously live under a human guise among people without revealing their identity until the correct time. Before society can accept the alien presence, its culture and organization must be changed, which is why they form an influential global network responsible for waves of UFO and alien phenomenon. These part alien/part-human individuals or hybrid-aliens are called 'Guardians.' The Guardians have been selectively bred with humans over the millenniums in order to produce spiritually evolved beings.

"Alien races have visited the earth for thousands of years for different purposes, but the explanation behind the majority of abductions is that alien beings based on earth are implementing a program to implant selected individuals with technically advanced information to condition, educate, and improve humanity. Across the globe and after an examination period, these people were chosen because of specific traits these beings were comfortable with. This microscopic implant, which lies dormant, is inserted into the brain through a condensed light source or manually using surgical instruments.

"The implant contains the foundation for understanding basic extraterrestrial knowledge, principles, and concepts. Very simple examples include: cures for diseases, undiscovered power sources, formulas for food processing and growing, applications of light, utilization of crystals, etc. The power of advanced knowledge will become second nature without ever affecting the implantees personality and memory. The process of learning has been condensed in a microscopic implant: all the chosen will have a sudden interest in an area that they never had previously as if an extraterrestrial course has been studied. This knowledge will be permanent, even if the implant is surgically removed. There are parts of the human brain that naturally becomes a 'biological storage area' for the information stored in the implants.

"A guide is free to scan any field of interest he or she desires. After implantation, these people are called 'Implantees.' After the implant is unlocked, these individuals are called 'Guides.' The Formation is the global event that will simultaneously notify and gather the selected implantees and activate or unlock the implants. The crucial conditioning period after The Formation is called 'The Convergence.'

"The implant also acts as a tracking device in order for the movements of each implantee to be occasionally monitored by a Guardian in close proximity. Through this means, each person will be protected and prevented from an unnatural death. Prior to The Formation, each implantee, regardless of what he or she is doing, will be confronted and informed in detail by a Guardian either verbally or telepathically of what is about to take place. Simultaneously across the world in different countries, all implantees will be transported by means of small crafts to larger crafts situated above the earth. Here the implantee is free to mingle and converse with others across the globe that have been selected, which may or may not include past acquaintances.

"Demonstrations will be given, there will be freedom to interact with hybrid-aliens, and virtually all questions will be answered. Sometime during this period, the implant will be activated or 'unlocked' via a harmless fine-tuning light directed at each person's head, leaving a small red mark for several days. The Convergence has now commenced...the great transformation and advancement these extraterrestrials have been guiding humanity toward. The Guides are now ready to introduce revolutionary and innovative ideas to mankind. For the first time in history, there will be a direct relationship between alien knowledge and society."

"LEAKED" DOCUMENTS REVEAL U.S./ALIEN SECRET ALLIANCE

An Iranian news agency reported in 2014 that documents leaked by NSA whistleblower Edward Snowden prove that the United States has been ruled by a race of tall, white space aliens who also assisted the rise of Nazi Germany in the 1930s.

Fars, Iran's English-language news service, revealed that Snowden, who has been given asylum in Russia, leaked documents that a race of extraterrestrial "tall whites" arrived on Earth, helped Nazi Germany build a fleet of advanced submarines in the 1930s, and then met in 1954 with President Dwight Eisenhower "where the 'secret regime' currently ruling over America was established."

"Most disturbingly," the leaked document continues, "is that the 'Tall White' agenda being implemented by the 'secret regime' ruling the United States calls for the creation of a global electronic surveillance system meant to hide all true information about their presence here on earth as they enter into what one of Snowden's documents calls the 'final phase' of their end plan for total assimilation and world rule."

More than likely this document had its origins with Russia's FSB spy agency and is part of a long-term propaganda campaign against the United States. However, any good disinformation campaign needs a mixture of truth and lies in order for it to be effective.

Some people, like self-styled prophet Marshall Vian Summers, may be unwittingly part of a world-wide disinformation campaign involving extraterrestrial visitors. Summers claims that a community of aliens "watching the Earth" told him through God that other malevolent alien races "regularly visit" our planet in UFOs and this alien presence is growing.

Summers, who heads a growing church called **The New Message From**

UFO HOSTILITIES AND THE EVIL ALIEN AGENDA

God (TNMFG) movement, in Boulder, Colorado, has transcribed the "alien messages" he claims to have received into a book entitled "***The Allies of Humanity Book One.***"

"Over 20 years ago," Summers said, "a group of individuals from several different worlds gathered at a discreet location in our solar system for the purpose of observing the alien intervention that is occurring in our world."

The benevolent aliens claim to have been sent by "unseen ones" to help us avoid a takeover by other species, who hope to "control" us and "interbreed with us."

"Your world is being 'visited' by several alien races and by several different organizations of races. There have been visitations throughout human history. What has brought these forces to your shores in such numbers with such intention are the resources of your world."

Summers claims that evil aliens were responsible for abducting people and may eventually "take over our planet."

"Each alliance represents several different racial groups who are collaborating for the purpose of gaining access to your world's resources and maintaining this access. They see your world as a great prize, something they want to have for themselves."

However, Summers says that we would not experience a War of the Worlds-style invasion, but a gradual takeover.

"The alien visitors do not come armed with great weapons or with armies or with armadas of vessels. They come in relatively small groups, but they possess considerable skill in influencing people via telepathy. To serve this purpose, they will create establishments here, though not in view. These establishments will be hidden, but they will be very powerful in casting a mental influence on human populations that are near them.

"Many more people are falling under its persuasion, losing their ability to know, becoming confused and distracted, believing in things that can only weaken them and make them impotent in the face of those who would seek to use them for their own purposes. Planet Earth could be overtaken without firing a shot, for violence is considered primitive and crude and is rarely employed in matters such as this.

"They will promise anything, offer anything and do anything to achieve this goal. Their influence is growing and their program of interbreeding, which has been underway for several generations, will eventually be effective. They are here to establish themselves. They want humanity to believe in them and to serve them."

CONTACTEES – OR ALIEN MIND CONTROL VICTIMS?

The UFO phenomena are baffling to those who have taken the time and ef-

UFO HOSTILITIES AND THE EVIL ALIEN AGENDA

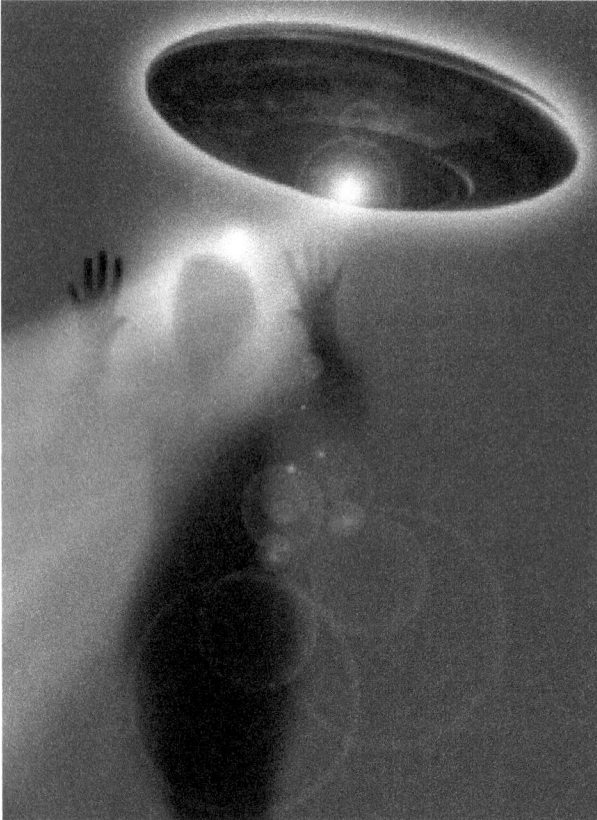

fort to study them on multiple levels. On one side, UFOs appear to be physical, constructed machines, flown by creatures who claim to be from other planets. On the other side is the unphysical nature of the phenomena, with UFOs and the strange beings associated with them manifesting like ghosts. People who are unlucky enough to get caught up in the confusing world of UFOs and their occupants are often subjected to weird forms of possession, behavioral changes and mind control. Victims of UFO abduction usually report periods of "missing time," which is almost certainly achieved with some kind of mental manipulation of the abductee.

The late John Keel speculated that the contactee syndrome is a fundamental reprogramming process. No matter what frame of reference is being used, the experience usually begins with either the sudden flash of light or a sound - a humming, buzzing or beeping. The subject's attention is riveted to a pulsing, flickering light of dazzling intensity. He finds he is unable to move a muscle and is rooted to the spot.

Next the flickering light goes through a series of color changes and a seemingly physical object begins to take form. The light diminishes, revealing a UFO or an entity of some sort. What is really happening is that the percipient is first entranced by the flickering light. From the moment he feels paralyzed, he loses touch with reality and begins to hallucinate. The light remains a light, but the contactee's mind is hypnotized to see a spaceship and/or a strange alien creature.

Keel writes in his book, **The Mothman Prophecies**, that he was concerned with the falsified memories of the contactees. "I wondered what happened to the bodies of these people while their minds were taking trips in flying saucers. Trips that often lasted for hours, even for days."

A young college professor in New York State was haunted by the same question in 1967. After investigating a UFO-related poltergeist case, he suffered possession and was led to believe that he had committed a daring jewel robbery while he was in a trance or possessed state. He abandoned Ufology and nearly suffered a total nervous breakdown in the aftermath.

UFO HOSTILITIES AND THE EVIL ALIEN AGENDA

Are contactees and abduction victims being used by exterior intelligences to carry out crimes, even murder? The answer is a disturbing yes. If you review the history of political assassinations, you will find that many were performed by so-called religious fanatics who were obeying the "voice of God," or were in an obvious state of possession when they committed their crime. Assassins, such as Sirhan Sirhan, who murdered Robert Kennedy, had a strange fascination with the occult and hypnosis. It is not unusual for them to say that they have no recollection of committing the crime...a telltale indication of mind control.

In contactee parlance, persons who perform involuntary acts are said to be "used." A contactee may feel a sudden impulse to go for a pointless late-night walk or drive. During that drive he encounters, he thinks, the space people and is abducted. Actually his body goes on to, say, Point A where he picks up a letter or object left there by another contactee. He carries the letter or object to Point B and deposits it. Later he has no memories of these actions.

ALIEN ABDUCTIONS OR MILITARY EXPERIMENTS?

According to Helmut Lammer Ph.D., UFO abductions are a complex of phenomena with no easy answers. For skeptics, journalists and the public, it is difficult to believe that abductions by alien beings have their basis in physical reality. However, well respected researchers have shown that the core of the UFO abduction phenomena cannot be explained psychologically as hallucinations or mass delusions. Recently, some UFO abductees have reported that they have also been kidnapped by military intelligence personnel and taken to hospitals and/or military facilities, some of which are described as being underground.

Very few books on the subject of UFO abductions have mentioned these experiences. Especially disconcerting is the fact that abductees recall seeing military intelligence personnel together with alien beings, working side by side in these secret facilities. Researchers in the field of mind control suggest that these cases are evidence that the whole UFO abduction phenomena are staged by the intelligence com-

munity as a cover for their illegal experiments. Could the whole abduction scenario be a carefully manipulated hypnotic cover for experimentation by government or military intelligence services?

The alleged military involvement in the abduction phenomena could be evidence that the military uses abductees for mind control experiments as test-targets for microwave weapons. Moreover, the military could be monitoring and even kidnapping abductees for information gathering purposes during, before and after a UFO abduction.

Lammer's research suggests that abductees are often harassed by dark, unmarked helicopters that fly around their houses. This mysterious helicopter activity goes back to the late sixties and early seventies, when they showed an apparent interest in animal mutilations, but not in alleged UFO abductees. However, UFO researcher Raymond E. Fowler reported some helicopter activity in connection with UFO witnesses during the 1970s.

Many abductees report interaction with military intelligence personnel after the helicopters begin to appear. Debbie Jordan reported in a side note of her book "***Abducted!***," while she was with a friend, she was kidnapped, drugged and taken to a kind of military hospital where she was examined by a medical doctor. This doctor told her he was going to remove a "bug" from her ear and proceeded to take out an implant that resembled a BB.

The abduction experiences of Leah Haley and Katharina Wilson also include military-type encounters. Some of Wilson's experiences are reminiscent of reported mind control experiments. For example, she writes of a flashback from her childhood where she remembers being forced into what appeared to be a Skinner Box that may have been used for behavior modification purposes. In some military abduction cases, military doctors searched for implants and sometimes even implanted the abductee with what appeared to be a man-made implant.

The technology does exist for small, radio frequency electronic implants. More than three million animals worldwide have been successfully implanted with

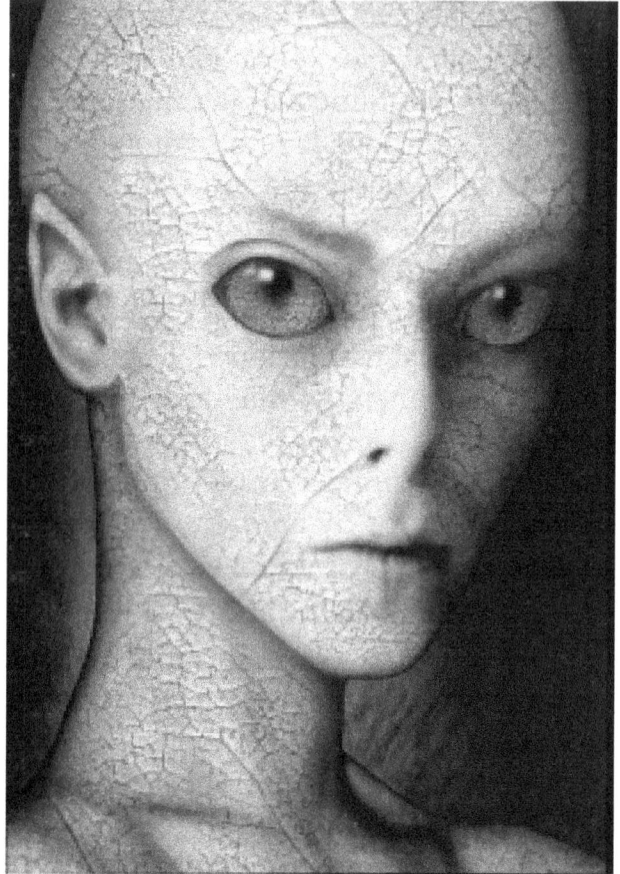

a transponder manufactured by Destron-Fearing. The transponder is a passive radio frequency identification tag, designed to work in conjunction with a compatible radio-frequency ID reading system.

The transponder is activated by a low-frequency radio signal. It then transmits the ID code to the reading system. The smallest transponder is about the size of an uncooked grain of rice. The transponder's tiny electronic circuit is energized by the low-power radio beam sent by a compatible reading device.

A similar bio-chip for humans was patented in 1989 by Dr. Daniel Man. The homing device, which can be implanted under the skin, was originally developed to locate missing children. This device is slightly larger than the Destron implant and a small surgical incision must be made for it to be implanted. Dr. Man claims that the best location for his implant may be behind the ear.

It is possible that some of the information received from abductees may be cover stories, induced by hypno-programming techniques of military psychiatrists. It is also possible that the military uses rubber alien masks and special effects during a supposed alien abduction. Katharina Wilson reported flashbacks where she remembered holding a rubber mask of an alien head in her hands. Facts such as these lead some mind control researchers to believe that all alien abductees are actually mind control and/or genetic experiments staged by a secret group within the government of the United States.

In a declassified memo dated February 17, 1994, former Naval Intelligence Commander Scott Jones, Ph.D. wrote to White House Presidential Science Advisor John Gibbons: "Whatever Roswell turns out to be, it is only the opening round. I urge you to take another look at the *UFO Matrix of Belief* that I provided you last year. My mention of mind-control technology at the February 4 meeting was quite deliberate. Please be careful about this. There are reasons to believe that some governmental group has interwoven research about this [mind-control] technology with alleged UFO phenomena. If that is correct, you can expect to run into early resistance when inquiring about UFOs, not because of the UFO subject, but because that has been used to cloak research and applications of mind-control activity."

UFO HOSTILITIES AND THE EVIL ALIEN AGENDA

WHAT IS EVIL?

The debate goes back and forth about the true nature of otherworldly visitors. Many see them as benevolent, almost angelic in nature. While others point out to incidents where human witnesses have come to physical harm and even death when interacting with UFOs and their occupants; thus, they reason, the visitors must be evil. However, are we projecting anthropomorphic ideals onto something that is not human?

Philosophers and theologians have argued for centuries on the nature of evil. The earliest concept of evil is often associated with the supernatural, especially in religious contexts. Satan, demons and witches were thought to be paragons of evil. These creatures possess powers and abilities that defy scientific explanation and, perhaps, even human understanding. It was thought that God was all-powerful, all-knowing and all-good, therefore incapable of evil. In order to explain this paradox, religious scholars invented Lucifer, a fallen angel who became the poster-child for everything bad that happens to humans.

Immanuel Kant, in his *"Religion Within the Limits of Reason Alone,"* was the first to offer a theory of evil that doesn't blame supernatural or divine entities. Kant's concern is to make sense of three apparently conflicting truths about human nature: 1. we are radically free, 2. we are by nature inclined toward goodness, 3. we are by nature inclined toward evil. According to Kant, we have a morally good will only if we choose to perform morally right actions because they are morally right. On Kant's view, anyone who does not have a morally good will has an evil will.

Naturally, these discussions are based on human concepts of good and evil. The farmer, who raises cattle, feeds them, provides shelter for them, nurses them when they are sick and finally slaughters them for their meat is not evil, at least not to the human viewpoint. However, if the cattle thought like humans, what would their perspective be? Would their loving caretaker who raises them from birth and then kills, dismembers and eats them be considered a loving God, or an evil demon?

Charles Hall, wrote a series of "fiction" books (*"Millennial Hospitality"*) based on his experiences with a race of human-like extraterrestrials – the "Tall Whites" – during a two year duty assignment at Nellis Air Force Base from 1965 to 1967. Hall described the Tall Whites as appearing quite human-looking in some respects and standing between six and nine feet tall. He said they had chalk white colored skin and their physique appeared frail and thin.

Hall says that the Tall Whites behaved arrogantly at times and they tended to value high ranking officials and people in higher ranking social structures over those of lower rank or those perceived to be from a lower socioeconomic class. The Tall Whites viewed humans with distain and had little restraint if someone got

Tall Whites

TS-SCI- S.A.M-422wxxy
Report prepared for S.A.A.L.M by xxxxxxxxxxxxxxx
A-C-Y-1-O-M_ACIO PINE GAP

Description
Extra-terrestrial
Species: SAM_Nordics
Aliases: Sweden, Tall whites, nordics
Height: 5 - 6.5 feet
Weight: 120 - 240 pounds (estimated)
Eyes: Human
Hair: Blonde
Skin: Pale white
Sex: Male and Female
Communication: Telepathic
Distinguishing Characteristics
Share common physical features with human beings
(especially Scandinavians)
Are taller than the average human
Have more of a muscular build than the average human

Charles Hall

"Charles Hall's "Millennium Hospitality" series has gained a foothold in American UFOlogy so quickly it's phenomenal. The 'Tall Whites' are on the fast track to becoming the most intriguing extra-terrestrials ever!"--Guy Vernon, BookFare.

in their way. Every Tall White adult carries a pencil-like weapon that can be set to stun, kill, immobilize, or "hypnotize" humans. It can also administer severe pain, and they frequently use it to discipline people who act in ways that annoy, frighten, or endanger them.

For example, one Tall White female wanted to kill a military servicemen who had unknowingly hit her child with a rock and broke its arm. Since the Tall Whites often stalked the servicemen, many serviceman thought they were wild animals and were quite frightened of them. The incident led to the Tall White threatening the servicemen with death if he didn't leave immediately and promise to never return.

This is how the serviceman in question described the incident to Hall: "I tried to reason with her. I told her that it wasn't my intention to break the little boy's arm and that I liked to play with kids. She wouldn't hear any of it. She told me that I was too stupid to know what I had done. Then she told me that the American generals had asked that I be given one warning before she and her friends killed me. She said that this was the warning. She said that if I ever came back out to the ranges alone, their Captain would kill me."

Other servicemen had similar negative encounters with the Tall Whites. Once a cook inadvertently went into the kitchen area on a day when the base was closed. He was threatened with death because he had accidently frightened some of the Tall White children who were playing in the kitchen.

Hall said the cook told him not to go back into the kitchen because he might "scare them like I did. They'll kill you if you scare them. They told me so. That tall one in the corner, he told me so. He said he'd kill me if I ever scared their children again."

Are the Tall Whites as referenced by Hall evil? They have weapons that can be used to discipline, torture and even kill people. Hall even admits that he had been attacked in the mid-desert by a group of the Tall Whites who had badly wounded him and then stood around and watched as he nearly bled to death.

Again, we may be anthropomorphizing creatures that, while they may superficially look like humans, are anything but. Chimpanzees and other primates are our closest genetic relatives, yet we treat them like all other animals, often in ways that are gruesome and, by human standards, "evil."

If we claim that we are the most highly evolved and intelligent species on this planet, yet treat all other species with little regard and disdain, how can we expect other intelligent species to treat us any differently? Evil, good...these are words that we use for our own benefit. Can they be properly used when someone mindlessly kicks over an anthill or accidently runs over a squirrel in the road?

We could be seen as nothing more than cockroaches running under a cosmic refrigerator when a more intelligent species turns their light onto our planet.

Time to get out the bug spray.

Chapter 36

ALIEN THOUGHT CONTROL
By Sean Casteel

In the New Millennium, as in decades past, the appearance of UFOs and their shadowy crew members may in fact be a political tool being used to reshape the patterns of our collective thought processes to the aliens' advantage, all of which forces us to ask questions like:

· Are the extraterrestrials here implementing plans to take control of the planet?

· Are the aliens offering solutions to mankind's problems as a kind of Trojan horse in order to get their foot in the door so they can lead us to a totalitarian form of world government?

· Do the ultra-terrestrials utilize religion—and fundamentalism in particular—to influence and even control the current program of terrorism?

· Will mankind one day be ruled over by hybrid human/aliens whose superior mental powers will completely subjugate us?

· Have we secretly achieved a new military technology that may one day provide us with effective weapons to use against an alien invasion?

· And finally, have the extraterrestrials found an ally in high-profile UFO researcher and advocate Dr. Steven Greer and his very popular Disclosure Project organization?

According to journalist and author Michael Brownlee, the answer to all of those questions is a resounding yes!

Brownlee recently published a two-part essay on what he sees as the potential peril certain UFO researchers such as Dr. Steven Greer may be placing our civilization in—either knowingly or unknowingly—by helping to popularize the unproven notion that the extraterrestrials are here to be helpful, benevolent friends of mankind. (Dr. Greer declined to be interviewed for this article, citing his busy schedule, but we extend a standing invitation to him to rebut any of the points made here by Brownlee.)

UFO HOSTILITIES AND THE EVIL ALIEN AGENDA

Brownlee initially viewed Greer as a hero.

"In 1997," Brownlee said, "which was a pivotal year in this field, I went to a conference in Phoenix, Arizona. It was a Prophets Conference, and one of the speakers there was Steven Greer. I'd heard of Steven. I had been reading some of his material and was very interested because he was the only one who, at least in the title of his organization [which is the Center for the Study of Extraterrestrial Intelligence], was talking about studying the extraterrestrials themselves. I thought his efforts toward disclosing the presence of extraterrestrials were very courageous and heroic.

"And I know he was suffering a lot of persecution at that time," Brownlee continued. "The person who was closest to him in that work died of cancer. [Dr. Greer also] suffered from cancer, and they felt very strongly that the cancers were implanted by someone who wanted to stop them. I was just very touched. I thought he was a very compassionate, brilliant man, passionate and doing something worthwhile."

But since then Brownlee says his opinion has changed.

"Whether for him it's knowingly or unknowingly," Brownlee said, "it would appear that Dr. Greer has been essentially 'co-opted' by the extraterrestrials who are visiting the Earth at this time."

In other words, the aliens have taken control of Greer, at least in terms of what he says publicly.

Meanwhile, Brownlee credits a reclusive spiritual leader named Marshall Summers with helping him to gain what he now feels is a more accurate viewpoint on both Greer and what the aliens really want.

"At a UFO Expo in California," Brownlee said, "I ran into people who had an underground version of a book that I had not heard of. It was called *The Allies of Humanity*, written by this man named Marshall Summers. The book's had a galvanizing effect on my life, certainly, and it has on a lot other people as well, because the perspective that it presents is just totally unprecedented in this field. It's not research. It's not about the phenomenon per se. It is a look at what's happening."

The "Allies" of the book's title are a group of friendly extraterrestrials who are sending a warning to mankind about the coming colonization of the Earth by less benevolent aliens intent on manipulating mankind into an ultimate submission.

"This book contains information," Brownlee said, "that just unravels the whole mystery of the extraterrestrial activities and presence here on this planet. And I say that not based upon something I just want to believe, because there's a lot of material in it that's damned uncomfortable."

Our prior attempts to understand the situation, Brownlee says, have been

UFO HOSTILITIES AND THE EVIL ALIEN AGENDA

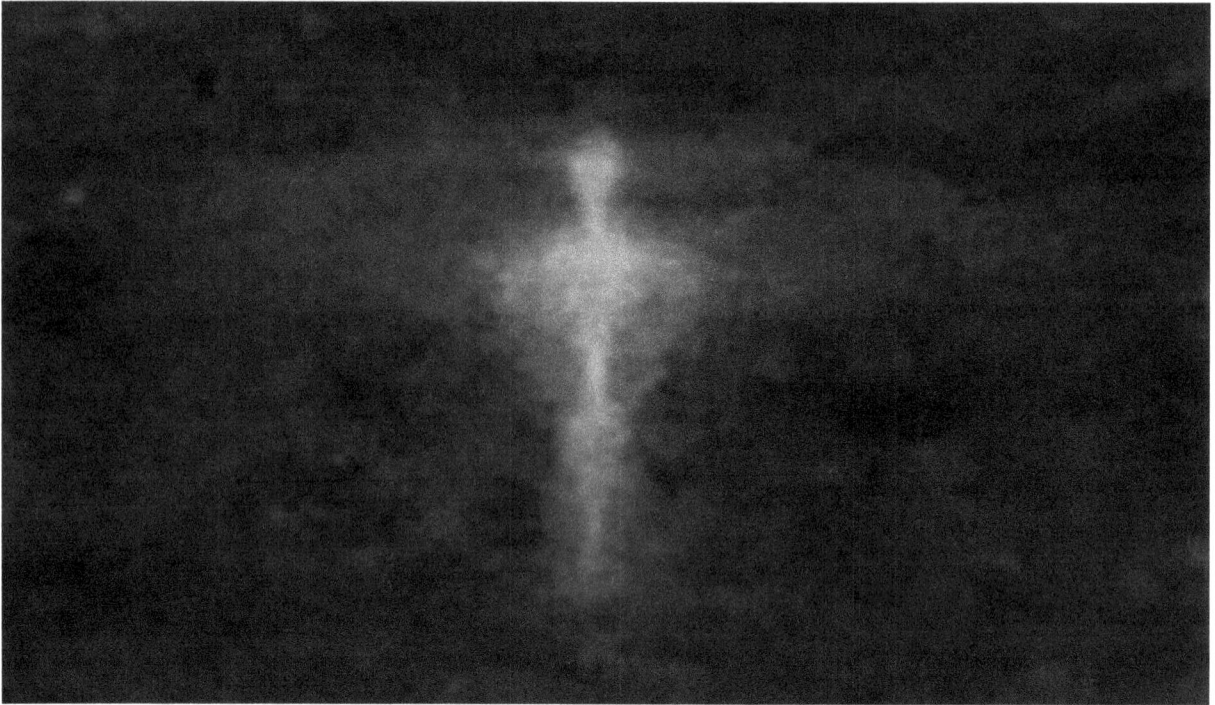

A "light being" photographed by Dr. Greer's team

woefully inadequate.

"The biggest problem we've had in the UFO/ET research community," he said, "is that this whole business of UFOs and extraterrestrials has been thoroughly enigmatic. It has just been a mystery. We haven't been able to penetrate it no matter how much research we do. After fifty years of hammering away at it, we haven't gotten much closer than we were in 1947. We have a lot of different ideas and theories and experiences, and they're all conflicting."

But Summers' book, while it is helpful to Brownlee in terms of getting a handle on the phenomenon, hasn't been welcomed with open arms by the UFO community.

"One of the things that I have noticed," Brownlee said, "is that the material presented in this little book is highly unpopular in the UFO community, because it presents a viewpoint that a lot of people would just not like to be true. The aliens are, just like [well-known UFO researcher] David Jacobs says, implementing plans to take control of this planet.

"Not to destroy humanity," he continued, "but to take control. And Steven Greer has become their most effective ally and ambassador with the human race. He's gone from responding to a deep passion within him to being used by the extraterrestrials. And as I say, I don't know whether he's aware of it or not. But, unfortunately, that's the position that he seems to be in. So in my eyes he's gone from being a hero for being so courageous to being a very, very dangerous force in the world today."

Brownlee vigorously contests certain opinions expressed by Greer, including Greer's claim that the majority of UFOs sighted are really manmade devices.

"Of course," Brownlee said, "the literature is filled with stories of how we have captured craft or downed spacecraft, and we have been working very hard to reverse-engineer them at Area 51 and other places. We've been hearing all that for years. And Greer was talking about that, of course, in his early years. But recently, in the last two or three years, he's started saying that, 'Well, what's really going on is that most of the UFOs in the sky are manmade.'

"Now, there is absolutely no evidence for that. It is an assertion on his part that is not backed up by evidence. He said, 'And we can prove it,' again and again. But he doesn't offer any proof. He doesn't offer any documentation whatsoever. I have no doubt that the government has been doing its best to reverse-engineer alien craft technology, but it's a long way from that to saying that the majority of UFOs are manmade."

Brownlee also disputes Greer's claim that the abduction phenomenon is again a human effort.

"This is called 'Escalation of Rhetoric,'" Brownlee explained. "Years ago, and I have heard Greer talk about this, he was saying that, based upon the work of Helmut Lammer, who wrote about MILAB abductions, he said, it looks like we're in the business of doing abductions. It's absolutely true that Helmut Lammer wrote about military-based abductions, human-based abductions. He says he doesn't know why it's happening. He just said it appears to be happening, but he also makes crystal clear that, based upon his research, this amounts to a very small fraction of abductions that are taking place.

Marshall Summers

"So Greer started using Helmut's research on MILAB abductions," Brownlee went on, "and he escalated the rhetoric in the last couple of years to start saying that all abductions are perpetrated by humans. All of them. Now, again, there is absolutely no evidence to that effect. There is no way he could even know that. If you look at the evidence, if you look at the studies that have been done, if you look at all of the abduction research, on balance you'd have to say that it is a completely ludicrous claim. Why is he making it? It's very odd. It's even odder that no one calls him on it."

But what if Dr. Greer is right, at least partially? How would a "hoaxed abduction," the kind put forth by Greer, typically play itself out?

"My understanding is," Brownlee replied, "that it follows much the same course. The one difference sometimes is that the abductees are taken to an underground facility. An underground, apparently *military* facility where they see lots of armed soldiers and things like that.

"Now, if you ask David Jacobs about this, he will say, well, it's pretty obvious now. We know from research on hundreds and hundreds if not thousands of abductees that human appearing beings often show up in abduction scenarios. And what he proposes is that those are not humans at all but are in fact hybrids. The hybrids are increasingly responsible for the abduction program. So to assert that they are human is a wild assertion, devoid of evidence."

However, in spite of what Brownlee considers to be Greer's false claims, Greer still provides an accurate barometer of what the aliens really want, says Brownlee.

"What I have learned about Steven Greer," Brownlee said, "is if you want to know what the extraterrestrials are concerned about, what they're worried about, what they're up to, pay attention to what Dr. Greer is saying and doing. He can give us the best insight into their strategy of almost anybody on the planet. But you have to look at it from the perspective that what's going on here is essentially a clandestine takeover of the planet. And, again, this is not about destroying humanity or destroying the planet. That's ludicrous. They want control. It's just that simple."

One method the aliens are using to manipulate our thinking is the time-honored ploy of making humanity feel guilty.

"The really stark thing that comes through everything else," Brownlee said, "is the statement that humans are the real problem in the universe. That's an outrageous statement. It is based upon a species-wide guilt, if you will. It's like, 'Oh, man, we're really bad. We're really screwed up.' And that's exactly where we're most vulnerable. So the extraterrestrials come and abduct our people and they tell them—and you can read the reports from [Harvard psychiatrist Dr. John] Mack or Jacobs or [seminal abduction researcher] Budd Hopkins, it doesn't matter, it's all the same—the point they make is, 'You guys are screwing up your planet.'

"Well, they don't say it this way, but the feeling is, 'You're bad. You're screwed up. You need us to come and straighten things out.' From a psychological perspective, that is the basis of a codependent relationship. That's exactly what they're trying to create. They want us dependent on them. They want us controlled. We're useful to them, because their numbers aren't very large. They've

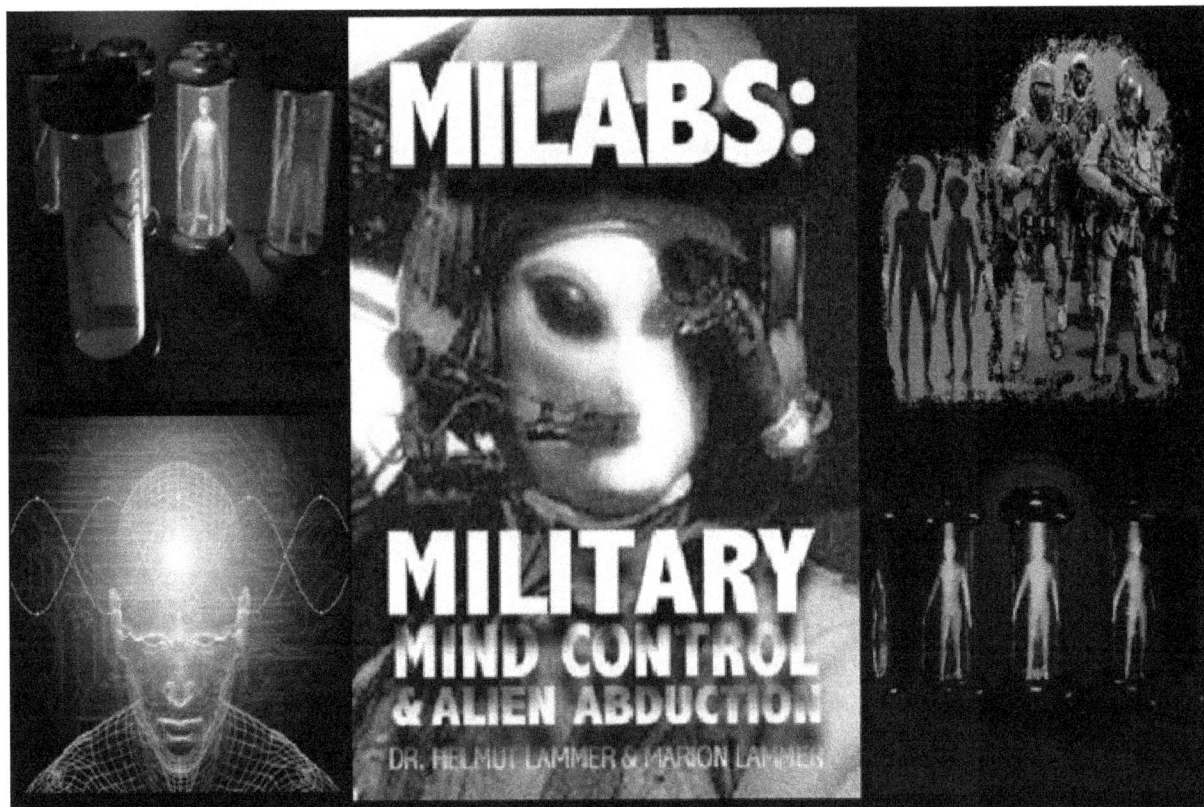

got a lot of work to do here, and they need our labor."

One primary motivation for the alien takeover may be simple, old-fashioned greed.

"The world holds incredible resources that are pretty rare in the universe," Brownlee theorizes. "Those resources include biological resources, they include mineral resources, and perhaps some other things. They don't want us, for instance, destroying our ecological system because it's bad business for them. It screws up their plans."

The aliens offer solutions to mankind's problems as a kind of Trojan horse in order to get their foot in our door so they can lead us to a totalitarian form of government.

"The message of fascism," Brownlee said, "is always 'This will solve all problems.' What we're looking at are the signs of a fascist takeover that, if it is not exposed and resisted, will wind up being not just a fascist state, but a fascist world.

"It's not like the aliens are going to come down as aliens and run things. No, they don't function that way. The Allies [the informants in Marshall Summers' aforementioned book] present a very interesting idea. And that is that the aliens are breeding hybrids to be the new rulers, the new leaders, of this world. They will look like humans, only they will be very, very powerful, particularly in what the

Allies refer to as the 'mental environment.'

"The particular aliens that are here now, while they're technologically advanced, they don't have great legions, they don't have great military forces. That's not how they work. You can only take physical technology to a certain point in the evolution of life in the universe. What's beyond physical technology is the technology of the mental realm. And that's where they excel."

Brownlee made reference to Harvard abduction researcher Dr. John Mack to drive his point home.

"Even John Mack," Brownlee said, "who takes a fairly positive view towards the alien presence here, he says the most common form of contact between humans and extraterrestrials is in the dream state. In other words, they can enter our dreams. They can plant images, ideas, emotions, thoughts, impulses. They're very good at that. And we're just so wide open that we don't even know it's going on, for the most part."

"In the mental realm," he continued, "they are very effective and we are unfortunately very naïve. So one of the ways that this plays out is in one of the key areas of alien activity—the manipulation of our religious and spiritual beliefs. They are very aware of our naivete and they are able to easily manipulate people who are particularly prone to fanatical beliefs, fundamentalism, in all of its varieties around the world. They can be very influential."

This has frightening implications on a global scale, according to Brownlee.

"One of the astonishing things that we ultimately learn out of this," he said, "is the extent to which the extraterrestrials are operating behind the scenes in world affairs today, influencing key people. And one of the great problems that we have in the world today, that's pushed us to the brink of global warfare, is this explosion of Islamic fundamentalism."

Brownlee again cited the Allies of Humanity as his source for this belief.

"The suggestion is that the extraterrestrials are meeting some resistance in the Western World," he said. "Meaning that we aren't just falling over for their program here. So they have decided to stir up some reprisals and to extract a little pain. Those are the roots, if you will, of the September 11 attack and the whole terrorist effort that has been building and escalating and is now pushing us literally to the brink of war. Now, I think that's a pretty extraordinary hypothesis, and if there's any basis to it at all, it's something that we should be looking at. I mean, talk about a conspiracy theory! That's about as big as it gets."

On top of that, the aliens have begun a form of warfare of their own.

"Craft have flown over military bases with nuclear missiles," Brownlee said, "and shut them down. They just shut the missiles down. People couldn't even restart them. There is no way that that could not be interpreted as a blatantly hostile act. If we did anything like that, to any other country, what would happen? It would be instant cause for war, right?

"It's a pretty universal principle. You enter my territory and you brandish weapons—or not even weapons, but just give a show of force, a show of power—that's throwing down the gauntlet. I mean, you've violated my territory, damn it! And what's interesting is our response to that. Our military forces, when events like that have happened, have found themselves to be completely powerless. There was nothing that they could do.

"Do you want the world to know that you were humiliated by someone like that? Absolutely not. You don't even have to talk about overflights. As Jacobs and others have done, you could simply talk about the abductions that are going on. They are a flagrant and fundamental violation of human rights and freedoms. You can't interpret it any other way. That's hostile.

"So the rhetorical claim that Steven Greer makes, that the aliens have no hostile intentions whatsoever, is preposterous. Again, it is an illogical, undocumented, unproven claim. Why is he making this claim? On whose behalf is he making such a claim?"

The aliens are also concerned that humanity may one day be able to defend itself militarily against an alien invasion.

"I know of people," Brownlee said, "who are involved in space defense projects who are very consciously aware that what they're doing is developing systems to defend against the extraterrestrials. It's happening. It's not just pie-in-the-sky or a rumor; it's actually happening. They're working on it.

"There are pretty good indications," he continued, "that the whole Star Wars missile defense effort has very little to do with terrorism and rogue states on this planet. I think it's much more likely that those systems are early attempts to design systems to build a shield against extraterrestrial incursion. I know that's not a politically popular viewpoint, but I think it's worth considering. And if that were true, it would kind of explain the way that some of those programs have been handled, the secrecy around them, the kind of skittishness that our government has had about them."

In spite of those possibly heroic machinations behind the scenes, there is still the *real* conspiracy to be dealt with, as Brownlee sees it.

"We talk about this dark conspiracy," he said, "of the wealthy and so on. It's not difficult to imagine that those people in positions of power are again either knowingly or unknowingly co-opted by the extraterrestrials. Yeah, there's a conspiracy, all right, but it's much larger than our conspiracy theorists have been talking about. We've been talking about human conspiracies here. Well, all human conspiracies are being manipulated by the extraterrestrials, essentially.

"We need to start understanding what they're up to. We need to start seeing the patterns of manipulation. We need to start seeing their strategies and their programs. We need to know what this hybridization program is about. We need to understand why they're manipulating religious and political and economic leaders throughout the world. We need to understand why they're acclimating us to their presence in every possible way they can, through the media, through entertainment, a variety of ways. They're getting us comfortable with the idea that aliens are among us."

All of which paints a dark picture indeed. But there is more.

"We are on the brink of losing our freedom," Brownlee cautioned, "of losing our sovereignty, and we're not aware of it. And we need to become aware. We need to help other people become aware. And that's a very big challenge."

BACKGROUND ON MICHAEL BROWNLEE AND DR. STEVEN GREER

Michael Brownlee is an author, publisher and journalist reporting on "the transformation of consciousness now unfolding on the planet." He cofounded Visibility Unlimited, LLC, a publishing venture committed to "communication that creates," and edited and published **Awakening World**, an online magazine.

Brownlee was the chief editor and coauthor of *Just In Case: Dispatches From the Front Lines of the Y2K Crisis*, a critically acclaimed anthology of essays published by Origin Press in 1999. Brownlee was a frequent public speaker, media interviewee and workshop presenter on this controversial topic. He is also the author of *The Creator's Workshop: A Complete Course In Creation*.

Brownlee has conducted workshops on Understanding the Extraterrestrial Presence in the World Today, an introduction to the mystery of "contact" and the growing challenge to human freedom; The Extraterrestrial Crisis: The Challenge of Our Lifetime, a wake-up call for those who suspect their destiny is somehow connected to humanity's emergence into the Greater Community; Awakening the Creator Within, a profound experiential journey into the Underlying Design of human evolution, from the personal to the universal; and The Cultural Creatives: An Awakening Culture, an introduction to the new paradigm of consciousness that is rapidly reshaping our world.

. . . . .

Steven Greer, M.D., is an emergency room physician and the founder of the Center for the Study of Extraterrestrial Intelligence, an international nonprofit scientific research and educational organization "dedicated to the furtherance of our understanding of extraterrestrial intelligence." CSETI was founded in 1990 by Dr. Greer, who is currently the organization's international director. Greer later founded The Disclosure Project, a nonprofit research project "working to fully disclose the facts about UFOs, extraterrestrial intelligence, and classified energy and propulsion systems."

UFO HOSTILITIES AND THE EVIL ALIEN AGENDA

According to the Disclosure Project website, "We have over 400 government, military and intelligence witnesses testifying to their direct, personal, first-hand experience with UFOs, extraterrestrials and extraterrestrial technology and the cover-up that keeps this information secret."

To learn more about the Center for the Study of Extraterrestrial Intelligence, visit their website at: "http://www.cseti.org"

For more on the Disclosure Project, go to:

"http://www.disclosureproject.org"

440

441

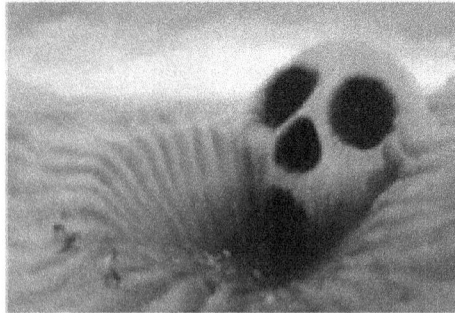

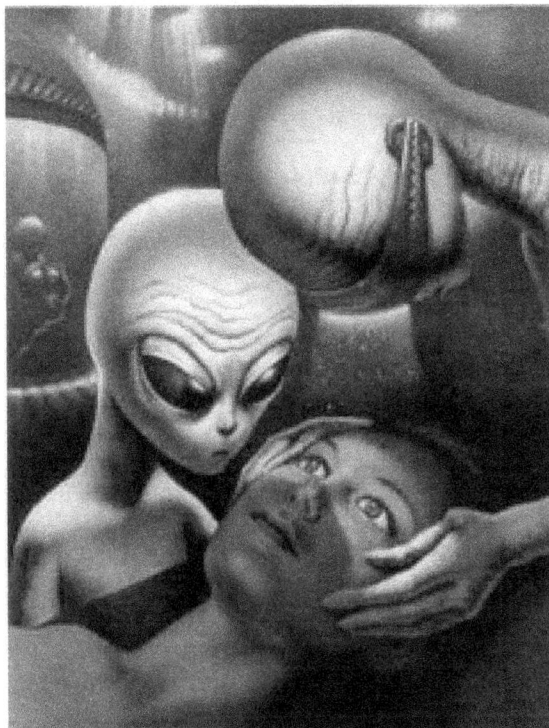